Applied Probability and Statistics

continued on back

Statistics

Statistics

A Biomedical Introduction

Byron Wm. Brown, Jr.

Stanford University

Myles Hollander

Florida State University

John Wiley & Sons

New York   London   Sydney   Toronto

Library of Congress Cataloging in Publication Data:

Brown, Byron Wm., Jr.
    Statistics.

    (Wiley series in probability and mathematical
statistics)
    Includes bibliographies and index.
    1. Medical statistics. 2. Biometry. I. Hollander,
Myles, joint author. II. Title. [DNLM: 1. Statistics.
2. Biometry. QH323.5 B877s]

RA409.B84   519.5'02'461   77-396
ISBN 0-471-11240-2

Printed in the United States of America

10 9 8 7 6 5 4 3

TO

JAN AND GLEE

Preface

Purpose. This book was written specifically as a text for an introductory course for persons interested in learning the fundamental concepts and techniques of statistical inference for application to the planning and evaluation of scientific studies in medicine and biology. The coverage includes the topics taught in most beginning statistics courses, but ours is a "biomedical introduction" because the chapters are centered on actual studies from the biomedical sciences.

The book evolved from lecture notes for a course that has been offered for many years at Stanford University, and, more recently, at Florida State University. The course is taken by medical students, by undergraduates in the natural sciences, and by graduate students in a variety of biomedical disciplines. The book contains enough material for about 65 one-hour lectures. Thus it can be used for a one-quarter (suggested coverage, Chapters 1–7) or two-quarter sequence. All of the material has been used in the classroom (at Stanford and at Florida State), much of it several times, in essentially the same form as it now appears.

Level. No mathematical knowledge beyond college algebra is needed for the learning of this material. The content can be used with assurance by the nonmathematically trained student or scientist. The applications, the rules of thumb, and the guidelines for the use of the various statistical procedures are based on the extensive experience of the authors and other professional statisticians. The examples and the problems are drawn from the experience of the authors and from the scientific literature. They can be taken to be a fair reflection of the application of statistical inference in the biomedical sciences today.

It should be stated here that although we have tried to treat the scientific studies used for illustration as realistically as space allowed, few serious scientific experiments can be dealt with as simply as our discussions might imply, and, in fact, detailed examination and qualification of certain of the problems might well alter certain of the analyses and conclusions. Nevertheless, we believe we have captured the flavor of the problems and done no great violence to the results.

Format. The chapters begin with the presentation of data sets arising from studies in the biomedical sciences. Statistical procedures are introduced, applied to the basic data sets, and conclusions are stated. Most chapters consist of two or three principal sections, with each section containing several unnumbered subsections.

Each principal section contains a subsection called comments and a subsection called problems.

The comments provide additional motivation, illustrate how the method under discussion can be applied in slightly different settings, furnish computational hints, relate the method to other procedures in different parts of the book, and call attention to certain references.

The problems test the reader's understanding of the lecture content and its applications to new situations. A relatively difficult or computationally tedious problem is marked by a •.

Interspersed within the chapters are discussion questions. They are designed to stimulate student participation in the classroom.

References at the end of each chapter provide the sources for the basic data sets, cite other introductory books where the reader can find supplementary material, and sometimes cite the articles where the statistical procedures were developed. A reference in the latter category may require more mathematics than college algebra for a careful reading. When this is the case, we mark the citation with a *.

Within each chapter equations, tables, figures, comments, and problems are numbered consecutively, without the chapter number, and they are referred to in this way within the chapter. For example, within Chapter 3 the fourth table of Chapter 3 is simply called Table 4. In other chapters this table is designated as Table 4 of Chapter 3. Sections, comments, figures, and problems are cited in the same manner.

Appendices.   Appendix A is a glossary that contains definitions of terms that appear in bold face in the text.

Appendix B gives answers to selected discussion questions and solutions to selected problems.

Appendix C contains the mathematical and statistical tables that are needed to implement the statistical procedures presented in the text.

Acknowledgments.   We are indebted to the Department of Statistics and the Department of Family, Community, and Preventive Medicine at Stanford University, and the Department of Statistics at Florida State University, for providing facilities which supported this effort. Byron Brown acknowledges Public Health Service Grant 1 R01 GM21215-01 and Myles Hollander acknowledges the Air Force Office of Scientific Research Grants AFOSR-76-3109 and 74-2581B for research support. Many friends contributed to this project. To all we are grateful, and in particular we mention here: Pi-Erh Lin—who gave the manuscript a hard reading and insisted on clear writing; Lynne Billard, Lincoln Moses, Frank Proschan, Jayaram Sethuraman, and Hans Ury—who made useful suggestions pertaining to certain chapters; William Blot, Duane Meeter, Rupert Miller, Jr.,

Ronald Thisted, Gerald van Belle, and Douglas Zahn—who helped locate examples; Gregory Campbell, Shih-Chuan Cheng, Edward Cooley, Robert Hannum, Alfred Ordonez, and Derek Wong—who checked final versions of various chapters; Jennifer Smith—who carefully typed the major portion of the manuscript; Kathy Bailey, Gail Lemmond, Annette Tishberg, and Beverly Weintraub—who did additional typing and provided secretarial assistance; Maria Jedd—who drew the illustrations; Beatrice Shube—our editor at John Wiley & Sons, who provided continuing cooperation.

To these workmates of high skill and dedication, we again say thank you.

BYRON WM. BROWN, JR
MYLES HOLLANDER

Stanford, California
Tallahassee, Florida
January 1977

Contents

Statistics

CHAPTER 1

Introduction

The following four problems are typical of the data analysis situations that arise in biomedical research; they are also typical of the problems the reader will encounter and learn to analyze in this book.

Medi-Cal Study

The Medi-Cal study was initiated by Cannon and Remen (1972) at the Stanford University Medical School. Medi-Cal is California's medical assistance program funded by federal, state, and county taxes. At the time of this writing, about 50% of the "sick visits" to Stanford's Pediatric Clinic are made by children covered under the Medi-Cal program and comprise the "Medi-Cal Group." The other 50% of patients are covered by other insurance programs (private insurance, military insurance, Stanford student dependents, etc.) or by private payment, and are called the "non-Medi-Cal Group."

Cannon and Remen wanted to study whether the health care given to Medi-Cal patients at the clinic differed from the health care given to the non-Medi-Cal patients. Specifically, they were interested in the physician's unconscious attitude towards the Medi-Cal patients, with the thought in mind that the physician might be biased in several ways:

1. The physician might be tempted to order *more* diagnostic tests (X-rays, lab tests, etc.) for a given illness, thinking it was of no financial cost to the patient under the Medi-Cal program.
2. The physician might view the Medi-Cal group as less "sophisticated" and rather disinterested in detailed diagnostic workup of a given illness, but more concerned with the fast and effective treatment of the symptoms. Therefore, the physician might be influenced to do *fewer* diagnostic tests in the Medi-Cal group compared with the non-Medi-Cal group.
3. The physician might view the Medi-Cal group as somewhat unreliable and less "compliant" and therefore treat a given illness differently in the two groups. For example, the physician might prefer to admit a severely sick (Medi-Cal) child rather than follow him closely as an outpatient. In the same spirit, the physician might prefer to treat a (Medi-Cal) child with a single injection of a long-acting drug rather than rely on daily oral treatment.

To reduce the number of variables, Cannon and Remen reviewed cases of an easily documented disease with a standardized diagnostic workup and therapy. They reviewed all the cases of pneumonia seen in the pediatric clinic at Stanford from July 1970 to August 1972. They took only first-episode pneumonias and eliminated all the cases in which some underlying disease could possibly affect the attitude of the physician towards his patient or affect the compliance of the parents.

Table 1, based on a subset of the Cannon–Remen data, shows the frequency of intramuscular injections versus oral administration of antibiotics for the Medi-Cal and non-Medi-Cal groups.

TABLE 1. Intramuscular (IM) versus Oral (PO) Administration of Antibiotics

	Medi-Cal	Non-Medi-Cal
IM	30	16
PO	26	40

Source: I. Cannon and N. Remen (1972).

At first glance, the data suggest that a Medi-Cal child is more likely to be treated via an intramuscular injection of antibiotics than is a non-Medi-Cal child. We get this impression by comparing the two proportions 30/56 and 16/56. Could the difference in these two proportions simply be due to chance, rather than differences in physicians' attitudes? This question can be answered by the statistical methods described in Chapter 8. This study forms the basis for Problems 2 and 5 of Chapter 8.

Study on Walking in Infants

Zelazo, Zelazo, and Kolb (1972) conducted a study that was designed to test the generality of the observation that stimulation of the walking and placing reflexes in the newborn promotes increased walking and placing. These investigators point out that "If a newborn infant is held under his arms and his bare feet are permitted to touch a flat surface, he will perform well-coordinated walking movements similar to those of an adult. If the dorsa of his feet are drawn against the edge of a flat surface, he will perform placing movements much like those of a kitten." Illingworth (1963) noted that the walking and placing reflexes disappear by about 8 weeks. The investigation of Zelazo, Zelazo, and Kolb was partially inspired by an earlier pilot experiment of the researchers, in which they noticed that the reflexes could be preserved beyond the second month through active exercise.

The subjects in Table 2 were about 1-week-old male infants from middle-class and upper-middle-class families. Each infant was assigned to one of four groups,

TABLE 2. Ages of Infants for Walking Alone (months)

1 Active-exercise Group	2 Passive-exercise Group	3 No-exercise Group	4 8-week Control Group
9.00	11.00	11.50	13.25
9.50	10.00	12.00	11.50
9.75	10.00	9.00	12.00
10.00	11.75	11.50	13.50
13.00	10.50	13.25	11.50
9.50	15.00	13.00	

Source: P. R. Zelazo, N. A. Zelazo, and S. Kolb (1972).

namely, an experimental group (active-exercise) and three control groups (passive-exercise, no-exercise, 8-week control).

Infants in the active-exercise group received stimulation of the walking and placing reflexes during four 3-minute sessions that were held each day from the beginning of the second week until the end of the eighth week. The infants in the passive-exercise group received equal amounts of gross motor and social stimulation as those who received active-exercise, but unlike the active-exercise group, these infants had neither the walking nor placing reflex exercised. Infants in the no-exercise group did not receive any special training, but were tested along with the active-exercise and passive-exercise subjects. The 8-week control group was tested only when they were 8 weeks of age; this group served as a control for the possible helpful effects of repeated examination.

If one casually looks at Table 2, one gets the impression that the ages for walking tend to be lower for the active-exercise group. Can this "eye-ball" impression be justified by a rigorous statistical analysis? Are numbers, such as those in Table 2, likely to arise when, in fact, there are no differences between groups? Basically, can we assert that active-exercise infants walk earlier than infants in the other three groups? This question is answered by the statistical methods in Section 1 of Chapter 10.

Study on Transfer of Information from Manual to Oculomotor Control System

Angel, Hollander, and Wesley (1973) report an experiment whose purpose was to study the mechanisms that coordinate movements of the eye with those of the hand. Whenever the hand is moved under visual guidance, several motor systems are employed.

1. The manual control system directs movement of the hand.
2. The oculomotor system, which moves the eyes, contains specialized subsystems for acquiring and tracking visual targets. (a) The saccadic system specializes in moving the eyes from one position to another very rapidly. (b) The smooth pursuit system matches the eye velocity to velocity of the visual target.

When a subject tries to keep his eyes on a continuously moving target, the smooth pursuit system tends to make the eyes follow the target, but occasionally the target "escapes." The saccadic system then produces a sudden jump of the eyes, bringing the target back to the center of the field. The more accurate the smooth pursuit, the fewer saccadic corrections needed. The number of "saccades" per unit time may thus be taken as a rough measure of the inaccuracy of smooth pursuit tracking.

Other workers have shown that visual tracking is more accurate (i.e., fewer saccades occur) when the target is moved by the subject rather than by an external agent. This is taken as evidence that the motor commands directed to the subject's hand are somehow made available to the oculomotor system. Eye movements can thus anticipate hand movements, without the delay involved in sensory feedback. If one assumes that information about hand movement is conveyed to the oculomotor system, the question arises: How long does this information persist in usable form? How long does the brain store information about motor commands directed to the hand? The following experiment was designed to answer questions of this nature.

Eye position was recorded while the subject tried to keep his eyes on a visual target. On each experimental run, the subject moved the target actively by means of a joy stick. Target motion was either synchronous with that of the hand or delayed by a fixed amount (.18, .36, .72, or 1.44 sec). On each control run, target motion was a tape-recorded copy of the pattern generated by the subject during the previous test run. (Regardless of the delay between joy stick and target motion, the frequency of saccades was generally smaller during the test run than during the matched control run. The results suggest that information about hand movement is available to the oculomotor system for at least 1 sec following manual performance.)

The experiment just described was repeated with ten normal volunteer subjects. The experimental conditions (different amounts of lag between hand and target movement) were presented in random order. After each run, the subject watched a tape-recorded copy of the target movement generated on the test run. Each run lasted 100 sec. The number of saccades (rapid, jumping eye movements) was determined for each test run and compared with the matched control run. The data are given in Table 3.

The variables (Y's, say) in Table 3 are defined as

$$Y = \frac{\text{saccades during test}}{\text{saccades during playback}}.$$

Do the Y's tend to increase as the lags increase? At (roughly) what time lag do the Y values tend to be close to 1, indicating the hand movement information is no longer available to the brain? Methods for analyzing data fitting the format of

TABLE 3.   Saccades During Test/Saccades During Playback

	Lags				
	1 (0 sec)	2 (0.18 sec)	3 (0.36 sec)	4 (0.72 sec)	5 (1.44 sec)
1	0.600	0.880	0.587	1.207	0.909
2	0.610	0.526	0.603	1.333	0.813
3	0.592	0.527	0.976	1.090	0.996
4	0.988	0.841	0.843	1.010	1.112
5	0.611	0.536	0.654	0.910	0.936
6	0.981	0.830	1.135	0.838	0.876
7	0.698	0.520	0.673	0.892	0.857
8	0.619	0.493	0.734	0.948	0.862
9	0.779	0.759	0.764	0.824	0.851
10	0.507	0.712	0.787	0.868	0.880

Source: R. W. Angel, M. Hollander, and M. Wesley (1973).

Table 3 are presented in Section 2 of Chapter 10. (The Table 3 data form the basis for Problem 6 of Chapter 10.)

Heart Transplant Study

Heart transplantation has held an extraordinary fascination for both layman and medical scientist ever since the first operation by Barnard in 1967. After the first flurry of operations, two programs in the United States continued to accumulate experience in transplantation, one at Baylor University and one at Stanford University. The Baylor program has been discontinued, but at the time of this writing, the program at Stanford continues, with one to three new transplants a month.

Turnbull, Brown, and Hu (1974) have investigated in some detail the question of whether the Stanford heart transplant program has prolonged lifetimes in the patients receiving new hearts. Their paper analyzed the available data on 82 heart transplant candidates, up to the closing date of March 1, 1973. Using the same data, Brown, Hollander, and Korwar (1974) considered which variables were correlated with the length of post-transplant survival.

We briefly describe some details of the Stanford heart transplant program. For each patient entering the program, there is a well-defined conference date on which he is designated officially as a heart transplant candidate. This selection means that it has been determined that he is gravely ill, that he will very likely benefit from a new heart, and that he and his family have consented and seem up to the trial. A donor is then sought, and the patient, given survival, receives a heart, usually within a matter of weeks, though occasionally the wait for a donor can be several months. A few candidates have survived for a time before transplant, shown noteworthy improvement, and then been "deselected."

The data for the 82 candidates admitted to the program, as of March 1, 1973, are given in Table 4. [The data of Table 4 appear in Brown, Hollander, and Korwar (1974) but were initially furnished by Dr. E. Dong (1973) of the Stanford University Medical Center.] For patients who died before a donor was found for them and for those still awaiting a heart, we give their birth date, date of declared acceptance into the program, sex, and the number of days of survival (to death or to the closing date for analysis, March 1, 1973). We also indicate whether the patient is alive or dead at the closing date. For patients receiving a new heart we give the birth date, date of acceptance, sex, days to transplant, days from transplant to death or closing date, and state (dead or alive) at closing date.

How does one estimate the probability that a patient who receives a transplant will survive at least one year, at least two years, and so forth? Does heart transplantation tend to increase the lifetimes of heart transplant candidates? Is the Stanford heart transplant program improving? Questions of this nature are difficult to answer and involve complicated statistical techniques, many of which are beyond the level of a first course in statistics. In Section 3 of Chapter 9 we investigate the question of whether there is a beneficial trend in the program. In Chapter 14 we describe a method for estimating survival probabilities, and the transplant data of Table 4 are the basis for Problems 4 and 5 and Supplementary Problem 1 of Chapter 14. For answers to some of the other questions mentioned, see Turnbull, Brown, and Hu (1974) and Brown, Hollander, and Korwar (1974).

Statistics and This Statistics Book

Statistical techniques can be extremely useful in describing and evaluating each of the foregoing studies. The principles and methods of statistics have been developed for the purpose of making sense from the results of scientific studies. In particular, statistical techniques allow the scientist to *measure* the strength of his conclusions, in terms of probability, and to do this in a language that is familiar to his peers around the world.

TABLE 4. Acceptance Dates, Birth Dates, Sex, Survival Times, and Time to Transplant for Stanford Program Patients

(Closing date: March 1, 1973)

Nontransplant Patients (30)

| Birth Date | | | | Acceptance Date | | | | alive = a |
Mo.	Day	Yr.	Sex	Mo.	Day	Yr.	$T_1{}^1$	dead = d
5	20	28	M	9	13	67	5	d
1	10	37	M	11	15	67	49	d
3	2	16	M	1	2	68	5	d
7	28	47	M	5	10	68	17	d
11	8	13	M	6	13	68	2	d
3	27	23	M	8	1	68	39	d
6	11	21	F	8	9	68	84	d
7	9	15	M	9	17	68	7	d
12	4	14	M	9	27	68	0	d
6	29	48	M	10	28	68	35	d
10	4	09	M	11	18	68	36	d
10	18	38	M	5	1	69	1400^2	a
2	6	19	F	7	14	69	34	d
10	4	14	M	8	23	69	15	d
8	4	26	M	1	21	70	11	d
3	13	34	M	8	21	70	2	d
6	1	27	F	10	22	70	1	d
5	2	28	M	11	30	70	39	d
1	23	15	M	2	5	71	8	d
1	24	30	M	4	25	71	101	d
9	16	23	M	7	2	71	2	d
6	8	30	M	9	13	71	148	d
8	19	42	M	11	1	71	427^3	a
5	12	19	M	12	4	71	1	d
8	1	32	M	12	9	71	68^4	d
1	2	19	M	3	20	72	31	d
7	25	20	M	9	29	72	1	d
8	27	31	M	10	6	72	20	d
2	20	24	F	11	3	72	118	a
2	18	19	M	11	30	72	91	a

Transplant Patients (52)

Birth Date Mo.	Day	Yr.	Sex	Acceptance Date Mo.	Day	Yr.	$T_2{}^1$	$T_3{}^1$	alive = a dead = d
9	19	13	M	1	6	68	0	15	d
12	23	27	M	3	28	68	35	3	d
8	29	17	M	7	12	68	50	624	d
2	9	26	M	8	11	68	11	46	d
8	22	20	F	8	15	68	25	127	d
2	22	14	F	9	19	68	16	61	d
9	16	14	M	9	20	68	36	1350	d
5	16	19	M	10	26	68	27	312	d
12	27	11	M	11	1	68	19	24	d
10	19	13	M	1	29	69	17	10	d
9	29	25	M	2	1	69	7	1024	d
6	5	26	M	3	18	69	11	39	d
12	2	10	M	4	11	69	2	730	d
7	7	17	M	4	25	69	82	136	d
2	6	36	M	4	28	69	24	1379	a
5	30	15	M	6	7	69	70	1	d
9	20	24	F	8	19	69	15	836	d
4	2	05	M	8	29	69	16	60	d
1	1	21	M	11	27	69	50	1140	a
5	24	29	M	12	12	69	22	1153	a
5	1	21	M	4	4	70	45	54	d
10	24	08	M	4	25	70	18	47	d
11	14	28	M	5	5	70	4	0	d
11	12	19	M	5	20	70	1	43	d
11	30	21	M	5	25	70	40	971	a
4	30	25	M	8	19	70	57	868	a
10	30	34	M	1	5	71	0	44	d
6	1	22	F	1	10	71	1	780	a
12	28	23	M	2	2	71	20	51	d
6	21	34	M	2	15	71	35	710	a
3	28	25	M	2	15	71	82	663	a
6	29	22	M	3	24	71	31	253	d
2	27	24	M	7	2	71	40	147	d
2	24	19	M	8	9	71	9	51	d
12	5	32	M	9	3	71	66	479	a

TABLE 4—continued

Transplant Patients (52)

| Birth Date | | | | Acceptance Date | | | | | alive = a |
Mo.	Day	Yr.	Sex	Mo.	Day	Yr.	$T_2{}^1$	$T_3{}^1$	dead = d
9	17	23	M	9	23	71	20	322	d
5	12	30	M	9	29	71	77	442	a
10	29	22	M	11	18	71	2	65	d
4	15	39	M	12	12	71	26	419	a
4	9	23	M	2	1	72	32	362	a
11	19	20	M	3	6	72	13	64	d
9	3	52	M	3	23	72	56	228	d
1	10	27	M	4	7	72	2	65	d
6	5	24	M	6	1	72	9	264	a
6	17	19	M	6	17	72	4	25	d
2	22	25	M	7	21	72	30	193	a
11	22	45	M	8	14	72	3	196	a
5	13	16	M	9	11	72	26	63	d
7	20	43	M	9	18	72	4	12	d
9	3	20	M	10	4	72	45	103	a
6	27	26	M	12	6	72	25	60	a
2	21	20	M	1	12	73	5	43	a

Source: B. Wm. Brown, Jr., M. Hollander, and R. M. Korwar (1974).
[1] T_1 = days to death or closing date.
T_2 = days to transplant.
T_3 = days from transplant to death or closing date.
[2] Deselected 8/21/69.
[3] Deselected 2/02/72 (also, for this patient only, survival experience is known only up to 1/1/73 so that the T_1 = 427 entry is measured from 11/1/71 to 1/1/73 rather than 3/1/73).
[4] Deselected 2/04/72.

In this book we concentrate on the statistical concepts and techniques most useful to the scientist who is doing research in the biomedical sciences. We address the problem of planning, executing, and evaluating studies on patients, carried out by the physician-scientist and typified by the four examples cited in this chapter.

Our approach is data based. The chapters revolve around actual experiments that are (or were) of interest to the scientific community. The approach of basing the chapters on scientific studies is seldom found in books on introductory biomedical statistics, but it has been used successfully in the teaching of introductory statistics (in particular, see the series of pamphlets edited by Mosteller *et al.*, 1973).

Other introductory biomedical statistics books include those of Armitage (1973), Bailey (1959), Bourke and McGilvray (1969), Daniel (1974), Dunn (1964), Goldstein (1964), Hill (1971), Huntsberger and Leaverton (1970), Lancaster (1974), Mainland (1963), and Zar (1974).

We close this chapter by calling the reader's attention to:

1. Appendix A: This appendix contains the glossary which gives definitions of the terms that are printed in bold face in the text.
2. Appendix B: In this appendix we give answers to selected discussion questions and solutions to selected problems.
3. Appendix C: This appendix contains the various mathematical and statistical tables needed to implement the statistical procedures advocated in the text.

REFERENCES

Angel, R. W., Hollander, M., and Wesley, M. (1973). Hand–eye coordination: The role of "motor memory." *Perception and Psychophysics* **14**, 506–510.

Armitage, P. (1973). *Statistical Methods in Medical Research*. Wiley, New York.

Bailey, N. T. J. (1959). *Statistical Methods in Biology*. English Universities Press, London.

Bourke, G. J. and McGilvray, J. (1969). *Interpretation and Uses of Medical Statistics*. Blackwell Scientific Publications, Oxford.

†*Brown, B. Wm. Jr., Hollander, M., and Korwar, R. M. (1974). Nonparametric tests of independence for censored data with applications to heart transplant studies. In *Reliability and Biometry*, F. Proschan and R. J. Serfling (Eds.). SIAM, Philadelphia, 327–354.

Cannon, I. and Remen, N. (1972). Personal communication.

Colquhoun, D. (1971). *Lectures on Biostatistics*. Clarendon Press, Oxford.

Daniel, W. W. (1974). *Biostatistics: A Foundation for Analysis in the Health Sciences*. Wiley, New York.

Dong, E. (1973). Personal communication.

Dunn, O. J. (1964). *Basic Statistics: A Primer for the Biomedical Sciences*. Wiley, New York.

Goldstein, A. (1964). *Biostatistics: An Introductory Text*. Macmillan, New York.

Hill, B. A. (1971). *Principles of Medical Statistics*, 9th edition. Oxford University Press, London.

Huntsberger, D. V. and Leaverton, P. E. (1970). *Statistical Inference in the Biomedical Sciences*. Allyn and Bacon, Boston.

Illingworth, R. (1963). *The Development of the Infant and Young Child*. Williams and Wilkins, Baltimore.

---

† Throughout the text, references marked with an asterisk may require more mathematics than college algebra for a careful reading.

Lancaster, H. O. (1974). *An Introduction to Medical Statistics*. Wiley, New York.

Mainland, D. (1963). *Elementary Medical Statistics*. Saunders, Philadelphia.

Mosteller, F., Kruskal, W. H., Link, R. F., Pieters, R. S., and Rising, G. R. (Eds.) (1973). *Statistics by Example*. Addison-Wesley, Reading.

*Turnbull, B. W., Brown, B. Wm. Jr., and Hu, M. (1974). Survivorship analysis of heart transplant data. *J. Amer. Statist. Assoc.* **69**, 74–80.

Zar, J. H. (1974). *Biostatistical Analysis*. Prentice-Hall, Englewood Cliffs.

Zelazo, P. R., Zelazo, N. A., and Kolb, S. (1972). Walking in the newborn. *Science* **176**, 314–315.

CHAPTER | 2

Elementary Rules of Probability

Chapter 2 introduces the language of probability and describes certain rules for calculating probabilities. In this chapter the calculations are relatively simple because they relate to relatively simple examples. In later chapters, in which the probability calculations are necessarily more difficult, the need for difficult calculations is removed by means of tables; the desired probability is obtained by entering an appropriate table and extracting a value.

Selecting a Digit at Random

Table 1 consists of 100 digits, 10 rows with 10 digits in each row.

TABLE 1. 100 Digits

0	7	0	4	8	5	2	8	4	1
5	4	9	6	9	8	7	0	5	7
3	0	5	7	0	5	0	4	9	4
2	9	9	3	6	9	3	9	6	7
1	0	6	4	1	7	9	8	7	1
0	9	1	6	5	5	6	9	2	6
1	7	2	9	4	0	3	8	0	3
3	1	7	5	5	1	1	3	2	1
3	3	6	8	1	1	2	9	9	7
1	7	6	2	5	2	5	9	5	4

If you take a few minutes to count the number of 0's, the number of 1's, and so forth, you will find that

$$\text{the number of 0's} = 10$$

$$\text{the number of 1's} = 13$$

$$\text{the number of 2's} = 8$$

$$\text{the number of 3's} = 9$$

$$\text{the number of 4's} = 8$$

$$\text{the number of 5's} = 12$$

$$\text{the number of 6's} = 9$$

$$\text{the number of 7's} = 11$$

$$\text{the number of 8's} = 6$$

$$\text{the number of 9's} = 14$$

Suppose we select a digit from Table 1 at random. By "at random" we mean in such a way that each of the 100 individual entries has the same chance of being selected. This might be done by writing each digit in Table 1 on a tag, mixing the tags in a bowl, and drawing one out blindfolded, as in a lottery. Then we can ask, for example, What is the chance or probability that the randomly selected digit is a 0? Since there are ten 0's, the probability is equal to 10/100 or .1.

More generally, an element is said to be randomly selected from a finite set of elements S if it is chosen in such a way that each element is equally likely to be the one that is picked. One may then wish to calculate the chance or probability that the randomly selected element is a member of a given subset of elements, A. This chance is written as $Pr(A)$ and read as "the probability of the event A." To calculate $Pr(A)$ we can use the formula

(1) $$Pr(A) = \frac{\text{number of elements in } A}{\text{number of elements in } S}.$$

To continue our Table 1 example, let A_0 denote the event that the randomly selected digit is a 0, let A_1 denote the event that the randomly selected digit is a 1, and so on. Then from formula (1) we find:

$$Pr(A_0) = \frac{10}{100} = .10$$

$$Pr(A_1) = \frac{13}{100} = .13$$

$$Pr(A_2) = \frac{8}{100} = .08$$

$$Pr(A_3) = \frac{9}{100} = .09$$

$$Pr(A_4) = \frac{8}{100} = .08$$

$$Pr(A_5) = \frac{12}{100} = .12$$

$$Pr(A_6) = \frac{9}{100} = .09$$

$$Pr(A_7) = \frac{11}{100} = .11$$

$$Pr(A_8) = \frac{6}{100} = .06$$

$$Pr(A_9) = \frac{14}{100} = .14.$$

Long-run Frequency Interpretation of Probabilities

The probabilities computed in the previous subsection can be interpreted in terms of long-run frequencies as follows. Suppose we perform a very large number (n, say) of independent repetitions of the experiment which consists of selecting a digit at random from Table 1. Let f_0 denote the number of times the digit 0 is selected in the sequence of n trials, f_1 denote the number of times the digit 1 is selected, and so forth. Then we would find that

$$\frac{f_0}{n} \approx Pr(A_0) = .10,$$

$$\frac{f_1}{n} \approx Pr(A_1) = .13,$$

$$\vdots$$

$$\frac{f_9}{n} \approx Pr(A_9) = .14.$$

The symbol $\approx$ means approximately equal to. We do not expect, for any particular number n of repetitions, that the observed relative frequencies of the events would be exactly equal to the "true" probabilities of the events. But the accuracy of the approximation can be trusted for large n and generally improves as n gets larger. [You might test this experimentally. First, perform the experiment yourself $n = 100$ times and compute the relative frequencies $f_0/100$ (of A_0), $f_1/100$, and so on. Next, combine your results with those of nine willing friends or classmates who have also made 100 random selections each. Compare the accuracy of the results for $n = 1000$ with those for $n = 100$.]

Selecting a Case at Random

Return to the Medi-Cal study of Chapter 1. Table 1 of Chapter 1, reproduced below as Table 2, shows the frequency of intramuscular injections versus oral administration of antibiotics for the Medi-Cal and non-Medi-Cal groups, where a total of 112 cases were considered.

TABLE 2. Intramuscular (IM) versus Oral (PO) Administration of Antibiotics

	Medi-Cal	Non-Medi-Cal
IM	30	16
PO	26	40

Suppose a case is selected at random from the 112 cases corresponding to the entries of Table 2. Let us define some events corresponding to some possible outcomes of the selection:

B_1: case is Medi-Cal

B_2: case is non-Medi-Cal

B_3: case is IM

B_4: case is PO

From formula (1) we then find:

$$Pr(B_1) = \frac{\text{number of Medi-Cal cases}}{\text{total number of cases}} = \frac{30+26}{112} = \frac{56}{112},$$

$$Pr(B_2) = \frac{\text{number of non-Medi-Cal cases}}{\text{total number of cases}} = \frac{16+40}{112} = \frac{56}{112},$$

$$Pr(B_3) = \frac{\text{number of cases that received intramuscular injections}}{\text{total number of cases}}$$

$$= \frac{30+16}{112} = \frac{46}{112},$$

$$Pr(B_4) = \frac{\text{number of cases that received oral administration of antibiotics}}{\text{total number of cases}}$$

$$= \frac{26+40}{112} = \frac{66}{112}.$$

Life Tables

Table 3 gives the survival experience of 270 male *Drosophila melanogaster* (fruit fly). The data, due to Miller and Thomas (1958), are also cited by Chiang (1968).

TABLE 3. Survival Experience of Male *Drosophila melanogaster*

Age Interval (in days)	Number Dying in Interval
0–10	6
10–20	10
20–30	6
30–40	82
40–50	90
50–60	63
60 and over	13

Source: R. S. Miller and J. L. Thomas (1958).

From Table 3 we can construct a life table giving the probability of death for each interval. Suppose a fruit fly is selected at random from the group of 270 represented by Table 3. Let us define the following events:

D_0: death in interval 0–10,

D_1: death in interval 10–20,

D_2: death in interval 20–30,

D_3: death in interval 30–40,

D_4: death in interval 40–50,

D_5: death in interval 50–60,

D_6: death in interval 60 and over.

From formula (1) we find:

$$Pr(D_0) = \frac{6}{270} = .022,$$

$$Pr(D_1) = \frac{10}{270} = .037,$$

$$Pr(D_2) = \frac{6}{270} = .022,$$

$$Pr(D_3) = \frac{82}{270} = .304,$$

$$Pr(D_4) = \frac{90}{270} = .333,$$

$$Pr(D_5) = \frac{63}{270} = .233,$$

$$Pr(D_6) = \frac{13}{270} = .048.$$

Putting these results in tabular form we obtain Table 4, a life table:

TABLE 4.   Life Table for Male *Drosophila melanogaster*

Age Interval (in days)	Probability of Death
0–10	$\dfrac{6}{270} = .022$
10–20	$\dfrac{10}{270} = .037$
20–30	$\dfrac{6}{270} = .022$
30–40	$\dfrac{82}{270} = .304$
40–50	$\dfrac{90}{270} = .333$
50–60	$\dfrac{63}{270} = .233$
60 and over	$\dfrac{13}{270} = .048$

Other relevant probabilities relating to survival can be calculated directly by formula (1) or by using some rules of probabilities to be introduced in the next subsection.

For example, by rule III of the next subsection, we can find
Pr(death by age 20)
$$= Pr(D_0 \text{ or } D_1) = Pr(D_0) + Pr(D_1) = .022 + .037 = .059.$$

Some Rules of Probability

The probability of an event is a number between 0 and 1 indicating how likely the event is to occur. We denote the probability of the event A by $Pr(A)$. Probabilities satisfy the following rules:

Rule I: $Pr(A) = 1$ if A is certain to occur

Rule II: $Pr(A) = 0$ if it is impossible for A to occur

Rule III: $Pr(A \text{ or } B) = Pr(A) + Pr(B)$ if the events A and B cannot occur simultaneously

Rule III': $Pr(A \text{ or } B) = Pr(A) + Pr(B) - Pr(A \text{ and } B)$

Note that Rule III is really a special case of Rule III'. When A and B cannot occur simultaneously, we have $Pr(A \text{ and } B) = 0$ and thus by Rule III' we obtain

$$Pr(A \text{ or } B) = Pr(A) + Pr(B) - 0 = Pr(A) + Pr(B).$$

Rule IV: $Pr(\text{not } A) = 1 - Pr(A)$.

Applications of the Rules of Probability

Selecting a Digit at Random

a.   Referring back to Table 1, what is the chance that the randomly selected digit is less than or equal to 9? Since the event "less than or equal to 9" is certain to occur, we have, by Rule I,

$$Pr\{\text{digit is } \leq 9\} = 1.$$

b.   What is the chance that the randomly selected digit is a 10? Since it is impossible to select a "10" from a table consisting of the digits 0, 1, 2, ..., 9, we have, by Rule II,

$$Pr\{\text{digit is a } 10\} = 0.$$

c.   What is the chance that the randomly selected digit is a 0 or a 1? That is, what is $Pr(A_0 \text{ or } A_1)$. Since events A_0 and A_1 cannot occur simultaneously (the selected digit can't be *both* a 0 and a 1), we can use Rule III to find

$$Pr(A_0 \text{ or } A_1) = Pr(A_0) + Pr(A_1) = \frac{10}{100} + \frac{13}{100} = \frac{23}{100} = .23.$$

d.   What is the probability that the randomly selected digit is greater than 0? Note that we are asking you to obtain $Pr(\text{not } A_0)$. By Rule IV we have

$$Pr(\text{not } A_0) = 1 - Pr(A_0) = 1 - .10 = .90.$$

The Multiplication Rule

If $Pr(B|A)$ denotes the probability of the event B, *given that the event A has occurred*, then

(2)
$$Pr(B|A) = \frac{Pr(A \text{ and } B)}{Pr(A)},$$

or equivalently,

(2') $$Pr(A \text{ and } B) = Pr(B|A) \cdot Pr(A).$$

The value $Pr(B|A)$ is called the probability of B, *conditional* on A. Equation (2') shows that $Pr(A \text{ and } B)$ can be calculated by finding the probability of A and then multiplying that probability by the probability of B, conditional on A.

Applications of the Multiplication Rule

Selecting a Case at Random

a.   Referring back to Table 2, what is the probability that a randomly selected case will be a Medi-Cal patient who received an oral treatment? Note that, in the notation of the "Selecting a Case at Random" subsection, we seek to determine $Pr\{B_1 \text{ and } B_4\}$. From formula (1) we have

$$Pr(B_1 \text{ and } B_4) = \frac{\text{number of cases that are both PO and Medi-Cal}}{112}$$

$$= \frac{26}{112}.$$

Alternatively, we could use the multiplication rule. From formula (2'), with B_4 playing the role of B and B_1 playing the role of A, we find that

$$Pr(B_1 \text{ and } B_4) = Pr(B_4|B_1) \cdot Pr(B_1).$$

Recall that $Pr(B_1)$ was calculated earlier; $Pr(B_1) = 56/112$. Thus

$$Pr(B_1 \text{ and } B_4) = Pr(B_4|B_1) \cdot \frac{56}{112}.$$

To determine $Pr(B_4|B_1)$, note that, *given* the case is a Medi-Cal case, 26 of the 56 total (Medi-Cal) cases are "PO" (i.e., received an oral treatment). Thus, by formula (1),

$$Pr(B_4|B_1) = \frac{26}{56},$$

and hence by the multiplication rule,

$$Pr(B_1 \text{ and } B_4) = \frac{26}{56} \cdot \frac{56}{112} = \frac{26}{112}.$$

b.   What is the probability that a randomly selected case will be a Medi-Cal patient or a patient who received an oral tratment? Note that now, in our notation, we are asking you to calculate $Pr(B_1$ or $B_4)$. Since the events B_1 and B_4 can occur simultaneously, we cannot calculate $Pr(B_1$ or $B_4)$ using Rule III. But we can use Rule III'. We find

$$Pr(B_1 \text{ or } B_4) = Pr(B_1) + Pr(B_4) - Pr(B_1 \text{ and } B_4).$$

Now, all the probabilities on the right-hand side of this equation are known to us. In particular, in part a of this subsection we found, via the multiplication rule, that $Pr(B_1$ and $B_4) = 26/112$. Thus

$$Pr(B_1 \text{ or } B_4) = \frac{56}{112} + \frac{66}{112} - \frac{26}{112} = \frac{96}{112}.$$

Equivalently, by direct enumeration in Table 2, we find that 30 patients were Medi-Cal and IM, 40 were PO and non-Medi-Cal, and 26 were both Medi-Cal and PO, giving $Pr(B_1$ or $B_4) = (30+40+26)/112 = 96/112$.

Independent Events

Two events A, B are said to be **independent** if

$$Pr(B|A) = Pr(B).$$

If this equality does not hold, A and B are said to be dependent. Roughly speaking, A and B are independent, if the occurrence of A does not prejudice for or against (that is, does not change the probability of) the occurrence of B.

Note that by the multiplication rule [see formula (2')], when A and B are independent we have

$$Pr(A \text{ and } B) = Pr(B) \cdot Pr(A).$$

We have already encountered dependent events without explicitly mentioning this property. Recall part a of the previous subsection. We found that

$$Pr(B_4|B_1) = \frac{26}{56} = \frac{52}{112},$$

but

$$Pr(B_4) = \frac{66}{112}.$$

Since

$$Pr(B_4|B_1) \neq Pr(B_4),$$

we know that the events B_1, B_4 are dependent. In other words, the probability that a randomly selected patient is treated orally is 66/112, whereas the probability that a Medi-Cal patient is treated orally is slightly less, namely, 52/112.

Death Months and Birth Months

Phillips (1972) considered the possibility that some people may be able to postpone their deaths until after their birthdays. Table 5 shows the number of deaths, classified by month of birth and month of death, for the people listed in *Four Hundred Notable Americans* (1965). Phillips examined notable Americans partly because it seemed reasonable that famous people would tend to look forward to their birthdays more eagerly than ordinary people (because famous people generally receive more publicity, attention, and gifts on their birthdays) and partly because it is easier to find birth and death dates of famous people than it

TABLE 5. Deaths of 348 Notable Americans by Death Month and Birth Month

Month of Death	Month of Birth												Row Total
	Jan.	Feb.	Mar.	Apr.	May	June	July	Aug.	Sept.	Oct.	Nov.	Dec.	
Jan.	1	1	2	1	2	2	4	3	1	4	2	4	27
Feb.	2	3	1	3	1	0	2	1	2	2	6	4	27
Mar.	5	6	5	3	1	0	5	1	2	5	3	1	37
Apr.	7	6	3	2	1	3	3	1	3	2	4	4	39
May	4	4	2	2	1	2	4	1	3	2	1	5	31
June	4	0	4	5	1	1	1	2	1	2	4	0	25
July	4	0	3	4	3	3	4	1	6	4	2	5	39
Aug.	4	4	4	4	2	2	3	3	1	1	2	0	30
Sept.	2	2	1	0	2	0	2	4	2	0	5	2	22
Oct.	4	2	2	3	2	2	2	3	3	1	4	5	33
Nov.	0	2	0	2	1	1	0	3	3	3	1	0	16
Dec.	1	2	2	1	2	1	4	1	4	0	2	2	22
Column total	38	32	29	30	19	17	34	24	31	26	36	32	348

Source: D. P. Phillips (1972).

is to find vital event dates for less famous persons. The classification of Table 5 is by birth month and death month, rather than birthday and death day, because a listing by day is tedious. The total number of deaths in Table 5 is only 348 because some of the 400 people had not yet died and for some the month of birth or death was not known.

Suppose we select a notable American at random from the set of 348 represented in Table 5. Let us denote by $B_{\text{Jan.}}$ the event that the randomly selected person was born in January; $D_{\text{Jan.}}$ the event that the randomly selected person died in January; $B_{\text{Feb.}}$, born in February; $D_{\text{Feb.}}$, died in February, and so forth. Then we find:

$$Pr(D_{\text{Jan.}}) = \frac{27}{348} = .078; \qquad Pr(D_{\text{Jan.}}|B_{\text{Feb.}}) = \frac{1}{32} = .031,$$

$$Pr(D_{\text{Feb.}}) = \frac{27}{348} = .078; \qquad Pr(D_{\text{Feb.}}|B_{\text{Mar.}}) = \frac{1}{29} = .034,$$

$$Pr(D_{\text{Mar.}}) = \frac{37}{348} = .106; \qquad Pr(D_{\text{Mar.}}|B_{\text{Apr.}}) = \frac{3}{30} = .100,$$

$$Pr(D_{\text{Apr.}}) = \frac{39}{348} = .112; \qquad Pr(D_{\text{Apr.}}|B_{\text{May}}) = \frac{1}{19} = .053,$$

$$Pr(D_{\text{May}}) = \frac{31}{348} = .089; \qquad Pr(D_{\text{May}}|B_{\text{June}}) = \frac{2}{17} = .118,$$

$$Pr(D_{\text{June}}) = \frac{25}{348} = .072; \qquad Pr(D_{\text{June}}|B_{\text{July}}) = \frac{1}{34} = .029,$$

$$Pr(D_{\text{July}}) = \frac{39}{348} = .112; \qquad Pr(D_{\text{July}}|B_{\text{Aug.}}) = \frac{1}{24} = .042,$$

$$Pr(D_{\text{Aug.}}) = \frac{30}{348} = .086; \qquad Pr(D_{\text{Aug.}}|B_{\text{Sept.}}) = \frac{1}{31} = .032,$$

$$Pr(D_{\text{Sept.}}) = \frac{22}{348} = .063; \qquad Pr(D_{\text{Sept.}}|B_{\text{Oct.}}) = \frac{0}{26} = 0,$$

$$Pr(D_{\text{Oct.}}) = \frac{33}{348} = .095; \qquad Pr(D_{\text{Oct.}}|B_{\text{Nov.}}) = \frac{4}{36} = .111,$$

$$Pr(D_{\text{Nov.}}) = \frac{16}{348} = .046; \qquad Pr(D_{\text{Nov.}}|B_{\text{Dec.}}) = \frac{0}{32} = 0,$$

$$Pr(D_{\text{Dec.}}) = \frac{22}{348} = .063; \qquad Pr(D_{\text{Dec.}}|B_{\text{Jan.}}) = \frac{1}{38} = .026.$$

Thus, for example, for the notable Americans we see that the conditional probability of death in January, given birth in February, is .031, and this is lower than the unconditional probability of death in January, the latter being .078. (In the terminology of the preceding subsection, for the experiment of selecting a person at random from the 348 people represented by Table 5, the events "death in January" and "birth in February" are not independent.) Furthermore, for 10 of the 12 months, namely January, February, March, April, June, July, August, September, November, and December, we see that the probability of death in the month is decreased for persons born in the following month.

Thus, at least for these 348 notable Americans, the data appear to support the notion that famous people tend to postpone their deaths until after their birthdays. (The question of an association between birth month and death month can also be treated by the chi-squared method of Section 9.2. See Problem 8 of Chapter 9.)

Discussion Question 1. How could one examine the data of Table 5 to see if there is an *increase* in the death rate *after* the birth month?

$$P(D_2 \mid B_1) \neq P(D_2 \mid B_2) \cdots$$

Detecting Cases of Disease

For some diseases there are tests available that detect the disease in a person before the person gets sick enough to be aware of the disease himself. For example, a chest X-ray can reveal some diseases of the lungs, such as tuberculosis and lung cancer, before the person has any symptoms of disease. Blood tests for syphilis and the Pap smear for cancer of the cervix are other examples. Diagnostic tests like these are often used to screen populations in order to detect and treat diseases early. Early detection often increases the chance of cure, and, in cases of infectious disease, decreases the contagious spread of the disease through the population.

If a diagnostic test is to be useful in detecting disease, it must give correct answers when it is used. There are two kinds of errors that a test can make. If the test is given to a person who really has the disease, the test might be negative, indicating erroneously that the person is healthy. This would be called a **false-negative result.** If the test is given to a healthy, disease-free person, the test might be positive, indicating erroneously that the person has the disease. This would be called a **false-positive result.**

The possibilities of error in a diagnostic test can be readily appreciated. In reading chest X-rays it is obvious that X-rays of diseased lungs may not always clearly reveal the disease, and that the interpretation of the X-ray depends on the skill of the X-ray reader. Further, it should be easy to believe that certain artifacts and transient lung abnormalities seen in an X-ray may be interpreted erroneously as disease, especially by the less-experienced reader.

Before a diagnostic test is used in practice, it is important to evaluate the rates of error that are experienced with the test. This is done by using the test on persons whose disease status is known by other means. For example, chest X-rays may be taken on several hundred people known to have tuberculosis and another several hundred known to be free of TB. The X-rays would be read by someone who was not aware of the disease status of the people; the X-rays would be presented to him in random order, so that he would not be reading all the X-rays in one group together. After the X-rays were read and a judgment of TB made on each, the results could be tabulated to see how many errors were made. Yerushalmy, Harkness, Cope, and Kennedy (1950) studied the diagnostic errors made by six readers in reading chest X-rays for TB. Table 6 contains the results for one of the six readers. Note that in 1790 X-rays of persons without disease, 51 false-positive errors were made, an error rate of approximately 3%. Among X-rays of persons with TB, the error rate was 8 in 30, a false-negative error rate of 27%.

TABLE 6.   False-positive and False-negative Errors in a Study of X-ray Readings

	Persons without TB	Persons with TB	
Negative X-ray reading	1739	8	1747
Positive X-ray reading	51	22	73
	1790	30	1820

Source: J. Yerushalmy, J. T. Harkness, J. H. Cope, and B. R. Kennedy (1950).

In interpreting the results of Table 6, it is helpful to define certain events. Suppose we select a person at random from the 1820 people represented by Table 6. Let B denote the event that the person tested has TB and let A denote the event that the chest X-ray "states" that the person X-rayed has TB. That is:

A: person tested has "positive" chest X-ray

B: person tested has TB.

From Table 6 we have

$$Pr(A|B) = \frac{22}{30} = .733,$$

$$Pr(\text{not } A|\text{not } B) = \frac{1739}{1790} = .972.$$

Note that $Pr(A|B)$ is the rate of correct decision on persons with TB. In diagnostic testing, the rate of correct decision on persons with disease is defined as the **sensitivity** of the test. For the data of Table 6, the sensitivity is 73%. The rate of correct decisions for persons without disease is defined as the **specificity** of the test. For the data of Table 6, the specificity is $Pr(\text{not } A|\text{not } B) = 97\%$.

Although it is informative to determine the sensitivity and specificity, these values must be interpreted cautiously as measures of the diagnostic test's perfor- mance. For example, direct inspection of Table 6 reveals that only 22 of the 73 persons with positive X-ray findings had TB:

$$Pr(B|A) = \frac{22}{73} = .30.$$

Of course, people who have positive tests will undergo further diagnostic evalua- tion, rectifying many of the false-positive errors. The eight persons whose TB was not detected would be less fortunate, missing an opportunity for treatment until the next test or until clinical symptoms become manifest.

More Difficult Probability Calculations—Bayes' Theorem

Bayes' theorem enables one to calculate $Pr(B|A)$ by the following formula:

(3) $$Pr(B|A) = \frac{Pr(A|B)Pr(B)}{Pr(A|B)\,Pr(B) + Pr(A|\text{not } B)\,Pr(\text{not } B)}.$$

Letting B denote the event "person has disease" and A denote the event "person's test is positive," we see that equation (3) yields a general formula for calculating the probability of disease for a person with a positive test reading. Similarly, the conditional probability of disease, for a person with a negative reading, is

(4) $$Pr(B|\text{not } A) = \frac{Pr(\text{not } A|B)Pr(B)}{Pr(\text{not } A|B)Pr(B) + Pr(\text{not } A|\text{not } B)Pr(\text{not } B)}.$$

For the TB example of the previous subsection we have

$$Pr(A|B) = .733,$$

$$Pr(B) = .016,$$

(5) $$Pr(A|\text{not } B) = 1 - Pr(\text{not } A|\text{not } B) = 1 - .972 = .028, \text{ (see Comment 5)},$$
$$Pr(\text{not } B) = 1 - Pr(B) = 1 - .016 = .984,$$

$$Pr(\text{not } A|B) = 1 - Pr(A|B) = .267,$$

$$Pr(\text{not } A|\text{not } B) = .972.$$

We note that $Pr(B)$ is the prevalence rate for TB in the group considered by Yerushalmy *et al.* In general, the percentage of persons in a group who have the disease (that the diagnostic test is attempting to detect) is called the **prevalence rate** for the disease in the group.

Substituting the specific values given by (5) into formulas (3) and (4) we obtain:

$$Pr(B|A) = \frac{(.733)(.016)}{(.733)(.016) + (.028)(.984)} = .30,$$

$$Pr(B|\text{not } A) = \frac{(.267)(.016)}{(.267)(.016) + (.972)(.984)} = .004.$$

Thus 30% of all positive X-rays might be expected to belong to persons actually having TB, and 0.4% of all negative X-rays might belong to persons actually having the disease.

Of course, formulas (3) and (4) can be applied to any disease. They can be used to calculate the probability of disease, conditional on the test result, if the prevalence rate of the disease and the false-positive and false-negative error rates of the test are known (equivalently, if the prevalence rate of the disease and the specificity and sensitivity of the test are known).

It should be emphasized that the study of Yerushalmy *et al.* was performed to estimate the specificity and sensitivity of a diagnostic procedure, namely, the chest X-ray. In such studies, the proportion of cases of disease may be higher than found in a natural population. This is purposely done to provide a better estimate of the sensitivity of the test. Hence the proportion of cases with the disease often does not furnish a realistic value for the prevalence rate of the disease in the population to be screened, and this rate must be obtained from other sources.

Discussion Question 2.   In the chest X-ray test for TB, how could the false-negative rate be decreased?

Comments

1.   Note that in our X-ray problem, $Pr(B|A)$, the conditional probability that the person has TB, given that the person's test is positive, is not equal to $Pr(B)$, the (unconditional) probability that the person has TB. The former value is .30, whereas the latter value is .016. Thus the events "person has TB" and "person's test result is positive" are *not* independent. This corresponds with our intuition. Knowing that a person's test is positive leads to a change (in this case an increase from .016 to .30) of the probability that the person has TB.

2.   Formulas (3) and (4) are special cases of a probabilistic result known as *Bayes' theorem*. If the reader is interested in more advanced concepts of the mathematical theory of probability, we suggest Parzen's (1960) standard text. In particular, see page 119 of Parzen's text for a formal statement of Bayes' theorem.

3.   See Chapter 1 of Fleiss (1973) for more discussion on diagnostic screening tests. [We note here that in his discussion of screening tests, Fleiss' terminology is different from ours. Letting B denote the event "person has disease" and A denote the event "person's test is positive," we call $Pr(A|\text{not } B)$ the false-positive rate. However, Fleiss defines $Pr(\text{not } B|A)$ to be the false-positive rate. Similarly, whereas we call $Pr(\text{not } A|B)$ the false-negative rate, Fleiss defines $Pr(B|\text{not } A)$ to be the false-negative rate. Our usage follows that of the World Health Organization as presented by Wilson and Junger (1968).]

4.   It is important to mention that the computed probabilities in the examples of this chapter correspond to probabilities based on a random selection from a given table of entries (digits, Table 1; cases, Table 2; fruit flies, Table 3; notable Americans, Table 5; persons, Table 6). If one wishes to interpret these probabilities for a larger group than the one represented by the given table, then the probabilities must be interpreted as *estimated* probabilities of *unknown* true probabilities. Thus, for example, whereas Table 3 represents the probabilities of death for various age intervals for a fly selected at random from a group of 270 males, if one wishes to view these probabilities in terms of the conceptual population of all male *Drosophila melanogaster*, then the probabilities of Table 3 are only *estimates* of true *unknown* probabilities corresponding to the larger conceptual population.

5.   In display (5) we calculated $Pr(A|\text{not } B)$, using the value of $Pr(\text{not } A|\text{not } B)$ and the equation

(6) $$Pr(A|\text{not } B) = 1 - Pr(\text{not } A|\text{not } B).$$

Equation (6) is analogous to our Rule IV: $Pr(\text{not } A) = 1 - Pr(A)$, but in (6) the reader will note the presence of the conditioning event "not B". In (6) we are using the following extension of Rule IV:

Rule IV*: For any event H,

$$Pr(\text{not } A|H) = 1 - Pr(A|H).$$

Rules I, II, III, and III′ have analogous conditional versions, namely:

Rule I*:   For any event H, $Pr(A|H) = 1$, if, given that the event H has occurred, A is certain to occur

Rule II*:   For any event H, $Pr(A|H) = 0$ if, given that the event H has occurred, it is impossible for A to occur

Rule III*: For any event H, $Pr(A \text{ or } B|H) = Pr(A|H) + Pr(B|H)$ if, given that the event H has occurred, events A and B cannot occur simultaneously

Rule III′*: For any event H,

$$Pr(A \text{ or } B|H) = Pr(A|H) + Pr(B|H) - Pr(A \text{ and } B|H).$$

Summary of Statistical Concepts

In this chapter the reader should have learned that:

(i)   The probability of an event is a number between 0 and 1 indicating how likely the event is to occur.

(ii) Probabilities can be interpreted in terms of long-run frequencies.

(iii) There are useful rules for calculating the probabilities of events.

(iv) The probability of an event B may change when it is known that event A has occurred; that is, $Pr(B|A)$ need not equal $Pr(B)$. When $Pr(B|A) = Pr(B)$, then A and B are said to be independent.

PROBLEMS

Problems 1 and 2 refer to Table 7, which gives the survival experience of 275 female *Drosophila melanogaster*.

1. Calculate the probability that a female fruit fly selected at random from the 275 represented by Table 7 will die by age 20.

2. Calculate the probability that a female fruit fly will survive beyond age 40, given that she has survived beyond age 30.

TABLE 7.   Survival Experience of Female
Drosophila melanogaster

Age Interval (in days)	Number Dying in Interval
0–10	11
10–20	10
20–30	35
30–40	99
40–50	89
50–60	31

Source: R. S. Miller and J. L. Thomas (1958).

3. Return to Table 1 and calculate the probability that a randomly selected digit will be a 0, given that it is in row 1 of the table.

4. Note that, for the "Selecting a digit at random" example, $Pr(A_0) + Pr(A_1) + \cdots + Pr(A_9) = 1.00$. Explain why this result is consistent with Rules I and III of probability.

5. A particular screening test for diabetes has been reported by Wilson and Jungner (1968) to have sensitivity of 52.9% and specificity of 99.4%. The prevalence of cases in the United States, for all ages, is approximately eight per thousand. For a positive test result, what is the probability that the person actually has diabetes? If the result is negative, what is the probability that the person does not have the disease?

6. Suppose that the interpretation of the diabetes test in Problem 5 can be changed so as to declare the test positive at a lower threshold. This would increase sensitivity but decrease the specificity. Suppose the new rates were 80 and 95%, respectively. How would this change the answers to the questions raised in Problem 5?

7. Suppose there exists a diagnostic test for disease D with the properties that $Pr(+|D) = .95$ and $Pr(-|\text{not } D) = .95$ where $+$ and $-$ denote positive and negative results for the test, respectively. Suppose the prevalence rate for disease D is $Pr(D) = .001$. Find $Pr(D|+)$ and $Pr(D|-)$.

•8.† Suppose, for a given diagnostic test for disease D, we have $Pr(+|D) = Pr(-|\text{not } D) = s$. Suppose also the prevalence rate for disease D is $Pr(D) = .001$. What value must s be in order that $Pr(D|+) = .90$?

•9. Recall that when events A and B are independent we have

$$Pr(A \text{ and } B) = Pr(A) \cdot Pr(B).$$

† In this text, problems marked by • are either relatively difficult or computationally tedious.

Show directly, or illustrate by an example, that when A and B are independent we also have

$$Pr(\text{not } A \text{ and not } B) = Pr(\text{not } A) \cdot Pr(\text{not } B).$$

That is, not A and not B are also independent.

10. In the screening test for diabetes (Problem 5), are the events "person has diabetes" and "person's test result is negative" independent events? Justify your answer.

11. Explain why, in the diagnostic test for TB, having X-rays read independently by several persons could lower the false-negative rate. How could one lower the false-positive rate?

12. Define false-negative result, false-positive result, sensitivity, specificity, prevalence rate, probability, conditional probability, and independent events.

Supplementary Problems

Questions 1–3 refer to Table 8 which contains counts of numbers of individuals as given in the 1960 census for the state of Florida, as reported on page 28 of the *Statistical Abstract of the United States 1966*.

TABLE 8.   1960 Population of Florida by Race and Sex

Sex Race	Male	Female
White	2,000,593	2,063,288
Negro	432,107	448,079
Other races	4,083	3,410

Source: *Statistical Abstract of the United States* (1966).

Suppose we select an individual at random from the population represented by Table 8.

1. What is the probability that the individual selected will be white? What is the probability that the individual selected will be white and female? What is the probability that the individual selected will be white or female?

2. Given that the selected individual is a female, what is the probability that she is white?

3. Are the events "individual is male" and "individual is white" independent? (Calculate probabilities to four decimal places.)

4. Suppose that for the events A and B we have $Pr(A) = .2$, $Pr(B) = .3$, and $Pr(B|A) = .1$. Find $Pr(A \text{ or } B)$.

5. Let A be the event a person is left-handed. Let B be the event a person is female. Express the statement "10% of all males are left-handed" in terms of a conditional probability.

6. Questions 6–10 refer to the data in Table 9, which are taken from Lindley (1965) and are also considered by Savage (1968).

In a routine eyesight examination of 8-year-old Glasgow schoolchildren in 1955 the children were divided into two categories, those who wore spectacles (A), and those who did not (not A). As a result, visual acuity was classified as good (G), fair (F), or bad (B). The children wearing spectacles were tested without them. The results are in Table 9.

TABLE 9. Visual Acuity of Glasgow Schoolchildren (1955)

	A		not A	
	Boys (*m*)	Girls (*f*)	Boys (*m*)	Girls (*f*)
Category				
G	90	81	5908	5630
F	232	222	1873	2010
B	219	211	576	612
Total:	541	514	8357	8252

Source: D. V. Lindley (1965).

Suppose a child is selected at random from the 17,664 represented in Table 9. Compute the probability that the child is a boy. Compute the probability that the child is a boy or wears spectacles.

7. To examine the relationship between spectacle usage and sex compute $Pr(A|m)$, $Pr(A|f)$, $Pr(\text{not } A|m)$, $Pr(\text{not } A|f)$. In particular, compare $Pr(A|m)$ with $Pr(A|f)$ and compare $Pr(\text{not } A|m)$ with $Pr(\text{not } A|f)$.

8. To consider the relationship between visual acuity and sex, compute $Pr(G)$, $Pr(F)$, $Pr(B)$, $Pr(m|G)$, $Pr(m|F)$, $Pr(m|B)$, $Pr(f|G)$, $Pr(f|F)$, $Pr(f|B)$. In particular, compare $Pr(m|G)$ with $Pr(f|G)$; compare $Pr(m|F)$ with $Pr(f|F)$; compare $Pr(m|B)$ with $Pr(f|B)$.

•9.   To consider the relationship of visual acuity and spectacle usage compare:

$$Pr(A|G) \text{ with } Pr(\text{not } A|G),$$

$$Pr(A|F) \text{ with } Pr(\text{not } A|F),$$

$$Pr(A|B) \text{ with } Pr(\text{not } A|B),$$

$$Pr(G|A) \text{ with } Pr(G|\text{not } A),$$

$$Pr(F|A) \text{ with } Pr(F|\text{not } A),$$

$$Pr(B|A) \text{ with } Pr(B|\text{not } A).$$

10.   Are the events "child is a boy" and "child wears spectacles" independent?

11.   Data concerning 45 workers in an industrial plant were punched on tabulating cards, with one card pertaining to each individual. The following table indicates the sex and skill classification of these workers:

Skill Classification	Sex	
	Male	Female
Unskilled	14	9
Semi-skilled	10	11
Skilled	1	0

a.   If a card is selected at random, what is the probability that it pertains to either a female or semi-skilled worker?

b.   What is the probability that the randomly selected card pertains to an unskilled worker?

c.   What is the probability that the randomly selected card pertains to a semi-skilled male?

d.   Given that the randomly selected card pertains to a female, what is the probability that it pertains to an unskilled worker?

e.   What is the probability that the randomly selected card does not pertain to a semi-skilled worker?

•12.   Urn I contains five white and seven black balls. Urn II contains four white and two black balls. An urn is selected at random, and a ball is drawn from it. Given that the ball drawn is white, what is the probability that urn I was chosen?

•13.   A device for testing radio tubes can always detect a bad tube, but 3% of the time it claims that a tube is bad, when, in fact, it is a good tube. If 95% of manufactured tubes are good, what is the probability that if a new tube is selected at random and the testing device claims that it is a bad tube, the tube actually is a good one?

REFERENCES

*Chiang, C. L. (1968). *Introduction to Stochastic Processes in Biostatistics*. Wiley, New York.

Fleiss, J. L. (1973). *Statistical Methods for Rates and Proportions*. Wiley, New York.

*Lindley, D. V. (1965). *Introduction to Probability and Statistics from a Bayesian Viewpoint, Part* 2: *Inference*. Cambridge University Press, Cambridge.

Miller, R. S. and Thomas, J. L. (1958). The effect of larval crowding and body size on the longevity of adult *Drosophila melanogaster*. *Ecology* **39,** 118–125.

Morris, R. B. (Ed.) (1965). *Four Hundred Notable Americans*. Harper & Row, New York.

*Parzen, E. (1960). *Modern Probability Theory and its Applications*. Wiley, New York.

Phillips, D. P. (1972). Deathday and birthday: An unexpected connection. In *Statistics: A Guide to the Unknown*, Tanur, J. M. *et al.* (Eds.) pp. 52–65. Holden-Day, San Francisco.

*Savage, I. R. (1968). *Statistics: Uncertainty and Behavior*. Houghton Mifflin, Boston.

U.S. Bureau of the Census (1966). *Statistical Abstract of the United States* 1966. (87th edition). Washington, D.C.

Wilson, J. M. G. and Jungner, G. (1968). *Principles and Practice of Screening for Disease*. World Health Organization, Geneva.

Yerushalmy, J., Harkness, J. T., Cope, J. H., and Kennedy, B. R. (1950). The role of dual reading in mass radiography. *Amer. Rev. Tuberc.* **61,** 443–464.

CHAPTER 3

Populations, Samples, and the Distribution of the Sample Mean

In this chapter the most fundamental concepts in the field of statistics are defined and discussed. In Section 1 the idea of a population of objects is introduced, with an associated distribution for measurements attached to the objects. The mean and standard deviation of a distribution are defined, an inequality due to Tchebycheff is presented, and the normal distribution is introduced.

In the second section the idea of random selection of a sample of objects from the population is developed and practical procedures for obtaining random samples are discussed.

Finally, in Section 3 the concept of the distribution of the sample mean is elaborated and the characteristics of this distribution are given, including the results of the famous *central limit theorem*.

1.	**POPULATIONS AND DISTRIBUTIONS**

Examples of Populations

Consider an existing group of objects, each with a measurement associated with it. For example, we might consider the file of all hospital bills mailed out by the Stanford Medical Center for a specified year, the measurement being the total amount of the bill. Another example would be the group of all adult males residing within the limits of some city, with their individual serum cholesterol values. It is immediately clear that our examples are strictly conceptual until operational definitions are supplied. Suppose that we decided to determine and record the measurement of each object in one of these populations, say, the first. Definitions of hospitalization, bill, date sent, and amount are necessary. None of these is as easily set down as it might seem. Further, there is the problem of actually finding each and every qualified bill in the files and recording its amount once and only once. The difficulties in defining and obtaining the serum cholesterol value for each and every adult male in a specific city are even more obvious. In particular, a person's serum cholesterol, like his weight, varies somewhat from one day to the next, and the definition has to acknowledge this, perhaps by specifying a date of measurement. Nevertheless, in both examples the concept of a group of objects, each with a measurement attached to it, is clear, and it is clear that we could set up the necessary definitions and procedures for actually enumerating the whole group of objects and obtaining the corresponding measurement of each object in the group, given the time and money to do the job. Such a conceptual group of objects is called a **population**. If the group consists of a finite number of objects, it is called a **finite population**, in the parlance of the sample survey statisticians.

The Frequency Table

Although, ordinarily, we can never know all the measurements in a large population, we consider now how we might describe these measurements if they were available to us. Suppose, then, that we do have available to us the serum cholesterol value for every adult male in a specified city, one measurement per man, some tens of thousands of measurements altogether.* Listing the men

---

* To our knowledge no one has measured the cholesterol levels of all the residents in a fair-sized city, thus the measurements we deal with are fictitious, though closely resembling sample data collected by Farquhar (1974) in three California communities.

alphabetically by name, each with his cholesterol value, would not be a helpful description, since the list would be too long to study easily. Listing the cholesterol values in their order of magnitude would be somewhat helpful, since the smallest values and the largest values would be apparent, and the more typical values in the middle could also be determined without too much trouble. Even more helpful would be a table showing the number of men, or the percentage of men, with values falling into certain rather coarse intervals on the serum cholesterol measurement scale. Table 1 illustrates this method of tabling, through the

TABLE 1. Hypothetical Results of the Measurement of Serum Cholesterol on All the Adult Males in a Medium Sized City

Cholesterol Measurement Interval (mgm%)	Frequency f (Number of Males with Measurements Falling in the Interval)	Relative Frequency	Cumulative Frequency c	Cumulative Relative Frequency
75–99	317	0.006	317	0.006
100–124	622	0.012	939	0.019[a]
125–149	1427	0.028	2,366	0.047
150–174	2681	0.053	5,047	0.100
175–199	3638	0.072	8,685	0.172
200–224	4929	0.097	13,614	0.269
225–249	6332	0.125	19,946	0.394
250–274	6521	0.129	26,467	0.523
275–299	5714	0.113	32,181	0.636
300–324	4892	0.097	37,073	0.733
325–349	4622	0.091	41,695	0.824
350–374	3317	0.066	45,012	0.890
375–399	2473	0.049	47,485	0.939
400–424	1620	0.032	49,105	0.971
425–449	493	0.010	49,598	0.981
450–549	980	0.019	50,578	1.000
	Total: 50,578			

[a] The cumulative relative frequency for a given interval is equal to the sum of the relative frequency for that interval and the relative frequencies for all preceding intervals. This is not always reflected in the column 3 and column 5 values of Table 1 because these values have been rounded to three decimal places. Thus, for example, the cumulative relative frequency for the interval 100–124 is 939/50,578, and this is equal to the sum of the relative frequencies 317/50,578 of interval 75–99 and 622/50,578 of interval 100–124. However, when these three fractions are rounded to three decimal places, we obtain 0.019 for the cumulative frequency, and 0.006 and 0.012 for the relative frequencies.

frequencies themselves are fictitious. Table 1 is called a **frequency table**. The extreme-left-hand column shows the measurement values that have been grouped together to form each interval. Measurements are taken to the nearest unit so that a value of 80 represents a measurement in the interval 79.5–80.5. The interval denoted by 75–99 represents measurements in the range of 74.5–99.5. The midpoint of the interval is 87.0.

The second column of Table 1 gives the **frequency** of each interval, that is, the number of males with cholesterol measurements that fall in the interval. Thus the entry in column 2, line 1, shows that 317 males (out of the total 50,578 males), have cholesterol values in the range of 74.5 to 99.5. We use the symbol f for frequency.

The third column of Table 1 gives, for each interval, the **relative frequency** of that interval. The relative frequency of an interval is defined to be the frequency of the interval divided by the total number of measurements. If we denote the total number of measurements by N we have

$$\text{relative frequency} = \frac{f}{N}.$$

Thus the relative frequency of the first interval is

$$\frac{317}{50,578} = .006,$$

or, in terms of percent, .6%. The relative frequency of the second interval is

$$\frac{622}{50,578} = .012,$$

or, in terms of percent, 1.2%.

The fourth column of Table 1 gives, for each interval, the **cumulative frequency**, the number of men having serum cholesterol values in that interval or a lower interval. For example, the cumulative frequency for the interval 125–149 is

$$317 + 622 + 1427 = 2366.$$

We use the symbol c for cumulative frequency.

The fifth, and final, column of Table 1 gives for each interval the **cumulative relative frequency**, the proportion of men having serum cholesterol values in that interval or a lower interval. For example, the cumulative relative frequency for the interval 125–149 is

$$\frac{317 + 622 + 1427}{50,578} = \frac{2366}{50,578} = .047.$$

That is, in the medium sized city, 4.7% of the men have cholesterol values between 74.5 and 149.5. Note that the cumulative relative frequency of an interval is simply the cumulative frequency of the interval divided by N. That is,

$$\text{cumulative relative frequency} = \frac{c}{N}.$$

The Histogram, Cumulative Frequency Polygon, and Frequency Polygon

Figures 1 and 2 show, in graphical form, the third and fifth columns of Table 1, respectively. For Figure 2, the values of column 5 are multiplied by 100 so that

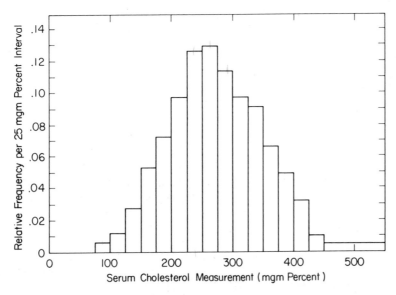

FIGURE 1.   Histogram of hypothetical results of the measurement of serum cholesterol for each of 50,578 adult males in a medium-sized city.

they are percentages. The graphs are called a **histogram** and a **cumulative relative frequency polygon**, respectively. (See Comment 1.) They are said to portray the **distribution** of cholesterol measurements in the population. Sometimes, for brevity, we refer to the distribution of measurements in the population as the population of measurements.

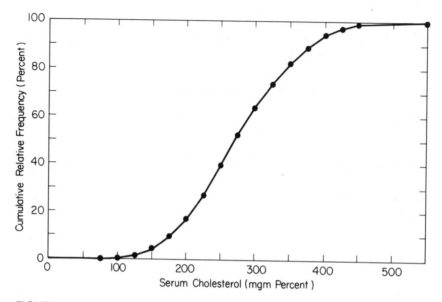

FIGURE 2. Cumulative relative frequency polygon showing the hypothetical results of the measurement of serum cholesterol on all the 50,578 adult males in a medium-sized city.

Figure 3 portrays a hypothetical histogram corresponding to a situation in which there are a large number of measurements finely grouped (so that the class intervals are very narrow). The smooth curve drawn through the tops of the bars in Figure 3 can be thought of as an idealized version of the histogram. From Figure 3 (or by sketching a smooth curve through the tops of the bars in the serum

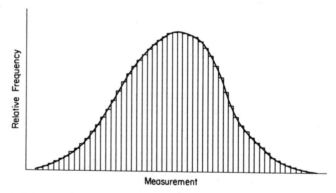

FIGURE 3. A smooth approximation to a hypothetical histogram that is based on a large number of finely grouped measurements.

cholesterol histogram of Figure 1) it is easy to see why the distribution of values among the objects in a large, finite population is often conceptualized as an infinite population represented by a rather smooth curve rising from zero on the left, attaining a maximum, and then declining smoothly to zero again on the right. Of course, the shape of the curve varies from one population to another, but the smoothness is observed in most instances where the measurement is made relatively precisely and the number of objects in the population is quite large.

Rather than simply "sketch" a curve through the tops of the bars of a histogram, most statisticians prefer the following procedure, which yields a specific curve—the **frequency polygon**—to represent the histogram. For each "bar top" of the histogram, determine the midpoint. Then connect adjacent midpoints, of bar tops, by straight lines. This is illustrated in Figure 4, using the histogram of Figure 1.

Probability Density Function

Note that the curve in Figure 3 is nonnegative (it never dips below the measurement axis) and that it includes a finite area between itself and the measurement

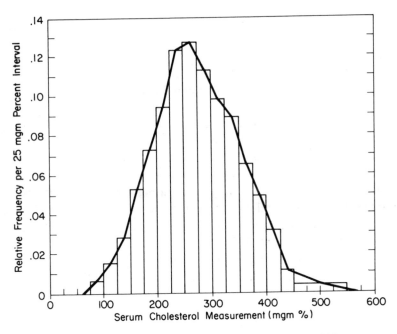

FIGURE 4.   Frequency polygon for the serum cholesterol histogram of Figure 1.

axis. Any curve having these properties can be viewed as an approximation to the histogram of a measurement (X, say) with a large number of possible values. Such a curve is called a **density**, or more technically, a probability density function. (Roughly speaking, it is a histogram for a conceptually infinite population.) For any two numbers a and b, the probability that the measurement X is between a and b is equal to the proportion of the area under the density between a and b. In symbols

$$Pr(a \leq X \leq b) = \frac{\text{area under curve between } a \text{ and } b}{\text{total area under curve}}.$$

Typically it is difficult to compute areas under curves. But for some densities the task is rather easy. In particular, consider the rectangular density given by Figure 5. The rectangular density could approximate the following situation. A group of

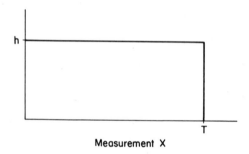

Measurement X

FIGURE 5.   Rectangular density.

patients who have had heart attacks undergo an exercise program. The program is scheduled to last one year, after which the patients are asked to continue to exercise, but not under supervision. The physician supervising the program has noticed that patients tend to drop out (stop coming to the scheduled exercise sessions) on a uniform basis. Letting the T in Figure 5 denote 365, we can ask, what is the chance that a patient will drop out between the 300th and 325th day? This chance is equal to the area under the curve between 300 and 325, divided by the total area between 0 and 365.

The total area of the rectangle with base of length 365 and height h in Figure 6 is $365 \times h$. The area of the shaded rectangle with base $325 - 300 = 25$ and height h is $25 \times h$. Thus, letting X denote the time measured from 0 at which the patient drops out, we have

$$Pr(300 \leq X \leq 325) = \frac{25 \times h}{365 \times h} = \frac{25}{365} = .068.$$

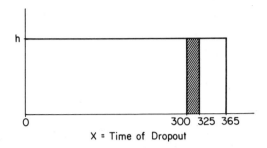

X = Time of Dropout

FIGURE 6. Rectangular density on the interval 0 to 365.

It is convenient to scale a probability density curve so that the total area under the curve equals 1. Then areas under the curve are exactly equal to probabilities, rather than just being proportional to probabilities. That is,

$$Pr(a \leq X \leq b) = \text{area under the curve between } a \text{ and } b.$$

From now on when we use the term probability density, it refers to a curve where the total area under the curve equals one.

As we have already mentioned, not all densities lend themselves to area calculations as easily as does the rectangular. For the important normal curve, to be encountered in a later subsection of this chapter, areas under the curve are obtained by entering an appropriate table.

Defining a Measure of Location for the Population

The distribution of measurements in a population could be presented in tabular or graphical form, but these forms suggest that certain characteristics of the distribution of measurements in the population might be abstracted as descriptive of the whole population. In particular, it is customary to think of a typical value for the group (usually some sort of average) that locates the center of the histogram. In Figure 1 it is clear that the center of the histogram is in the middle of the 200–300 range. The two most popular measures of location are the **population mean** and the **population median.**

The familiar mean is the balancing point or center of gravity of the histogram. Suppose the histogram were drawn on some stiff material of uniform density, such as plywood or metal, cut out, and balanced on a knife edge arranged perpendicular to the measurement scale, as shown in Figure 7. The X value corresponding to the point of balance is the mean.

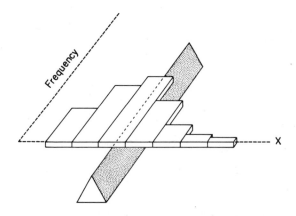

FIGURE 7.   The mean as a balancing point.

The median is the value that divides the area or frequency in the histogram into equal halves. For grouped data, the median can be approximated by the point on the horizontal axis of the graph of the cumulative relative frequency polygon at which the curve crosses the 50% level.

A value such as the mean or median of a population distribution is called a **parameter** of the population. A parameter is a constant associated with the conceptual population distribution and, since the whole set of population measurements is usually unknown, the parameter is unknown also. The business of statistical methodology is to make inferences about population parameters based on measurements of only a sample of objects from the whole population. The ideas of inference are explored in this chapter and the next.

For the moment, however, let us assume again that we do know all the measurements in the population. The population mean, denoted by μ (Greek mu), could then be computed. If the measurements are denoted by X's, and there are N measurements, then μ is obtained by summing the N values of X and dividing the total by N. We write

(1) $$\mu = \frac{\sum X}{N}.$$

[The summation symbol, $\sum$ (Greek upper case sigma), in (1) is used throughout this book. If you are not familiar with its use, or you need to refresh your memory, refer back to a college algebra book, such as the one by Hart (1967).] For example, if there are only five objects in the whole population and their measurements are 4, 19, 23, 38, 500, then

$$\mu = \frac{4+19+23+38+500}{5} = \frac{584}{5} = 116.8.$$

If the population measurements have been grouped for presentation in a frequency table, as in Table 1, the mean of the distribution can be approximated from the table by assigning to each individual in the population the value of the midpoint of the class interval in which his measurement lies. Letting m denote the midpoint of the interval and f the frequency of the interval, then the population mean, approximated by using grouped data, is

$$(2) \qquad \mu \approx \frac{\Sigma(f \cdot m)}{N} = \frac{\Sigma(f \cdot m)}{\Sigma f},$$

where the summation Σ is over the number of intervals.

We now illustrate the use of equation (2) with the cholesterol data. Table 2 breaks the calculation into the steps of multiplying m by f, for each interval. The second column of Table 2 gives for each interval the midpoint m. The third column gives for each interval the frequency f. Column 4 contains the values of $f \cdot m$, and column 5 contains the values of $f \cdot m^2$. Summing the values in column 4 gives the numerator of the right-hand-side of (2). Note that the denominator of (2) is the

TABLE 2.   Interval Midpoints and Frequencies for the Data of Table 1

Cholesterol Measurement Interval	Midpoint m	Frequency f	$f \cdot m$	$f \cdot m^2$
75–99	87	317	27,579	2,399,373
100–124	112	622	69,664	7,802,368
125–149	137	1427	195,499	26,783,363
150–174	162	2681	434,322	70,360,164
175–199	187	3638	680,306	127,217,222
200–224	212	4929	1,044,948	221,528,976
225–249	237	6332	1,500,684	355,662,108
250–274	262	6521	1,708,502	447,627,524
275–299	287	5714	1,639,918	470,656,466
300–324	312	4892	1,526,304	476,206,848
325–349	337	4622	1,557,614	524,915,918
350–374	362	3317	1,200,754	434,672,948
375–399	387	2473	957,051	370,378,737
400–424	412	1620	667,440	274,985,280
425–449	437	493	215,441	94,147,717
450–549	499.5	980	489,510	244,510,245
	Totals:	50,578	13,915,536	

sum of the values in column (3). We do not need the column 5 values to use equation (2), but we use them later (see Problem 4) when approximating the standard deviation; the latter is a parameter that remains to be introduced.

Summing the $f \cdot m$ entries of column 4 of Table 2 we have

$$\Sigma(f \cdot m) = 27{,}579 + 69{,}664 + \cdots + 489{,}510 = 13{,}915{,}536.$$

Thus, from equation (2),

$$\mu \approx \frac{13{,}915{,}536}{50{,}578} = 275.1.$$

See Problem 4 for a method (coding) that can reduce the computational work when the numbers are large, as in Table 2.

The median of a distribution can be calculated in several ways. If the number of individuals in the population is small, their measurements can be listed in increasing order and the value that has half the measurements below and half above can be selected by inspection. For example, if there were only five individuals involved and their measurements were 180, 180, 195, 223 and 286, the median would be 195, since two individuals have measurements below and two have measurements above this value. If there are an even number of individuals, with values, say of 9, 11, 18, 30, 32 and 50, any value between 18 and 30 would satisfy the definition. In this case, the average of the middle two values may be used, that is, $\frac{1}{2}(18 + 30) = 24$. If the number of measurements is large and the data have been grouped, the median can be approximated by inspecting the cumulative relative frequency polygon. For example, from Figure 2 it can be seen that the curve crosses the 50% cumulative relative frequency at a cholesterol level of approximately 269. A numerical procedure for interpolation is described in Problem 5.

Many scientists prefer the median to the mean as a measure of central tendency, because the median is much less sensitive to wildly large or small measurements. To illustrate this, consider the following example. Suppose the finite population consists of nine individuals with measurements 4, 19, 23, 38, 57, 58, 61, 63, 505. The median is 57, because four individuals have measurements below 57 and four have measurements above 57. The mean is

$$\frac{4 + 19 + 23 + 38 + 57 + 58 + 61 + 63 + 505}{9} = 92.$$

Note how the unusually large 505 value has more of an effect on the mean than it does on the median. In this simplified example, many users feel more comfortable with the value 57 rather than 92, as a measure of the center of the population.

Defining a Measure of Variation for the Population

An important characteristic of any distribution is the variation among the individual values, that is, the spread or range of the individual values. Either of the two methods of picturing the distribution, with a histogram or cumulative frequency polygon, will give a very satisfactory idea of the variation. For example, from Figure 1 it can be appreciated immediately that cholesterol values for males do vary from man to man, with most men having values between 150 and 400, and values under 100 or over 500 being very unusual. It is more difficult to arrive at some numerical description. The simplest solution is to state the minimum and maximum values among all the objects in the population. However, these values are not satisfactory without some idea of the shape of the curve in between, since the histogram might have very thin or very fat "tails" extending out to the minimum and maximum values. A much more satisfactory solution is to give several percentiles of the distribution. We define a *p*th **percentile** of a population to be a value such that *p* percent of the objects in the population have measurements below this value. (Note that this definition does not force a *p*th percentile to be unique—there may be an interval of values having the desired property. Note also that a median is a 50th percentile of the distribution.) Then it is clear that quotation of 5th, 25th, 75th, and 95th percentiles, for example, would reveal quite a lot about the variation in the population. In fact, from these few values one could sketch the cumulative frequency polygon. We use the notation $x_{p/100}$ to denote a *p*th percentile of the distribution of a variable X. For example, $x_{.45}$ will denote a 45th percentile. Often, we speak of a high percentile, such as $x_{.95}$, as an upper percentile, and we call it an upper fifth percentile. More generally, an **upper *p*th percentile** of a population is a value such that *p* percent of the objects in the population have measurements greater than or equal to this value.

The use of a set of percentiles to describe the variation in a distribution can be quite effective. However, for purposes of convenience and theoretical development, a single numerical measure of the variation of a population of values has been devised and is universally used. This measure of variation is called the standard deviation. It is based on the average squared deviation of the individual values from the population mean. The **population standard deviation** is usually denoted by σ (Greek lower case sigma), and it can be defined, for a finite population consisting of N objects, as

(3)
$$\sigma = \sqrt{\frac{\sum(X-\mu)^2}{N}},$$

where the parameter μ, defined by equation (1), is the mean of the finite population.

To compute σ proceed as follows:

1. Compute the mean μ.
2. Subtract μ from each measurement X, obtaining the difference $X - \mu$.
3. Square each $X - \mu$ difference to obtain $(X - \mu)^2$.
4. Sum the N squared differences and call the sum $\sum(X - \mu)^2$.
5. Divide the sum by N.
6. Now take the square root; this value is σ.

We illustrate these steps for a finite population of four objects with measurements 2, 11, 15, 8.

$$\mu = \frac{2 + 11 + 15 + 8}{4} = 9$$

X	$X - \mu$	$(X - \mu)^2$
2	-7	49
11	2	4
15	6	36
8	-1	1

$$\sum(X - \mu)^2 = 49 + 4 + 36 + 1 = 90.$$

Thus

$$\sigma = \sqrt{\frac{90}{4}} = \sqrt{22.5} = 4.74,$$

where the square root of 22.5 can be obtained from Table C1 of Appendix C.

An alternative way of computing σ, more suitable for use with certain calculators, is given by equation (4). (See Problem 3.)

The square of σ is also a parameter that is a measure of population variation. The parameter σ^2 is called the **population variance.**

Note that σ is in the same units as the original measurements X, and can never be negative since each term in the summation $\sum(X - \mu)^2$, being the square of a number, must be nonnegative.

If all objects in a population had identical values, then there would be no variation; that is, all values would be identical to the mean, and this would be reflected by a value of 0 for σ. On the other hand, it is clear that if the measurements in the distribution were widely spread, then some of the deviations from the mean would have to be large, and the standard deviation would have to be large. Thus in some sense the average squared deviation from the mean does measure the amount of variation in a distribution. Although the standard deviation seems much less informative than a set of percentiles for the distribution, there are ways of inferring much about the variation in a distribution from this single value, and these ways are described in the next subsection.

The Tchebycheff Inequality

The famous *Tchebycheff inequality*, an elementary but extremely useful inequality in the theory of mathematical probability, helps in the interpretation of the value of a standard deviation.

> *Tchebycheff's inequality.* *The area, under a histogram or probability density curve, that lies more than k standard deviations (that is, more than $k\sigma$) away from the mean cannot exceed $1/k^2$, where k is any positive constant.*

Thus, for example, if you are told that the population mean for serum cholesterol is 225 mgm% and the standard deviation is 75 mgm%, you know that no more than 25% (that is, $1/2^2$) of the individuals in the population have values more than 150 units (i.e., 2σ) away from the mean. In other words, at least 75% of the persons in the population have serum cholesterol values within the range 75–375 mgm%. Similarly, choosing $k = 3$, the inequality says that at least 8/9 of all persons have measurements within 225 units of the mean.

We can also interpret Tchebycheff's inequality in a probabilistic manner. Suppose again that the population mean for serum cholesterol is 225 mgm% and the standard deviation is 75 mgm%. Suppose a person is selected at random from the population, and we ask for the chance, or probability, that this person's serum cholesterol value X (say) will land within the range 75–375 mgm%. Then Tchebycheff's inequality states that this chance is at least .75. We write

$$Pr\{75 \leq X \leq 375\} \geq .75.$$

(This probability is to be interpreted in terms of long-run frequency, as indicated in Chapter 2.)

Tchebycheff's inequality is illustrated graphically in Figure 8.

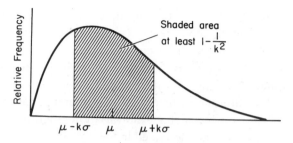

FIGURE 8. An illustration of the meaning of Tchebycheff's inequality. By Tchebycheff's inequality, the area under the curve (frequency), further than $k\sigma$ from μ, cannot be more than $1/k^2$ of the total area. Equivalently, the shaded area between $\mu - k\sigma$ and $\mu + k\sigma$ is at least $1 - 1/k^2$ of the total area. In terms of probabilities, if X is a randomly selected value from the population represented by the relative frequency curve, then $Pr\{\mu - k\sigma \leq X \leq \mu + k\sigma\} \geq 1 - 1/k^2$.

The Tchebycheff inequality gives bounds in terms of the standard deviation that are true for *any* distribution, and therefore, these bounds must necessarily be conservative for some distributions. That is, the actual area that lies more than k standard deviations from the mean may be substantially less than the upper bound $1/k^2$. In certain cases in which the shape of the distribution is known or can be well approximated, knowledge of the standard deviation can yield full knowledge of the variation in the distribution and one need not resort to what may be a conservative bound. Thus another way of inferring details about the variation of a distribution, from the value of σ alone, is based on knowledge of the shape of the distribution. The most familiar case is the normal (or Gaussian) distribution, which is bell-shaped.

The Normal Distribution

Many distributions encountered in practice have a symmetrical, bell-shaped look about them. This is true of the frequency polygon, for serum cholesterol, plotted in Figure 4. Figure 9, adapted from Hodges, Krech, and Crutchfield (1975), displays frequency polygons of three actual populations. These polygons, and the Figure 4 polygon, can be well approximated by a specific curve called the **normal**

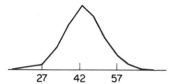

27   42   57

3830 marriages in California
with bride of age 40 to 44 (1961),
distributed by age of groom.

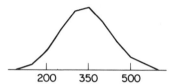

200   350   500

37,885 dairy cows (1949),
distributed by yearly pounds of
butterfat.

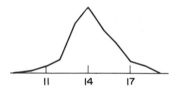

11   14   17

100 consecutive years (1841–
1940), distributed by mean June
temperature (Celsius) at Stockholm.

**FIGURE 9. Some distributions
that have a symmetrical, bell-
shaped look.**

(or Gaussian) **distribution**. We do not write out the mathematical function, which
specifies the curve, here. It is sufficient to know that the curve is a smooth,
bell-shaped symmetrical curve, as depicted in Figure 10, with some very special
properties. A normal curve depends on two parameters, the mean μ and the
standard deviation σ. The mean μ can be any number, and the standard deviation
σ can be any positive number. The curve has points of inflection (points where the
curve changes from cupping upward to cupping downward) exactly one standard
deviation σ distant from the mean μ, on either side, and the interval within one
standard deviation of the mean contains 68% of the total area (frequency) for the
distribution. Ninety-five percent of the total area lies within two (more precisely,

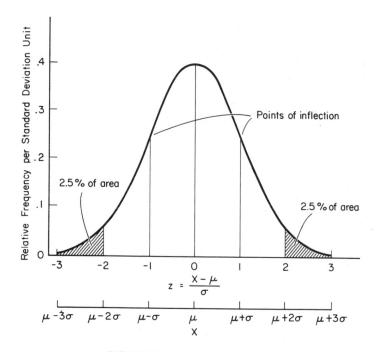

FIGURE 10.   Normal distribution.

1.96) standard deviations of the mean, and practically all (99.7%) lies within three standard deviations.

The normal distribution is symmetric about its mean μ. That is, for any positive value a, the area under the curve to the right of $\mu + a$ is equal to the area under the curve to the left of $\mu - a$.

Ideally, a normal distribution would be used to represent the distribution of a **continuous variable,** that is, a variable that assumes a continuum of values, as opposed to a variable that assumes only "separated" values (the latter is called a **discrete variable**). Actually, continuous variables exist only in theory; in practice, the values that are recorded are discrete, due to the limits of measurement precision, the rounding off of observations, and the grouping of measurements. Nevertheless, in certain cases in which measurements are made fairly precisely and the resulting histogram is reasonably symmetrical and bell-shaped, quite satisfactory approximations to the characteristics of the actual distribution can be obtained by assuming it is a normal distribution.

Table C2 (Appendix C) contains a detailed description of the distribution of areas for a standard normal distribution, that is, a normal distribution with mean $\mu = 0$ and standard deviation $\sigma = 1$. The point z_α is defined to be the point having the property that the area under the curve to the right of z_α is equal to α. Thus, for example, we write $z_{.0934} = 1.32$, because from Table C2 we find that the area under the curve to the right of 1.32 is equal to .0934.

The 95th percentile for any normal distribution is 1.645 standard deviations above the mean, the $97\frac{1}{2}$th percentile is 1.96 standard deviations above the mean, and the $99\frac{1}{2}$th percentile is 2.58 standard deviations above the mean.

From Table C2 it is possible to obtain the area under a normal curve in any specified interval if the mean μ and standard deviation σ of the distribution are specified. For example, suppose it is known that the distribution of cholesterol measurements in a population is approximately normal in form, with mean 225 and standard deviation 75. Then we can use Table C2 to calculate the percentage of people with cholesterol measurements between 200 and 350, or equivalently, to calculate the probability that a randomly selected person from the population has a cholesterol value X that lies between 200 and 350. We write this percentage or probability as

$$Pr\{200 \le X \le 350\}.$$

First we compute the **standardized deviates** (sometimes called standard scores) or Z values,

$$Z = \frac{X - \mu}{\sigma},$$

for the X values of 200 and 350. The standardized deviate for a given value is the distance of X from μ, measured in units of the standard deviation. Using the notation $X_1 = 200$ and $X_2 = 350$, we have

$$Z_1 = \frac{X_1 - \mu}{\sigma} = \frac{200 - 225}{75} = -.33,$$

$$Z_2 = \frac{X_2 - \mu}{\sigma} = \frac{350 - 225}{75} = 1.67.$$

Area = .3707 or 37.07 %   Area = .0475 or 4.75 %

-.33 0   1.67

Table C2 shows the area under the standard normal curve to the right of 1.67 is .0475. Since the total area under the standard normal curve is equal to 1, the area to the left of the point 1.67 is $1 - .0475 = .9525$. Also, by the symmetry of the normal distribution, we know that the area (under the curve) to the left of $-.33$ is equal to the area (under the curve) to the right of .33. The latter area is found from Table C2 to be .3707. The area between Z_1 and Z_2 is found by subtraction to be $.9525 - .3707 = .5818$. Thus approximately 58% of the people would have cholesterol values between 200 and 350. In terms of a probability statement, we have found

$$Pr\{200 \le X \le 350\} \approx .58.$$

Examples Illustrating the Use of the Standard Normal Distribution Table

1.   Find the area between 0 and .7 under a standard normal curve: From Table C2 we find that the area from .7 to $+\infty$ is .2420. Since the area from 0 to $+\infty$ is .5, we have

area between 0 and .7 = [area between 0 and $+\infty$]$-$[area between
.7 and $+\infty$]
$= .5000 - .2420 = .2580.$

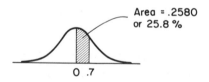

Area = .2580
or 25.8 %

0 .7

If Z is a measurement whose distribution is standard normal, then the probability that Z is between 0 and .7 is the area under the curve between 0 and .7. We write

$$Pr\{0 \le Z \le .7\} = .2580.$$

2.   Find the area between $-\infty$ and -1.7 under a standard normal curve: By the symmetry of the normal curve, the area between $-\infty$ and -1.7 is equal to the area between 1.7 and $+\infty$. The latter is found from Table C2 to be .0446.

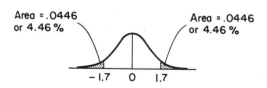

Area = .0446
or 4.46 %

Area = .0446
or 4.46 %

-1.7 0   1.7

If Z is a measurement whose distribution is standard normal, then the probability that Z is less than -1.7 is the area under the curve between $-\infty$ and -1.7. We write

$$Pr\{Z \le -1.7\} = .0446.$$

3.   Find the area between $-.2$ and $.7$ under a standard normal curve: The area (A, say) between $-.2$ and $.7$ is equal to $A_1 + A_2$ where A_1 is the area between $-.2$ and 0 and A_2 is the area between 0 and $.7$. In Example 1 of this subsection we found $A_2 = .2580$. Now

$$A_1 = [\text{area between } -\infty \text{ and } 0] - [\text{area between } -\infty \text{ and } -.2]$$

$$= .5000 - .4207 = .0793.$$

Thus,

$$A = A_1 + A_2 = .0793 + .2580 = .3373$$

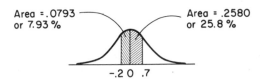

Area $=.0793$ or 7.93 %

Area $=.2580$ or 25.8 %

$-.2\ 0\ .7$

If Z is a measurement whose distribution is standard normal, then the probability that Z is between $-.2$ and $.7$ is the area under the curve between $-.2$ and $.7$. We write

$$Pr\{-.2 \le Z \le .7\} = .3373.$$

4.   Suppose white blood cell counts in a population of persons without discernible disease ("healthy" people) are known to have a mean of 8000 and a standard deviation of 1500. Suppose further that "healthy" counts have a distribution well-approximated by a normal curve. To determine, for example, the proportion of counts to be expected in the interval 10,000–12,000, proceed as follows. First, convert the raw values 10,000 and 12,000 to standardized deviates. We have

$$Z_1 = \frac{10,000 - 8000}{1500} = 1.33,$$

and

$$Z_2 = \frac{12,000 - 8000}{1500} = 2.67.$$

We need to determine the area under a standard normal curve between Z_1 and Z_2. This area (A, say) is equal to $A_1 - A_2$ where

$$A_1 = \text{area between 1.33 and } \infty,$$

$$A_2 = \text{area between 2.67 and } \infty.$$

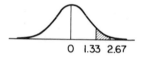

O 1.33 2.67

From Table C2 we find

$$A_1 = .0918, \qquad A_2 = .0038,$$

and

$$A = A_1 - A_2 = .0918 - .0038 = .0880.$$

Hence approximately 8.8% of the counts fall in the interval 10,000–12,000. Expressed in terms of probabilities, if X is the white blood count of a randomly selected person from the "healthy" population, then

$$Pr\{10,000 \leq X \leq 12,000\} \approx .0880.$$

The Coefficient of Variation

For distributions of values that are all positive (e.g., height, weight, heart rate), the amount of variation is sometimes expressed relative to the magnitude of the mean of the distribution, usually in terms of a percentage. If the variation is described by giving a value for the standard deviation, then the **coefficient of variation**, C.V. $= 100(\sigma/\mu)\%$, is often quoted. Since the coefficient of variation is dimensionless, it is useful for comparing variations of distributions measured in different units. For example, Farquhar (1974) found the following values in a population of adults between the ages of 35 and 60 in a California community south of Stanford University:

Variable	Mean	Standard Deviation	Coefficient of Variation(%)
Serum triglyceride	153.6	104.1	67.8
Serum cholesterol	214.8	34.2	15.9
Systolic blood pressure	135.6	17.6	13.0
Diastolic blood pressure	86.4	10.2	11.8

Note that systolic blood pressure has a much greater standard deviation than diastolic pressure, but the variations of the two measures are quite comparable when compared as percentage variations relative to their respective means. The variation of serum cholesterol is only slightly greater than the variation in blood pressure, but serum triglyceride is highly variable, because it depends so much on what has been eaten within hours prior to the blood sampling.

Discussion Question 1. What difficulties might be encountered if, when constructing a histogram, a large number of class intervals is used?

Discussion Question 2. Can you think of some populations of measurements that would not be well-approximated by a normal curve?

Comments

1. In the histogram, it is important to realize that the *area* is proportional to the frequency. This means that the height of each bar represents the frequency *per specified unit*. If all bars in the histogram are of equal width, the distinction is not important, but if the intervals for the bars in the histogram are unequal, the heights of the bars must be adjusted to make the *areas* proportional to *frequency*.

In Figure 1 the last interval has width 100, whereas the other intervals are of width 25. If the areas are to represent frequency and the heights are expressed as frequency per 25 unit interval, then the height for the last interval should be $(.019/100) \times 25$. Since the interval has width 100 units and the frequencies are expressed per 25-unit interval, the area is $(.019/100) \times 25 \times (100/25)$ or .019 for the whole 100-unit interval.

In general, if the widths of the intervals in a frequency table are L and the frequencies are f, then the heights of the bars in the histogram should be $(f/L) \cdot L^*$, where L^* is a conveniently chosen standard width.

The cumulative relative frequency polygon is graphed as follows. At the right-hand end point of each class interval, a dot is placed at a height equal to 100 times the cumulative relative frequency of that interval. Thus one plots cumulative relative frequency percentages. Then the successive dots are connected by straight line segments to form the cumulative relative frequency polygon. Note that technical accuracy would demand that the right-hand end point of each interval be determined by the measurement accuracy and grouping of the measurements. For example, in Figure 2, the right-hand end point of the interval 75–99 would be 99.5, since the interval contains all measurements between 74.5 and 99.5.

2.   Graphical display of data is an art in itself. Here we are interested only in the basic concepts of the histogram and cumulative frequency polygon as bases for thinking about population distributions. Dunn (1964) and Lancaster (1974), for example, contain more detail on the construction of both the histogram and the cumulative frequency polygon, and also on other methods of graphical and tabular display.

3.   Note that Tchebycheff's inequality gives rather weak results when compared to results one can obtain from knowledge of the population distribution. For example, if we know that the mean cholesterol value in a population is 225, the standard deviation is 75, and the distribution is approximately normal in shape, then we know quite a bit about the distribution. In particular, we know that (roughly) 95% of the values of the distribution are in the interval 75 to 375 $[\mu \pm 2\sigma]$, a much more informative statement than the lower bound of 75% based on the Tchebycheff inequality. In probabilistic notation, if X is a random value from the population, the assumption of (approximate) normality yields

$$Pr\{\mu - 2\sigma \leq X \leq \mu + 2\sigma\} \approx .95,$$

whereas Tchebycheff's inequality yields the more conservative probability statement

$$Pr\{\mu - 2\sigma \leq X \leq \mu + 2\sigma\} \geq .75.$$

Our discussions of the Tchebycheff inequality and the normal distribution have been intended to give understanding of the meaning of these concepts without formal mathematical statements or proof. For mathematical treatment the reader is referred to Parzen (1960).

PROBLEMS

1. Draw a histogram and also a cumulative relative frequency polygon, showing the distribution of the population of measurements tabled in Table 3.

TABLE 3. Hypothetical Results of the Measurement of Serum Cholesterol on the Adult Males in a Very Small City

Cholesterol Measurement Class	Frequency (number of males with measurements in the indicated class interval)
75–99	3
100–124	6
125–149	14
150–174	26
175–199	36
200–224	49
225–249	63
250–274	65
275–299	57
300–324	48
325–349	46
350–374	33
375–399	24
400–424	16
425–449	4
450–549	10
	500

2. If the first two lines of Table 3 had been compressed into a single line reading

75–124

how would you have modified your histogram? What would be the height of the bar representing the frequency for that interval? (See Comment 1.)

3. Suppose you have a very small population, containing only $N = 10$ objects, with the following measurements: 4, 6, 3, 2, 1, 10, 16, 3, 2, 23. Compute the population mean, median, and standard deviation. Compute the standard deviation by the defining equation

(3) and also by the following, equivalent formula, which is sometimes more convenient for use with a calculating machine (but more prone to rounding errors):

$$(4) \qquad \sigma = \sqrt{\dfrac{\sum X^2 - \dfrac{(\sum X)^2}{N}}{N}} = \sqrt{\dfrac{N(\sum X^2) - (\sum X)^2}{N^2}}.$$

4. Approximate the standard deviation for the population of measurements tabled in Table 1. For **grouped data** like this, the exact measurement for a given member of the population is not known, and it is customary to assign to each member the midpoint of its measurement interval when computing the mean and standard deviation. Denote the midpoints of the intervals by m and the corresponding frequencies of members in the intervals by f. Table 2 contains the values of $f \cdot m$ that were used to approximate μ via equation (2). Similarly, σ can be approximated by

$$(5) \qquad \sigma \approx \sqrt{\dfrac{N \cdot (\sum fm^2) - (\sum fm)^2}{N^2}}.$$

The value $\sum fm$ is obtained by summing the entries in column 4 of Table 2. The value $\sum fm^2$ is obtained by summing the entries in column 5 of Table 2.

When the values for the midpoints have three or more digits, as in Table 1, where they have values of 87.0 to 499.5, the calculations can be tedious. If the calculations are to be done with pencil and paper, or with a hand calculator, the work can be simplified by coding the midpoints with new values, calculating the mean and standard deviation on the coded scale, and then decoding the results to obtain the mean and standard deviation on the original scale. It is as if data in degrees Fahrenheit were converted into degrees Centigrade, and the mean and standard deviation then converted back into Fahrenheit values. The original midpoint values are converted into values that are easier to work with. In Table 2, where the midpoints are given in column 2, we might subtract 87.0 and divide by 25 to get the following coded values for the midpoints:

original values: 87, 112, ..., 499.5

coded values: 0, 1, ..., 16.5.

The coded values are integers, except for the last value, 16.5, which jumps 2.5 units from the next to last value 14 because of the long last interval.

Compute the mean and standard deviation of the coded values (using the same f values for the frequencies) and denote them by μ' and σ', respectively. The mean is then decoded by multiplying μ' by 25 and adding 87.0; the standard deviation is then decoded by multiplying σ' by 25.

In general, if coded values m' are obtained from original values m by subtracting a convenient constant A and dividing by a convenient constant B; that is,

$$m' = \dfrac{m - A}{B},$$

then the mean and standard deviation of the coded values, μ' and σ', can be decoded as follows:

$$\mu = B\mu' + A,$$

$$\sigma = B\sigma'.$$

The decoded values μ and σ do not depend on the original choices of A and B.

Now calculate μ and σ for Table 1 by using the coded values, $m' = (m - 87)/25$, and then decoding. Check your results with the results obtained without coding.

5. From the cumulative relative frequency polygon you drew for Problem 1, read off the median, the 5th, 25th, 75th, and 95th percentiles. Now compute, by linear interpolation, the median, the 5th, the 25th, the 75th, and the 95th percentiles for the data tabled in Table 3. For example, the 60th percentile would be calculated as follows. The person at the 60th percentile would be the 300th person in the population of 500. There are 262 persons with values up to the class 275–299 and 57 people in that class. Hence the 300th person would have a value in that class interval, and we might estimate his value as:

$$x_{.60} = 274.5 + \frac{300 - 262}{57}(25) = 291.2.$$

The 274.5 in the preceding expression is the lower-class boundary, that is, the lowest possible measurement, for the interval containing persons with measurements 275 through 299, measured to the nearest unit. The length of the interval is $299.5 - 274.5 = 25$, and $(300 - 262)/57$ is the fractional part of the total frequency, 57, needed to achieve a cumulative frequency of 300, starting with 262 at the beginning of the class interval.

Compare your calculated results with the values read from the graph.

6. Suppose the ages of a pediatric population served by a particular medical clinic fall into the following age intervals.

Age interval (yr).	Number of Children
0–1	245
1–2	213
2–3	208
3–4	173
4–5	167
5–10	625
10–18	349

Construct a histogram showing the population distribution. Show the class boundaries accurately, and label the vertical scale accurately. Draw a cumulative relative frequency distribution for the population.

Look at your graphs and estimate, by eye, the mean, median, 5th percentile, 25th percentile, 75th percentile, 95th percentile, and the standard deviation of the distribution. Then compute each of these population parameters.

Compare the bounds on the percent of distribution outside one and two standard deviations from the mean, obtained from the Tchebycheff inequality, with the actual percentages as read from your graph of the cumulative distribution.

7. Suppose that I.Q.'s are distributed normally in a defined population, with mean 100 and standard deviation 15. What proportion of the population will have I.Q.'s less than 90? What proportion will have I.Q.'s greater than 145? What proportion will have I.Q.'s between 120 and 140?

2.   SAMPLING FROM FINITE POPULATIONS

Random Sampling

In the case of large populations of objects it is usually not feasible to obtain and measure each object in the population in order to determine the nature of the distribution of the measurements. Instead, a subset or **sample** of objects is obtained and measured. We use n to denote the number of objects in the sample and N to denote the number in the population. On the basis of the sample, inferences or conclusions are drawn regarding the distribution of measurements in the whole population. Since the distribution of measurements in the sample of objects only approximates the distribution in the whole population, it is clear that inferences about the population distribution, based on the sample, are subject to some uncertainty. The discipline called **statistics** consists of principles and techniques directed to the questions of how to measure the uncertainty of such inferences, and how to minimize this uncertainty.

If statistical logic is to be employed in drawing a conclusion about a population from a sample of objects, the sample must be obtained from the population randomly. The concept of randomness is a mathematical concept, an ideal concept; it can be only imperfectly defined and realized in practice. Statistical logic can be used in drawing inferences from samples only to the extent that the actual sampling procedures approximate random sampling ideals.

The word "random" refers to the procedure used in drawing the sample and not to the sample itself. This is analogous to the notion of a fair deal in a game of cards. The "fairness" has to do with the procedure of shuffling and dealing the cards; once the deal is completed, there is no way of judging whether the deal was fair, since any particular distribution of the cards *could* arise from a fair deal.

There are a number of types of random procedures for selecting a sample of n objects from an existing population of N objects. We concentrate on **simple random sampling**, and only briefly mention other types of random sampling.

Prior to giving a formal definition of simple random sampling, it is useful to consider the following special case. The Statistics Department at a small university consists of five faculty members. All five members wish to attend an annual meeting of statistical societies in New York City. Unfortunately, because of a lack of funds, it is deemed by the department chairperson (who is, of course, one of the five faculty members) that only two members of the faculty will be granted travel money. What is the number of different groups of size two that could be granted travel money? Let us denote the five faculty members by the letters A, B, C, D, and E. Then the number of possible two-person groups can be denoted as pairs of letters, as given in Table 4.

TABLE 4.   The Samples of Size 2 that can be Formed from a Population of Five Objects

AB	AC	AD	AE	BC	BD	BE	CD	CE	DE

Table 4 shows that there are 10 different samples of size two that can be formed from a population of five objects. The department chairperson, being a fair-minded person, wants to select a group, from the total of 10 possibilities, in a way that gives each of the 10 samples the same chance of being the group selected. Thus if we let $Pr(AB)$ denote the probability that the sample consisting of faculty members A and B is selected, the chairperson seeks a scheme that has the property that

$$Pr(AB) = 1/10, \quad Pr(AC) = 1/10,$$

$$Pr(AD) = 1/10, \quad Pr(AE) = 1/10,$$

$$Pr(BC) = 1/10, \quad Pr(BD) = 1/10,$$

$$Pr(BE) = 1/10, \quad Pr(CD) = 1/10,$$

$$Pr(CE) = 1/10, \quad Pr(DE) = 1/10.$$

We are now ready to generalize the travel money example and give a formal definition of simple random sampling. Let $\binom{N}{n}$ denote the number of different samples of size n that can be formed from a population of N objects.

Simple random sampling, of n objects from a population of N objects, is any procedure that assures each possible sample of size n an equal chance or probability of being the sample selected, that is, any procedure that assures each possible sample has probability $1 \Big/ \binom{N}{n}$ *of being the sample selected.*

The listing in Table 4 shows that $\binom{5}{2} = 10$. The reader can also check that $\binom{5}{3} = 10$. Rather than actually listing the 10 possible samples of size three that can be formed from a population of five objects, a direct and simple verification that $\binom{5}{3} = 10$ can be obtained from Table 4 as follows. Every time a group of size two is listed in Table 4, the remaining three objects form a group of size three. Thus when we list the two-object group AB as a sample of size two, we are implicitly also listing CDE (the three objects left over) as a sample of size three. Thus for every two-person sample listed in Table 4, there corresponds a three-person sample. This argument shows that

$$\binom{5}{2} = \binom{5}{3}.$$

More generally, the same reasoning establishes the equation

(6) $$\binom{N}{n} = \binom{N}{N-n}.$$

The values $\binom{N}{n}$ are known as the number of combinations of N objects taken n at a time. These quantities are also known as binomial coefficients and we encounter them again in Chapter 7. (In particular, Comment 7, in Chapter 7, presents a formula for computing these coefficients.) Table C3 of Appendix C gives values of $\binom{N}{n}$ for all n, N pairs with $2 \le N \le 30$, $2 \le n \le 15$, and $N \ge n$. It is easy to verify that $\binom{N}{1} = N$, and thus Table C3 starts with $n = 2$ rather than $n = 1$. Furthermore, values of $\binom{N}{n}$ where $N \le 30$ but $n > 15$ can be retrieved from the Table C3 entries by using equation (6). For example, to find $\binom{20}{17}$, note that by equation (6)

$$\binom{20}{17} = \binom{20}{20-17} = \binom{20}{3},$$

and from Table C3 entered at $N = 20$, $n = 3$, we find $\binom{20}{3} = 1140$.

A glance at Table C3 shows that for even moderate values of n and N, a procedure for simple random sampling that required a listing of all possible $\binom{N}{n}$ samples would be tedious indeed. In the next subsection we see that we can, in fact, perform simple random sampling without listing all possible samples.

How to Obtain a Simple Random Sample

We continue the discussion of the $n = 2$, $N = 5$ example of the previous subsection in order to make the definition of simple random sampling clear and to demonstrate how a procedure might be developed that would approximate the ideal of random selection. Consider again a population of only five objects, named A, B, C, D, and E. Now suppose that we want to obtain a simple random sample of two objects from this population of five objects. There are $\binom{5}{2} = 10$ possible samples of two objects that might be selected, and we have listed them in Table 4. According to the definition of simple random sampling, we must select one sample from among the 10 in such a way that all of the 10 have equal probabilities of being selected. This can be done by writing the designation of each pair on a ping pong ball, throwing the 10 balls into a large container, thoroughly shaking the container, and then reaching in and selecting one of the balls blindly. This procedure should lead to the same chance of selection for each of the balls, hence each of the 10 possible samples. If there were any doubt, this could be checked by repeatedly sampling from the container (replacing the ball after each draw), recording the occurrences of the various outcomes in a long series of sampling results, and checking to see that the outcomes did indeed occur with approximately equal frequencies.

Another method of selecting a simple random sample of two from the population of five would circumvent the listing of all possible samples. We could put the labels of the objects in the population, namely, A through E, on five ping pong balls and put these in the container. After shaking, we would reach into the container and draw out two balls blindly. The two could be drawn out together or consecutively. Note that if the two balls are drawn out consecutively, the first should not be thrown back before the second is drawn out or the number of possible combinations will be augmented by the unwanted AA, BB, CC, DD, EE, combinations.

A still faster way of selecting a random sample of two from the population depends on the use of a table of random numbers, such as Table C4 of Appendix C. First, the objects in the population must be numbered, in any order, for example, $A = 1$, $B = 2, \ldots, E = 5$. Then, starting at a place in the table that is chosen blindly, (e.g., by putting a pencil point down on the page, or by choosing a row and column without looking at the page), we can proceed in a prespecified direction, say columnwise, and read the first two numbers. These are the numbers of the objects to be selected in the random sample. If an unusable digit is encountered, that is, 6, 7, 8, 9, or 0, discard it and go on to the next.

More Complex Random Sampling Procedures

We have defined simple random sampling. There are other types of random sampling that also provide a valid basis for statistical inference. For example, in **stratified random sampling**, the population of units is first divided into mutually exclusive groups or strata. Then a simple random sample is obtained from each of the strata. In our example, in which the population has only five objects, we might separate the population objects into two strata, one containing objects A and B and the other containing objects C, D, and E. Then we might obtain a simple random sample of one object from each stratum. It is clear that this two-stage procedure does not yield a simple random sample of size two from the whole population. For example, the sample AB consisting of A and B has no chance of being the selected sample of size two. Thus it is obvious that the simple random sampling requirement—that all samples are equally likely to be selected—is not satisfied. Nevertheless, statistical techniques can be used for this more complicated type of random selection. As one might suspect, these techniques are more complicated than the techniques appropriate to inference from simple random samples. Slonim (1966) has written a very good, simple introduction to the ideas of sampling, covering the techniques and types of random sampling.

Applications and Cautions Concerning Random Sampling

Note that although we have started with consideration of simple random sampling from an existent population of objects, this type of sampling is not common in medical research. We choose it as a starting point because of its simplicity, not because it is of most general use. However, some examples that do occur are (1) sampling the total of all hospital charts belonging to patients admitted to a hospital in a specified period, having a specified type of confirmed disease, for example, stomach cancer, and (2) selecting a sample of households in a village or small town for the purpose of measuring the frequency of occurrence of some infection in the population. Most laboratory experiments and clinical studies do not involve randomly selected animals or patients from some existing population, and thus the application of statistical procedures to such experimental data must be justified on some basis other than random selection of the experimental subjects. (This matter is discussed in Chapter 6.)

Before leaving the topic of random selection, several notes of caution are in order:

1. If you intend to use a table of random numbers to select a sample from a population, you must have a list of the names of all units in the population so that you can associate a number with each unit in the population. This list is called the **frame**. Frequently, a frame is not easily obtained.

2. If you do not have a frame and cannot afford to obtain one, then it is possible to handle the problem in several ways. For small populations of objects that are arranged or listed in some sort of "pseudorandom" manner (e.g., alphabetically in a file), it may be acceptable to sample systematically by selecting every kth object in the file or list. This will certainly *not* yield a simple random sample, but the procedure may approximate a simple random sampling procedure. This is a matter for consultation with an experienced sampling expert familiar with the content area.

3. If a random sample of human beings is to be drawn from a population residing in a specified area, it is usually not feasible to try to obtain a simple random sample. Rather, some sort of **multiple stage sampling** must be used. For example, in drawing a sample of persons from the whole of the United States one might first obtain a frame or listing of the counties in the United States and obtain a random sample of these counties. Then each of the selected counties could be divided into sections, (e.g., townships and cities) and the sections could be sampled. In each selected section further divisions could be made and sampled, finally coming down to housing units, and then the people residing within these units. A good example is the regularly recurring U.S. National Health Examination Survey, a sampling of the United States population. The techniques of such **area sampling** constitute a large body of methodology and if you are ever faced with such a problem it will be most important for you to study this part of applied statistics and to consult specialists in area sampling.

4. Defects in the design or execution of a random sampling effort can easily occur and remain undetected. Such defects can lead to systematic deviations of sample from population and serious errors in inference. The things to watch for are opportunities for the sampler to depart from sampling protocol and to accept or reject an object on some other basis (convenience, curiosity, etc.), or for the object to exclude itself. For example, we suppose that a naive sampler sets about obtaining a simple random sample of 100 hospital charts. From experience he knows that when he calls for a hospital chart by chart number in the record room there is about a 20% chance that the chart will be signed out and not available when he calls for it. He allows for this by carefully selecting a simple random sample of 125 charts, so that he will have about 100 charts at hand for his sample. It is clear that the 100 or so charts he finally reports as a random sample are a random sample not of all charts but of the charts available in the record room, and it is quite conceivable, even likely, that the charts that are not available are materially different from those that are available.

In conclusion, it should be emphasized again that random sampling is a mathematically defined ideal concept, but it can be well approximated in practice if sufficient care is taken. Random sampling does *not* mean careless selection, haphazard selection, or willy-nilly selection of units from a population. There are two main reasons for taking the care to obtain a random sample instead of a haphazard sample from a population. First, the sample is not subject to the systematic selection or biases that might affect a sample drawn in another manner. Second, random sampling furnishes a basis for the use of statistical logic in drawing inferences from sample to population. This second point is amplified in subsequent chapters.

Discussion Question 3.   Recall the heart-transplant study described in Chapter 1. Do you think the candidates who actually receive transplants can be viewed as a random sample of the heart transplant candidates?

Discussion Question 4.   Suppose that out of 500 students registered in Stanford's Medical School, 100 are to be selected for an interview concerning their nutrition habits. How could you select a simple random sample of 100?

Comments

4.   Random number tables are constructed by some sort of gaming device or by the use of a high-speed electronic computer. The numbers are generated in

such a way that all 10 digits appear with the same frequency and in no recognizable pattern when examined in small groups. After a population of objects is numbered, the table can be used to select a simple random sample by proceeding down the rows of the table and selecting units corresponding to the numbers. If the population numbers in the hundreds, three columns must be used to allow the selection of three-digit numbers. If a number is encountered that does not correspond to a population unit, it is discarded and the next number is considered. If a number is encountered that has already been used, it also is discarded. To save time, by allocating a larger proportion of numbers, each object can be assigned several numbers. For example, if the population contains 163 objects, the objects can be assigned five numbers each, using the series 1–163, 201–363, 401–563, 601–763, and 801–963. Thus, the second object in the population is assigned the numbers 2, 202, 402, 602, 802. If any of these numbers is encountered in proceeding through the random number table, the second object is selected for the sample. A good source of random digits is *A Million Random Digits with 100,000 Normal Deviates*, The RAND Corporation (see References).

5. This section has emphasized simple random sampling from a finite population. However, we also must explain the concept and properties of a random sample from an *infinite population*. Suppose the infinite population is represented by a density, as described in Section 1. Recall that such a density is a curve (on a graph) that never lies below the horizontal axis. Furthermore, it is customary to scale the curve so that the total area between the curve and the horizontal axis is one. When this is done, the area between the curve and the horizontal axis represents probability. Then if one measurement is randomly selected from the population, the probability or chance that the measurement falls in a given interval is given by the area between the curve and the horizontal axis over that interval. If a **random sample of n measurements** is selected from the infinite population, then for any given interval, each measurement has an equal and independent chance of falling in the interval, and the chance is given by the area between the curve and the horizontal axis over that interval.

PROBLEMS

8. From the 365 dates of a regular calendar year, select a simple random sample of five dates. Use Table C4, and describe your method in detail, including your results. If all Stanford Medical School students born on these dates are selected as a sample of the 500 students registered in Stanford's Medical School, would you consider this to be a simple random sample of the Medical School student body? If not, why not?

3.	THE DISTRIBUTION OF THE SAMPLE MEAN

Generating the Distribution of the Sample Mean

The purpose of sampling is to obtain information about the population. We say that we draw inferences about the population from the sample. Our inference concerns the population distribution (histogram), or some characteristic of it—for example, its mean or standard deviation, or a percentile. Recall that such population characteristics are called parameters. We start by studying the problem of making inferences about the population mean. It is natural to base such inferences about the population mean on the **sample mean**. Suppose that we obtain a random sample from the population, compute the sample mean, and then take this value as an estimate of the population mean. How good is this estimate likely to be? Clearly, we can be far wrong—but it seems likely that the sample mean will be close to the unknown population mean. Is there some way to make these ideas more precise?

We start by examining the variation in the sample mean in repeated simple random samples from a specified population. Think of drawing a simple random sample of four objects from a population and computing the sample mean. We denote a sample mean by $\bar{X}$ and continue to denote the mean of the population by μ. If the four objects in the sample have values 8, 15, 3, 9, then

$$\bar{X} = \frac{8+15+3+9}{4} = 8.75.$$

Then think of putting the objects back into the population and repeating the procedure again and again. Figure 11 shows a sketch of a hypothetical distribution for an infinite population, and also a sketch of the kind of distribution that might result from tabulating and picturing the distribution of means for many samples of size $n = 4$ drawn from the population. This distribution of means is called the distribution of the sample mean in samples of size four from the specified parent distribution. We frequently use the abbreviated phrase "the distribution of the sample mean," or alternatively, "the sampling distribution of $\bar{X}$."

If you had the graph of the distribution of the sample mean before you, it would be easy to see how far off an estimate of the population mean might be if it were based on a single sample. Some samples yield good estimates (i.e., estimates close to the population mean) and some do not, and the histogram can tell us the proportion of samples with means far from μ, the population mean. Of course, only in an occasional statistics course exercise does one generate the distribution

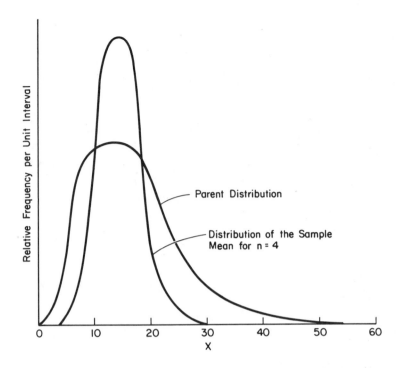

FIGURE 11. **The distribution of the sample mean in samples of size 4 from a specified parent population.**

of the sample mean, with all possible samples. The reason is that we have no need for this. Mathematical results tell us all we need to know about the distribution of the sample mean, in repeated samples from (practically) any parent distribution. The following subsections discuss the nature of the distribution of the sample mean in random samples from a population.

The Mean and Standard Deviation of the Distribution of the Sample Mean

Regardless of the shape of the parent distribution, if the parent distribution has a (finite) mean and (finite) standard deviation denoted by μ and σ, then the mean and standard deviation of the distribution of the sample mean* ($\mu_{\bar{X}}$ and $\sigma_{\bar{X}}$,

---

* For brevity, we often simply say the mean and standard deviation of the sample mean.

respectively) are given by the following equations:

(7)
$$\mu_{\bar{X}} = \mu,$$

(8)
$$\sigma_{\bar{X}} = \frac{\sigma}{\sqrt{n}}, \quad \text{(see Comment 6)},$$

where n is the sample size. Stated in words instead of mathematical notation: the mean of the distribution of the sample mean is equal to the mean of the original or parent population; the standard deviation of the distribution of the sample mean is given by the standard deviation of the original population, divided by the square root of the number of observations in the sample. Equation (8) is exact when the parent population is infinite. If the parent population is finite, with N objects, say, then equation (8) can only be regarded as an approximation. The closeness of the approximation is discussed in Comment 6.

The result for $\sigma_{\bar{X}}$ is simple and quite reasonable. It says that variation in the sample mean, $\bar{X}$, from one random sample to the next, depends on two factors: the variation from unit to unit in the parent population (as measured by σ) and the number of values that are averaged in each sample (i.e., n). Note that $\sigma_{\bar{X}}$ is directly proportional to σ, and that it is proportional to the inverse of the square root of n.

The Central Limit Theorem

The preceding results concerning the mean and standard deviation of the distribution of the sample mean are useful but not very surprising. On the other hand, you may be surprised to know that under mild restrictions on the original population, the form of the distribution of the sample mean is approximately normal, whether or not the distribution of the original population is well approximated by a normal distribution. If the original population has a distribution that is exactly normal, then the distribution of the sample mean is exactly normal for any sample size n.

Of course, no real distribution can be exactly normal in form, since the process of measurement, in practice, groups the measurements to some specified limit of precision, making the population histogram appear to be a series of bars rather than a smooth curve. But when the number of units in the population is large and the measurements are made to a fine degree of precision relative to the amount of variation from unit to unit in the population, then the population histogram may be nearly smooth and, in some cases, nearly normal. Perhaps it is not surprising that in this case the distribution of the sample mean is also normal in form.

The surprising result is that even if the original population distribution is not normal, the distribution of the sample mean tends to be normal in form and, if the sample size n is sufficiently large, the form of the distribution of the sample mean is almost exactly normal. This remarkable normality of the distribution of the sample mean is the substance of the famous central limit theorem, a theorem with many specific and general versions, applicable to almost all parent populations encountered in practice. Because of the central limit theorem, the normal distribution assumes far greater importance than it would have as merely an approximation to certain population distributions found in nature. In fact, the normal distribution can be useful in many problems of inference where the population sampled has a distribution that is known to be far from normal.

The Central Limit Theorem.   If random samples of n observations are taken from a population with mean μ and standard deviation σ (where μ and σ are assumed finite), then the distribution of the sample mean $\bar{X}$ is approximately normal with mean μ and standard deviation σ/√n. The accuracy of the approximation generally increases as n becomes large.

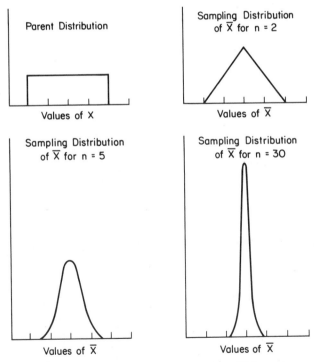

FIGURE 12.   Sampling distributions of $\bar{X}$ for various sample sizes when the parent population is rectangular.

Figures 12–15 illustrate the central limit theorem for four different (infinite) parent populations. Note that in each of Figures 12, 13, and 14, the distribution of the sample mean "approaches" a normal distribution as n, the sample size, increases. Note also that at $n = 30$, the normal approximation to the distribution of the sample mean is well justified. Figure 15 differs from Figures 12–14 in that in Figure 15 the parent population is that of a normal distribution, whereas in Figures 12–14 the parent populations are not normal distributions. Figure 15 illustrates that when the parent population is that of a normal distribution, then the distribution of $\bar{X}$ is *exactly* normal for any sample size n. Figures 12–14 show that when the parent population is not that of a normal distribution, then the distribution of $\bar{X}$ is only *approximately* normal—but the approximation improves as n gets larger.

In conclusion, we can say much about the distribution of the sample mean and how far the mean of a single sample might stray from the population mean,

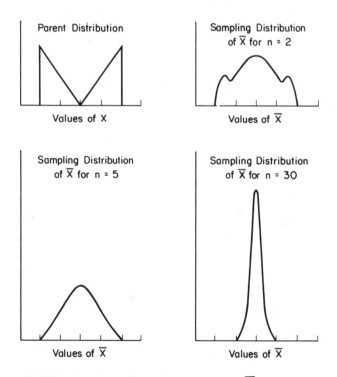

FIGURE 13.   Sampling distributions of $\bar{X}$ for various sample sizes when the parent population is described by oppositely facing right triangles.

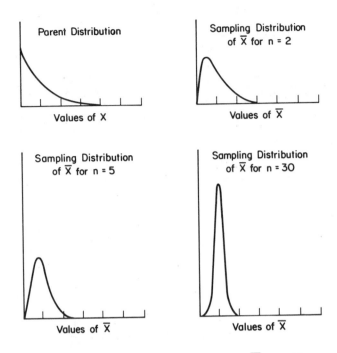

FIGURE 14. Sampling distributions of $\overline{X}$ for various sample sizes when the parent population is described by an exponential distribution.

without actually generating the distribution for a particular situation. Now, how can this information be used in evaluating an inference about an *unknown* population mean from the mean of a random sample from the population? This is considered in Chapter 4.

A Probability Calculation Based on the Central Limit Theorem

Consider a situation in which a large batch of ampules of penicillin is manufactured. Suppose it is known that the mean penicillin content of the ampules in the batch is 10,000 international units. However, the content varies from ampule to ampule, the standard deviation being $\sigma = 500$. If 25 ampules are randomly selected from the large batch of ampules, the average content in the 25 ampules can be viewed as one sample mean from the distribution of averages from all

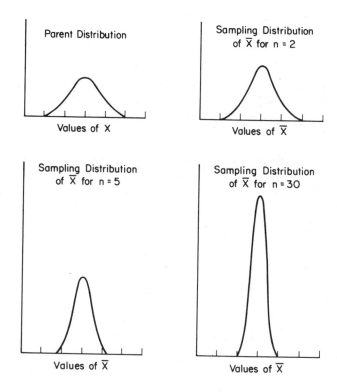

FIGURE 15.   Sampling distributions of $\bar{X}$ for various sample sizes when the parent population is that of a normal distribution.

possible samples of size 25. This distribution will be approximately normal in form, with mean and standard deviation as follows:

$$\mu_{\bar{X}} = \mu = 10{,}000,$$

$$\sigma_{\bar{X}} = \frac{\sigma}{\sqrt{25}} = 100.$$

Thus, using Table C2, we find there is approximately a 68% chance that a given sample mean is within ± 100 (that is, $\pm \sigma_{\bar{X}}$) of the population mean of 10,000. There is (approximately) a 95% chance that the sample mean is within ± 196 (that is, $\pm 1.96\,\sigma_{\bar{X}}$) of the population mean. In fact, using the values for the mean and standard deviation of the distribution of the sample mean, together with the normal table (Table C2), we can calculate the chance that the sample mean lies in any specified interval.

For example, suppose we wish to calculate the chance that the sample mean $\bar{X}$ is between 9930 and 10,020. We write this chance formally as $Pr\{9930 \le \bar{X} \le 10,020\}$. (We read this as "the probability that $\bar{X}$ is between 9930 and 10,020.") Since $\mu_{\bar{X}} = 10,000$ and $\sigma_{\bar{X}} = 100$, the desired probability is the area, under the standard normal curve, between the points

$$Z_1 = \frac{9930 - 10,000}{100} = -.7,$$

and

$$Z_2 = \frac{10,020 - 10,000}{100} = .2.$$

From Table C2 we find the area to the right of .7 is .2420, and thus (by symmetry) the area to the left of $-.7$ is .2420. Hence the area to the right of $-.7$ is $1 - .2420 = .7580$. Again, from Table C2, the area to the right of .2 is .4207. Thus the area between $-.7$ and .2 is $.7580 - .4207 = .3373$. We thus have found that $Pr\{9930 \le \bar{X} \le 10,020\} = .3373$.

Comments

6. When the underlying population is finite, equation (8), for the standard deviation of the distribution of the sample mean, is only an approximation. The approximation is good when the sample size n is a small fraction of the population. The exact expression for $\sigma_{\bar{X}}$, in the finite population case

$$(9) \qquad \sigma_{\bar{X}} = \sqrt{\frac{N-n}{N-1}} \cdot \frac{\sigma}{\sqrt{n}},$$

where, if the elements of the finite population are denoted by X's, σ is given by equation (3). See Cochran (1963) for mathematical development.

The factor under the first radical sign in (9) is called the **finite population correction.** Note that when N is much larger than n, $\sqrt{\dfrac{N-n}{N-1}}$ is close to 1, and the right-hand side of (9) is close to the right-hand side of (8). Note also that when $n = 1$, $\sigma_{\bar{X}}$, as given by (9), is equal to σ itself. Explain.

7. For those with more than a passing interest in the mathematical aspects of central limit theorems, it should be remarked that the theorems require fairly sophisticated proofs. The theorem we have discussed here is usually not proved in

an introductory course. See Chapter 10 of the text by Parzen (1960) for a rigorous mathematical proof.

Summary of Statistical Concepts

In this chapter the reader should have learned that:

(i)   There is variation in a measured characteristic from one object to another in a population.

(ii) Histograms, cumulative relative frequency polygons, and frequency polygons are very useful for graphically describing a distribution of measurements.

(iii) The mean, standard deviation, and percentiles of a distribution convey much information about the distribution.

(iv) If statistical logic is to be employed in drawing a conclusion about a population from a sample of objects, the sample must be obtained from the population randomly.

(v)   Simple random sampling of n objects from a population of N objects is any procedure that assures each possible sample of size n an equal chance of being the sample selected.

(vi) The normal curve resembles many frequency polygons encountered in practice. Even if the parent population is not well approximated by a normal population, the central limit theorem assures that the distribution of the sample mean $\bar{X}$ can be approximated by a normal distribution. The approximation improves as the sample size n increases.

PROBLEMS

9. Suppose the mean diastolic blood pressure in a certain age group is 78 and the standard deviation is 9. Approximate the probability that in a sample of size 16, the sample mean is greater than 81.

10. What are undesirable characteristics of the sample mean when used as an estimator of the population mean? [Hint: Suppose one of the sample values is erroneously recorded and that the recorded value is much larger than the actual value observed, and also much larger than the other sample values. What effect would such a gross error have on $\bar{X}$?]

11. Assume that it is known that the mean survival time (measured from date of diagnosis) for a certain type of cancer is 1577 days with a standard deviation of 500 days. Consider a random sample of 16 patients suffering from the disease. Use the central limit theorem to approximate the probability that the sample mean of the 16 patients will be greater than 1700 days.

12. Use equation (9) to calculate the standard deviation of the sample mean when the sample size is two and the finite population consists of the elements 4, 9, 2, 13, 5.

13. Suppose that the standard deviation of a sample mean based on a sample of size four, drawn from a finite population of size 10, is 2.4. What is the standard deviation of the population?

14. Define population, finite population, frequency table, frequency, relative frequency, cumulative frequency, cumulative relative frequency, histogram, cumulative relative frequency polygon, frequency polygon, distribution, population mean, population median, parameter, pth percentile, upper pth percentile, population standard deviation, population variance, normal distribution, continuous variable, discrete variable, standardized deviate, coefficient of variation, grouped data, sample, simple random sampling, stratified random sampling, frame, multiple stage sampling, area sampling, random sample of n objects, sample mean, and finite population correction.

Supplementary Problems

Questions 1–4 refer to the data in Table 5. These data, due to Gershanik, Brooks, and Little (1974), can be viewed as the birth weights of a finite population: the population of newborn infants delivered between March 1 and September 30, 1972, at Confederate Memorial Medical Center.

TABLE 5. Birth Weights

Birth Weight (gm)	Number of Babies
1000–1499	3
1500–1999	5
2000–2499	12
2500–2999	58
3000–3499	82
3500–3999	39
>3999	19

Source: J. J. Gershanik, G. G. Brooks, and J. A. Little (1974).

1. Construct a histogram and a cumulative relative frequency polygon for the data of Table 5.

2. For the data of Table 5, use formula (2) to approximate μ.

3. For the data of Table 5, use formula (5) to approximate σ.

4. From the cumulative relative frequency polygon that you drew for Supplementary Problem 1, approximate the 25th and 75th percentiles.

5. Television commercials on station WXYZ average 30 sec in length, with a standard deviation of 5 sec. Using Tchebycheff's inequality, we may say that at least three-fourths of the commercials on WXYZ have lengths between ——— and ——— seconds.

6. Let X be a normal variate with mean 1 and variance 1. Find $Pr(-1 \leq X \leq 2)$.

7. Consider the following finite population of measurements

$$-1, -2, -4, 6, 4, 5, 3.$$

Find the mean, median, and standard deviation.

8. The distribution of seed sizes X utilized in the diet of a particular sparrow species is closely approximated by a normal distribution with mean $\mu = 1.5$ mm and variance $\sigma^2 = .99$. Calculate $Pr(X \geq 1.9)$.

9. List the samples of size three that can be formed from a population of six objects.

•10. An investigator plans to take a sample from a normal population with unknown mean μ and known standard deviation 2. The investigator wants the sample to be large enough to assure, with probability .95, that $|\bar{X} - \mu| \leq 1$. How should the investigator select n, the sample size?

11. Let X represent the height of an individual selected at random from a population of adult males. Assume that X possesses a normal distribution with mean $\mu = 68$ inches and standard deviation $\sigma = 3$ inches. Suppose a random sample of size $n = 25$ is taken from this population.

   a. What is the distribution of the sample mean $\bar{X}$, and what is its mean and standard deviation?

   b. Calculate the probability that the sample mean $\bar{X}$ will differ from the population mean by less than 1 inch.

12. a. Show that the total number of ways of selecting a committee of n from a group of N and naming one as chairperson is $\binom{N}{n} \times \binom{n}{1}$. (Hint: Select first the committee; then from the committee members a chairperson.)

   b. Show that an alternative answer to part (a) is $\binom{N}{1} \times \binom{N-1}{n-1}$. (Hint: Select first the chairperson from the whole group; then select the $n-1$ ordinary members.)

   c. By comparing the answers to (a) and (b) show that

$$\binom{N}{n} = \frac{N}{n} \times \binom{N-1}{n-1}.$$

   d. Use the result given in part (c) and Table C3 to determine the value of $\binom{31}{16}$.

13. Use the result in part (c) of Supplementary Problem 12 to show that the probability that a specified object is included in a simple random sample of size n from a population of N objects is

$$\frac{\binom{N-1}{n-1}}{\binom{N}{n}} = \frac{n}{N}.$$

14. A finite population consists of the values 1, 2, 3, 4, 5, 6, 7. Calculate the population mean and the population variance.

15. Scores on a given civil service examination are assumed to be normally distributed with mean 50 and standard deviation 10.
   a. Calculate the probability a person taking the examination will score higher than 60.
   b. If 100 persons take the examination, calculate the probability that their average grade will exceed 51.6.

16. The length of life of a Junetag automatic washer is approximately normally distributed with mean 3.8 years and standard deviation 1.4 years.
   a. Calculate the probability that a randomly selected Junetag washer will have a life length that exceeds 5.2 years.
   b. If Junetag washers are guaranteed for 1 year, what fraction of original sales will require replacement during the warranty period?

17. The mean and standard deviation of the distribution of claims against an auto insurance company are $500 and $100, respectively. Use Tchebycheff's inequality to determine an interval such that at least 75% of the claims lie in that interval.

REFERENCES

A Million Random Digits with 100,000 *Normal Deviates* (1955). The RAND Corporation. Free Press of Glencoe, Glencoe, Illinois.

Cochran, W. G. (1963). Sampling Techniques, 2nd edition. Wiley, New York.

Dunn, O. J. (1964). *Basic Statistics: A Primer for the Biomedical Sciences*. Wiley, New York.

Farquhar, J. W. (1974). Personal communication.

Gershanik, J. J., Brooks, G. G., and Little, J. A. (1974). Blood lead values in pregnant women and their offspring. *Amer. J. Obstet. Gynecol.* **119**, 508–511.

Hart, W. L. (1967). *Contemporary College Algebra and Trigonometry*. Heath, Boston.

Hodges, J. L., Jr., Krech, D., and Crutchfield, R. S. (1975). *Statlab: An Empirical Introduction to Statistics*. McGraw-Hill, New York.

Lancaster, H. O. (1974). *An Introduction to Medical Statistics*. Wiley, New York.

Parzen, E. (1960). *Modern Probability Theory and its Applications*. Wiley, New York.

Slonim, M. J. (1966). *Sampling, A Quick, Reliable Guide to Practical Statistics*. Simon and Schuster, New York.

CHAPTER 4

Analysis of Matched Pairs Using Sample Means

In this chapter we analyze paired data. The data consist of two samples of equal size, and each member of one sample is paired with one (and only one) specific member of the other sample. This type of data arises, for example, in experiments designed to investigate the effect of a treatment. A treated subject is compared with a control subject in each of n pairs of subjects that are matched on factors such as age, sex, and severity of disease, factors that may affect the measured response. Matched pairs also arise naturally when, on each of n subjects, there are two measurements—say a pretreatment value and a posttreatment value. We wish to know if there is a treatment effect, and each subject serves as his own control.

Let d be the amount by which the response of a treated subject in a matched pair exceeds the response of the control subject of the pair. (Note the excess may be negative; that is, d may be a positive or a negative value.) With n matched pairs, we view the d values as a random sample of size n from a population with mean μ. The parameter μ is called the treatment effect. In Section 1 we obtain a confidence interval for μ. Section 2 presents tests of hypotheses concerning μ. These tests are directly related to the confidence interval of Section 1. Section 3 considers the determination of sample size—that is, how large should n be?

1.	A CONFIDENCE INTERVAL FOR THE TREATMENT EFFECT

Changes in Blood Samples During Storage

When blood samples or plasma samples are stored, does the concentration of cholesterol, triglycerides, and other substances in the blood change? If not, then blood samples can be taken from an individual at monthly or yearly intervals, stored, and then all measured at the same time to study changes in the individual through time. This could be a great convenience to medical scientists in the organization of laboratory work time and resources in certain population studies. More important, it would obviate the need for extraordinary care in eliminating variations in laboratory techniques over the course of the months or years of blood sampling.

A Stanford University team randomly selected hundreds of individuals from several small California communities, for a study of methods of preventing heart disease. One of their objectives was to help the subjects lower their plasma triglyceride levels through changes in diet, more exercise, and other changes in living habits. To measure changes in triglyceride level, they wanted to store plasma and measure the triglyceride levels later. A small study of storage effects was done, and the results are presented in Table 1 (Wood, 1973).

In Table 1 we view the data as consisting of 30 matched pairs, corresponding to 30 different subjects. In each pair there are two measurements, the triglyceride level of the fresh blood plasma and the triglyceride level after 8 months of storage. In this case, the question of whether there is a "treatment" effect is equivalent to asking whether the two measurements are in close agreement.

A glance at the values in Table 1 indicates that the determination made on the fresh blood plasma and on the plasma after 8 months of storage are in fairly close agreement. Is this impression statistically sound? We need a statistical summary of the results, and Table 1 is helpful for obtaining such a summary. Thus we now go through the calculations in Table 1 in detail and interpret the results.

First, it is helpful to compute the mean of the triglyceride measurements obtained in September of 1972 on the fresh blood plasma samples, then compare

TABLE 1. Plasma Triglyceride Measurements (mg/100 ml) Made on 30 Fresh Blood Samples in 1972 and Repeated in 1973 after 8 Months of Frozen Storage

Plasma Sample	First Measurement (1972)	Second Measurement (1973)	Difference d	Deviation $d - \bar{d}$	Squared Deviation $(d - \bar{d})^2$
1	74	66	−8	−9.1	82.81
2	80	85	+5	3.9	15.21
3	75	71	−4	−5.1	26.01
4	136	132	−4	−5.1	26.01
5	104	103	−1	−2.1	4.41
6	102	103	+1	−0.1	0.01
7	177	185	+8	6.9	47.61
8	88	96	+8	6.9	47.61
9	85	76	−9	−10.1	102.01
10	267	273	+6	4.9	24.01
11	71	73	+2	0.9	0.81
12	174	172	−2	−3.1	9.61
13	126	133	+7	5.9	34.81
14	72	69	−3	−4.1	16.81
15	301	302	+1	−0.1	0.01
16	99	106	+7	5.9	34.81
17	97	94	−3	−4.1	16.81
18	71	67	−4	−5.1	26.01
19	83	81	−2	−3.1	9.61
20	79	74	−5	−6.1	37.21
21	194	192	−2	−3.1	9.61
22	124	129	+5	3.9	15.21
23	42	48	+6	4.9	24.01
24	145	148	+3	1.9	3.61
25	131	127	−4	−5.1	26.01
26	228	227	−1	−2.1	4.41
27	115	129	+14	12.9	166.41
28	83	81	−2	−3.1	9.61
29	211	212	+1	−0.1	0.01
30	169	182	+13	11.9	141.61
Totals:	3803	3836	33	0	962.70

| Sample Means: | $\dfrac{3803}{30} = 126.8$ | $\dfrac{3836}{30} = 127.9$ | $\dfrac{33}{30} = 1.10$ | | |

Source: P. D. Wood (1973).

this 1972 mean with the mean of the measurements made on the plasma after storage. The means are given in Table 1, 126.8 mg/100 ml in 1972 and 127.9 in 1973. If there were any change in triglyceride concentration at all during storage, one estimate of that change can be obtained by subtracting the means. Thus we estimate that the concentration increased by an average of 1.10 mg/100 ml, or less than 1% of the 1972 mean value. (A different estimate of the change can be obtained by using the estimator $\tilde{\mu}$ defined in Comment 10 of Chapter 12.)

If the 1972 and 1973 measurements on the individual plasma samples are examined, however, it is clear that the changes vary from one blood sample to another. The differences are shown in Table 1. The 1972 measurement is subtracted from the 1973 measurement in each individual case, so that a positive difference indicates an increase in triglyceride during storage, a negative difference a decrease. Note that the differences range from an apparent decrease of 9 for one blood sample to an apparent increase of 14 for another. A histogram showing the distribution of the 30 differences is given in Figure 1. The individual differences do cluster about zero; in fact, glancing down the column of differences, exactly half of them are positive and half are negative. The mean of the

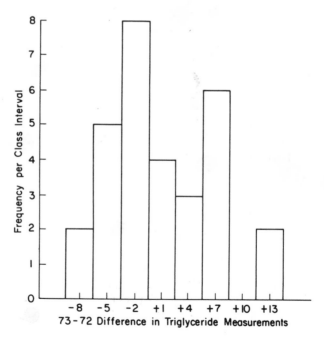

FIGURE 1. A histogram of the 30 triglyceride differences.

differences, $+1.10$ mg/100 ml, is the same as the difference in the means, a number already noted.

Although the mean of the 30 differences is small, it is disturbing that the differences for the individual samples vary quite widely, from -9 to $+14$. These differences may be due in part to real changes in the triglyceride concentration during storage. In addition, the technique for measuring triglyceride concentration is not perfect. Every effort was made to standardize and calibrate the technique for the two sets of measurements, but it is known that, as with any laboratory procedure, repeated measurements on the same sample, even if made on the same day, vary to some extent. In the light of these variations in the individual measurements, it is clear that if a different set of 30 blood samples had been measured, the average change would not have been the same as the 1.10 mg/100 ml actually obtained. How reliable is this mean, based on this set of 30 blood samples? To what extent can we take the 1.10 as a reliable average for all blood samples?

Reliability of the Mean

The mean difference $\bar{d}$ (say) of 1.10 mg/100 ml in Table 1 should be regarded only as an estimate of true difference μ (say) that might be obtained as the mean of the population of all differences. The reliability of $\bar{d}$ can be computed by first estimating the standard deviation of $\bar{d}$ as follows:

1. Compute each individual difference d. Note that $d = $ [second measurement] $-$ [first measurement].
2. Sum the n differences (in our case $n = 30$) and call the sum Σd. Compute the sample mean of the differences, $\bar{d} = \Sigma d / n$.
3. Subtract $\bar{d}$ from each d. The $d - \bar{d}$ values are given in column 5 of Table 1.
4. Square each $d - \bar{d}$ value. The squares are given in column 6 of Table 1.
5. Sum the $(d - \bar{d})^2$ values and call the sum $\Sigma(d - \bar{d})^2$.
6. Compute the **sample standard deviation** s:

(1)
$$s = \sqrt{\frac{\Sigma(d - \bar{d})^2}{n - 1}}.$$

7. Compute the estimate of the standard deviation of $\bar{d}$:

(2)
$$\widehat{SD} = \frac{s}{\sqrt{n}}.$$

The quantity $\widehat{SD}$ estimates the standard deviation of $\bar{d}$. The standard deviation of $\bar{d}$ is equal to $\sigma/\sqrt{n}$, where σ is the population standard deviation. Since we do not usually know the value of σ, we estimate it by s. Note that s is calculated with $n-1$ rather than n in the denominator (see Comment 1).

Many writers call $\widehat{SD}$ the **standard error** of $\bar{d}$; they use the phrase "standard error of the mean" and denote it by SE rather than $\widehat{SD}$. By omitting the circumflex, these writers are not distinguishing between the true standard deviation of the mean, $\sigma/\sqrt{n}$, and its estimate, $s/\sqrt{n}$.

In our example we find, from Table 1,

$$\sum (d-\bar{d})^2 = 962.70,$$

and thus, from equation (1),

$$s = \sqrt{\frac{962.70}{29}} = 5.76.$$

From (2), our estimate $\widehat{SD}$ of the standard deviation of $\bar{d}$ is found to be

$$\widehat{SD} = \frac{5.76}{\sqrt{30}} = 1.05.$$

The statistic $\widehat{SD}$ has a straightforward interpretation. The true storage effect μ, that is, the average effect that would be found in a very large series, can be estimated to lie in the range:

(3) $\bar{d} - t_{\alpha/2, n-1} \cdot \widehat{SD}, \qquad \bar{d} + t_{\alpha/2, n-1} \cdot \widehat{SD}.$

In the sequel, we often use the following shorthand notation to denote the interval: $\bar{d} \pm t_{\alpha/2, n-1} \cdot \widehat{SD}$. The interval defined by (3) is called a **confidence interval** for μ with **confidence coefficient** equal to $100(1-\alpha)$ percent. (See Comment 2.) The value $t_{\alpha/2, n-1}$ is the upper $\alpha/2$ percentile point of Student's t **distribution** with $n-1$ **degrees of freedom**. There is a family of mathematical functions, called Student's t distributions. There is one such function for each positive integer (called its degrees of freedom). The t distributions appear here because, when the underlying population of the d's is normal with mean μ, the probability distribution of $(\bar{d}-\mu)/\widehat{SD}$ is the t distribution with $n-1$ degrees of freedom. Thus the t distribution arises here (and in most situations) as a *derived* distribution, rather than a distribution that is assumed for the basic data.

The t distribution is symmetric about 0. Thus for any positive value a, the area under the t curve to right of a is equal to the area under the curve to the left of $-a$. Because of this symmetry, it is necessary to table only upper (rather than upper and lower) percentiles of the distribution. Some upper percentiles are given in Table C5.

Except for cases in which the number of degrees of freedom is small, a t distribution looks much like a normal distribution. Furthermore, as the degrees of freedom increase, the t distribution tends to the standard normal distribution. Figure 2 shows the t distribution with five degrees of freedom and the standard normal distribution.

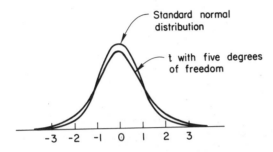

FIGURE 2.   Standard normal distribution and t distribution with five degrees of freedom.

Let us now return to the interval given by (3). It is a random interval that depends on the particular d values we observe. The lower endpoint of the interval is $\bar{d} - (t_{\alpha/2,n-1})(\widehat{SD})$, and the upper end point of the interval is $\bar{d} + (t_{\alpha/2,n-1})(\widehat{SD})$. (The lower end point of a confidence interval is often called the lower confidence limit, and the upper end point, the upper confidence limit.)

Consider, in particular, the plasma data where $n = 30$ and suppose 99% confidence is desired. Then $\alpha = .01$, $(\alpha/2) = .005$, and from Table C5 we find $t_{.005,29} = 2.756$. Thus a 99% confidence interval for μ is

$$1.10 \pm (2.756) \cdot (1.05) = 1.10 \pm 2.89$$

$$= -1.79 \text{ and } 3.99.$$

The confidence interval defined by (3) has the following interpretation. The chance that the interval contains the unknown value μ is equal to $1 - \alpha$. This chance should be interpreted as a long-run probability as follows. Imagine a sequence of independent repeated experiments, each producing a set of n d-values. For each experiment, the interval defined by (3) is calculated. Assuming that the d-values are from a normal population, in the long run $100(1 - \alpha)\%$ of these confidence intervals will contain the unknown value μ and $100(\alpha)\%$ of these intervals will not contain the unknown value μ. However, for any particular interval that is calculated, such as the interval $(-1.79, 3.99)$ computed for our blood storage data, μ is either in that particular interval or not. It is *not correct* to say $\Pr(-1.79 < \mu < 3.99) = .99$. This probability is either 1 or 0, depending on whether or not the true value of μ is indeed between -1.79 and 3.99. We simply must be content with the knowledge that this confidence interval procedure, with confidence coefficient 99%, in the long run yields confidence intervals that contain the true value μ 99% of the time.

The confidence interval defined by (3) is derived under the assumption that the d's are a random sample from a normal population. For more information on this interval, see Hoel (1971a, pp. 146–149). Also see Comment 3.

A **nonparametric** confidence interval, which does not require the assumption of normality, is described in Chapter 12.

Discussion Question 1.   If 95% confidence were sufficient for the inference, what would be the confidence limits for the true or long-run average μ in the blood storage problem? Do you expect the limits to be narrower or wider than the 99% confidence limits?

Comments

1.   Although the population standard deviation is defined as the root average squared deviation, using the population size N in the denominator [see equation (3) of Chapter 3], the sample standard deviation is usually calculated with $n - 1$ as the denominator. The reason for this is that the deviations in the sample must be measured from the sample mean rather than from the unknown population mean; in repeated samples this causes the squared deviations to be slightly smaller than if they had been measured from the population mean. Division by $n - 1$ instead of n tends to correct for this underestimation. See Hoel (1971b, p. 192) for a mathematical proof.

2.   The interval estimate for μ given in expression (3) is called a $100(1 - \alpha)\%$ confidence interval. (Sometimes we simply call it a $1 - \alpha$ confidence interval.) In constructing the interval the sample mean is taken to lie within $t_{\alpha/2, n-1} \cdot \widehat{SD}$ of the population mean. Since the probability that this will be true of a randomly selected sample is simply $1 - \alpha$, $100(1 - \alpha)\%$ of all confidence intervals will be correct, that is, will indeed contain μ. In our example, a 99% confidence interval was computed; since 99% of such intervals will be correct, the confidence interval is said to have a confidence coefficient of 99% or .99, though there is no way to know if the calculated interval itself is one of the large majority of correct statements

about μ, or one of those few (1%) erroneous intervals based on an erratic sample mean.

It is instructive to see how the length of the interval given by (3) depends on α and n. The length of the interval, the upper end point minus the lower end point, is

$$\bar{d}+t_{\alpha/2,n-1}\cdot\widehat{SD}-(\bar{d}-t_{\alpha/2,n-1}\cdot\widehat{SD})=(2)(t_{\alpha/2,n-1})(\widehat{SD}).$$

Fix n and let α get smaller. Then $1-\alpha$ gets larger, and we are insisting on more confidence. But as α gets smaller, the percentile point $t_{\alpha/2,n-1}$ increases, and thus the length of the interval increases. The price for a higher confidence coefficient is a longer confidence interval.

Now fix α and let n get larger. The percentile point $t_{\alpha/2,n-1}$ decreases (as can be seen from Table C5). Furthermore, recall that $\widehat{SD}$ is an estimator of $\sigma/\sqrt{n}$, where σ is the standard deviation of the d population. Since $\sigma/\sqrt{n}$ decreases as n increases, $\widehat{SD}$ tends to decrease as n increases. Thus we see that for a fixed value of α, increasing the sample size n tends to decrease the length of the confidence interval. This corresponds to our intuition. As we increase the sample size we obtain more information, and thus we should be able to obtain a better interval estimate of the unknown value of μ.

3.   The confidence interval defined by (3) is exact (i.e., the true probability that the random interval contains μ is equal to the desired coverage probability $1-\alpha$) when the d's are a random sample from a normal population. An *approximate* $1-\alpha$ confidence interval for μ can be obtained without the normality assumption, by replacing the t percentile $t_{\alpha/2,n-1}$ in (3) with the normal percentile $z_{\alpha/2}$. The approximate interval has the property that its coverage probability is approximately equal to the desired value $1-\alpha$ and this coverage probability tends to $1-\alpha$ as n gets large. (The theoretical basis is provided by a slight modification of the central limit theorem.) With $\alpha=.01$ this approximate interval, for the data of Table 1, becomes

$$1.10\pm(2.58)\cdot(1.05)=1.10\pm2.71=-1.61 \text{ and } 3.81.$$

An exact nonparametric confidence interval, which does not require that the underlying population be a normal population, is described in Chapter 12.

4.   The estimator $\bar{d}$ and the confidence interval based on $\bar{d}$ are very sensitive to wildly outlying observations—**outliers**. In Chapter 12 we present another interval estimator of μ that is less sensitive to outliers than the interval estimator based on $\bar{d}$.

5. The sample standard deviation s, defined by equation (1), can also be computed using the following equivalent formula;

$$s = \sqrt{\frac{\sum d^2 - \frac{(\sum d)^2}{n}}{n-1}} = \sqrt{\frac{n\sum d^2 - (\sum d)^2}{n(n-1)}},$$

where $\sum d^2$ is the sum of the n d^2-values. Note that with this alternate formula we do not need to subtract $\bar{d}$ from each d. Direct calculations with the data of Table 1 show

$$\sum d^2 = (-8)^2 + (5)^2 + \cdots + (13)^2 = 999.$$

Thus

$$s = \sqrt{\frac{999 - \frac{(33)^2}{30}}{29}} = \sqrt{33.20} = 5.76,$$

agreeing with the result obtained with equation (1) earlier.

PROBLEMS

1. In the same blood storage study analyzed in this section, the triglyceride concentrations were measured in 1973 by two different methods. The data in Table 1 were obtained by a Model I autoanalyzer, but measurements were also obtained using a Model II autoanalyzer. The measurements obtained by both instruments are given in Table 2. Compute 95% confidence limits for the long-run average difference between the two methods. (Save this result for comparison with the 95% nonparametric interval to be obtained in Problem 8 of Chapter 12.)

2. Consider the operant conditioning data of Table 1, Chapter 12. Determine a 90% confidence interval for μ, the operant conditioning effect, using interval (3) of this section. (Save this result for comparison with the 89% nonparametric interval obtained in Section 2 of Chapter 12.)

3. Consider the tensile strength data of Table 3, Chapter 12. Determine a 95% confidence interval for μ using interval (3) of this section. (Save this result for comparison with the interval to be obtained in Problem 5 of Chapter 12.)

4. Construct an example that illustrates the sensitivity of the estimator $\bar{d}$ to outlying observations. Why is this sensitivity undesirable?

TABLE 2. Triglyceride Measurements (mg/100 ml) Made by Two Different Autoanalyzers on 30 Blood Samples after 8 Months of Frozen Storage

Plasma Sample	Autoanalyzer I	Autoanalyzer II
1	66	71
2	85	88
3	71	75
4	132	133
5	103	104
6	103	108
7	185	185
8	96	96
9	76	79
10	273	277
11	73	73
12	172	171
13	133	133
14	69	73
15	302	307
16	106	100
17	94	95
18	67	70
19	81	83
20	74	75
21	192	195
22	129	130
23	48	51
24	148	152
25	127	131
26	227	236
27	129	133
28	81	85
29	212	216
30	182	185

Source: P. D. Wood (1973).

2.	A HYPOTHESIS TEST FOR THE TREATMENT EFFECT

Does Storage Have an Effect on Triglyceride Concentration in the Blood?

The confidence limits calculated in Section 1 provide a very satisfactory kind of conclusion for the study results in Table 1. In particular, since the confidence interval includes the value 0, the data are consistent with the **null hypothesis**, the hypothesis that there is no storage effect. (See Comment 6.) In other cases, the confidence limits may not include zero, and in such cases, this would suggest a departure from the null hypothesis. The sample mean, $\bar{d}$, is then said to be significantly different from zero, at the α level, and the null hypothesis can be rejected as a tenable hypothesis. In Section 1 we used $\alpha = .01$.

A formal test of the null hypothesis,

$$H_0: \text{"storage has no effect on triglyceride concentration,"}$$

can be quickly done by noting whether $\bar{d}$ is more than $t_{\alpha/2,n-1} \cdot \widehat{SD}$ from zero. In other words, the **test statistic**, T can be computed:

(4)
$$T = \bar{d}/\widehat{SD},$$

and if T is larger than $t_{\alpha/2,n-1}$ or smaller than $-t_{\alpha/2,n-1}$, the null hypothesis can be rejected. If not, the null hypothesis is accepted. For the data of Table 1,

$$T = 1.10/1.05 = 1.05.$$

Since T does not exceed $t_{.005,29} = 2.76$ in absolute value, the null hypothesis cannot be rejected at $\alpha = .01$, as has already been noted from the 99% confidence limits.

The theoretical basis for the significance test based on the statistic T is that when the d's are a random sample from a normal population, with mean zero, the distribution of T is the t distribution with $n-1$ degrees of freedom.

Discussion Question 2.   What is the lowest α value at which you could reject the null hypothesis using the data of Table 1? (You can only approximate this value, rather than determine it precisely, because Table C5 contains percentiles for only certain selected values of α.) The lowest α value at which you can reject the null hypothesis, with the observed value of your test statistic, is called the **P value**. (Some writers call the P value the **significance probability**.) The smaller the value of P, the more surprising the outcome of the experiment should to be a believer of the null hypothesis. Explain why it is more informative to give a P value rather than, for example, to state simply that the null hypothesis was rejected (or accepted) at the $\alpha = .05$ level. [Hint: You can reject the null hypothesis at all α values that are as large or larger than the P value, and you cannot reject the null hypothesis at α values that are smaller than the P value.]

Discussion Question 3.   Refer to Table 1 and suppose, for the moment, that the 1973 measurement on plasma sample 3 (the 71 value) was not available. That would leave 29 "complete" pairs and one "pair" where an observation was missing. Can you suggest some ways to test $\mu = 0$ using the available observations? [A formal approach to the missing data problem in paired replicates studies is given by Lin and Stivers (1974).]

Comments

6.   Hypothesis testing procedures can be described in a very formal way, as follows. The null hypothesis to be tested is denoted by H_0. [The adjective "null" is used by statisticians because the basic hypothesis to be tested is often one which asserts that certain claims of novelty are not true.] In our storage example, we have

$$H_0: \mu = 0.$$

A test statistic, T, is calculated from the sample values, and if T falls into the **critical region** (the set of values for which H_0 is to be rejected) we reject the null hypothesis. In our storage example, the critical region consisted of the two regions $\{T > 2.76\}$ and $\{T < -2.76\}$.

The test is designed so that if H_0 is actually true, in our case if the population mean μ is actually equal to the hypothesized value, zero, then the probability of rejection is equal to α. Rejecting H_0 when it is true is called a **Type-I error**. Thus, if H_0 is tested at the α level, the probability of a Type-I error is equal to α.

Another type of error is also possible. We may accept H_0 when it is false. This is called a **Type-II error**, and the probability of a Type-II error is denoted by β. The value of $1 - \beta$ is called the **power** of the test. It is equal to the probability of rejecting H_0 when the alternative is true, a probability we want to keep high. The actual value of $1 - \beta$ depends on which particular alternative to the null hypothesis is true. Table 3 summarizes the notions of Type-I and Type-II errors in terms of the two possible states of the null hypothesis, true or false, and the two decisions for the experimenter, reject H_0 or accept H_0.

TABLE 3.   Errors with Hypothesis Tests

	Null Hypothesis H_0	
Decision	True	False
Reject H_0	Type-I error	Correct decision
Accept H_0	Correct decision	Type-II error

It is nice to have tests where both α and β are small, but these two goals are antagonistic in the sense that, for a fixed sample size, lowering α raises β and lowering β raises α. However, increasing the sample size would enable one to lower both α and β.

7.   If the rejection rule allows for rejection of H_0 for very large values or for very small values of the test statistic, the test is called a **two-tailed test** (sometimes called a two-sided test). In our example the alternatives to the null hypothesis $\mu = 0$, were that either $\mu < 0$ or $\mu > 0$ (i.e., a **two-sided alternative**) and so rejecting for large and small values of T was appropriate. In some applications the alternative hypothesis (H_a say) is a **one-sided alternative**

$$H_a : \mu > 0.$$

In this case, the test statistic T causes rejection only if T is too large. For the matched pairs problem of this chapter, in order to control the probability of a Type-I error when $\mu = 0$ at α, the null hypothesis is rejected only if T exceeds $t_{\alpha,n-1}$ rather than $t_{\alpha/2,n-1}$, and the test is called a **one-tailed test**. A one-sided α level test in the opposite direction, designed for the alternative $\mu < 0$, is: reject H_0 if T is less than $-t_{\alpha,n-1}$; accept H_0 otherwise.

8.   In the storage example and in the discussion of hypothesis testing thus far, the null hypothesis has been that the population mean is zero. If, for example, in the blood storage problem we wished to test the hypothesis that $\mu = 2$, we simply redefine the test statistic T to be $T = (\bar{d} - 2)/\widehat{SD}$, and refer this value of T to the percentiles of the t distribution with $n - 1$ degrees of freedom. More generally, to test

$$H_0 : \mu = \mu_0,$$

where μ_0 is any specified value, use the test statistic

$$T = \frac{\bar{d} - \mu_0}{\widehat{SD}}.$$

Significantly large values of T indicate $\mu > \mu_0$; significantly small values of T indicate $\mu < \mu_0$.

9. Consider a situation where, rather than paired data, we have a single random sample from a population, and we wish to use the sample to make inferences about the unknown population mean μ. For example, in a study of body temperature it may be hypothesized that the average body temperature of subjects in a specified population is $98.6°$ F. The null hypothesis may be stated:

$$H_0: \mu = 98.6.$$

Suppose our sample consists of the values

$$98.3, 98.6, 99.0, 98.1.$$

The test procedure of this section is applicable with the four sample values playing the roles of the d's. We need to compute (see Comment 8)

$$T = \frac{\bar{d} - 98.6}{\widehat{SD}}.$$

We find

$$\bar{d} = 98.5,$$

$$\sum (d - \bar{d})^2 = .46,$$

$$s = \sqrt{\frac{.46}{3}} = .392,$$

$$\widehat{SD} = \frac{.392}{\sqrt{4}} = .196,$$

and thus,

$$T = \frac{98.5 - 98.6}{.196} = -.51.$$

With $\alpha = .05$, the two-sided test rejects if T exceeds $t_{.025,3}$ or if T is less than $-t_{.025,3}$. From Table C5 we find $t_{.025,3} = 3.182$. Thus our observed value of $T = -.51$ is not in the critical region, and we accept H_0.

The test based on T also yields, as in Section 1, a confidence interval for μ. A $1 - \alpha$ confidence interval for μ is given by equation (3). For $\alpha = .05$ and $n = 4$, the 95% interval is

$$\bar{d} \pm (t_{.025,3})(\widehat{SD}) = 98.5 \pm (3.182) \cdot (.196)$$

$$= 97.9 \text{ and } 99.1.$$

To test $H_0: \mu = \mu_0$, the test statistic (discussed in Comment 8 and this comment)

$$T = \frac{\bar{d} - \mu_0}{\widehat{SD}}$$

is referred to a t distribution with $n-1$ degrees of freedom. Note that the denominator of T is $\widehat{SD}$, our estimate of the standard deviation of $\bar{d}$. However, if the standard deviation, σ, of the d population is known, then we do not have to estimate the standard deviation of $\bar{d}$, for in such a case we can compute it directly as $\sigma/\sqrt{n}$. In such cases where σ is known, the test statistic to be used is

$$T_\sigma = \frac{\bar{d} - \mu_0}{\sigma/\sqrt{n}}$$

and T_σ should be referred to the standard normal distribution (Table C2). More specifically, the two-sided α level test of $H_0: \mu = \mu_0$ versus $\mu \neq \mu_0$ rejects H_0 when T_σ is larger than $z_{\alpha/2}$ or smaller than $-z_{\alpha/2}$; if not H_0 is accepted. The one-sided α level test of H_0 versus $\mu > \mu_0$ rejects H_0 if $T_\sigma > z_\alpha$; if not H_0 is accepted. The one-sided α level test of H_0 versus $\mu < \mu_0$ rejects H_0 if $T_\sigma < -z_\alpha$ and accepts H_0 otherwise. Here z_α is the upper α percentile point of the standard normal distribution.

In the case where σ is known, the $1-\alpha$ confidence interval for μ is

$$\bar{d} - \left(z_{\alpha/2} \cdot \frac{\sigma}{\sqrt{n}} \right), \qquad \bar{d} + \left(z_{\alpha/2} \cdot \frac{\sigma}{\sqrt{n}} \right).$$

This is analogous to the confidence interval when σ is unknown but here $\sigma/\sqrt{n}$ replaces $\widehat{SD}$ and the normal percentile $z_{\alpha/2}$ replaces the t percentile $t_{\alpha/2, n-1}$.

10.　The tests of this section, based on referring T to percentiles of the t distribution with $n-1$ degrees of freedom, are exact when the d's are a random sample from a normal population. Furthermore, for any population with a finite variance, these tests are nearly exact for large n. Competing nonparametric tests, which do not require the assumption of a normal population, are presented in Chapter 12.

11.　Investigators who report their studies in the scientific journals rarely practice hypothesis testing in the rigorous decision-making mode suggested by mathematical statistics texts. The common method is to report the P value (see Discussion Question 2) and then interpret this value in their own verbal discussion of their results. If the P value is small, they regard their data as inconsistent with the null hypothesis. If the P value is not strikingly small, they are more hesistant, less insistent on rejection of the null hypothesis. If the P value is large, they conclude that the data are consistent with the null hypothesis. Note that when the

P value is large, that is, when the null hypothesis cannot be rejected, *this does not mean that the null hypothesis is accepted without further question.* In the blood storage example the data did not lead to rejection of the hypothesis of no storage effect. Yet *there may have been a storage effect,* in the range of -2% to $+4\%$, over the 8 month period. Further investigation is appropriate.

12.   Scientists often ask just how small the significance test P value must be before the null hypothesis can be rejected as inconsistent with the sample data. The question does not have an answer that is both easy and satisfying. The following answers are ranked in order with regard to both criteria, starting with the answer that is easiest and least satisfying:

1.   A common practice among scientists and editors is to regard 5%, sometimes 1%, as the dividing line between consistency and inconsistency of the data with the null hypothesis. The development of this custom is difficult to trace; it has little justification in statistical logic, and it is regularly condemned by statisticians and knowledgeable scientists.

2.   It is true that if you were to follow some set rule, such as rejecting H_0 when P is less than some specified cutoff value, denoted by α (for example, α might be .05), then, in only $100\alpha\%$ of the cases in which H_0 is actually true would you make the error of rejection. Thus, by specifying a cutoff point beforehand, you can control your error rate for this type of error, called an error of the first kind or Type-I error (see Comment 6). J. Neyman and E. S. Pearson suggested that if the comparison of $\bar{d}$ with 0 were regarded as a decision problem, in which the hypothesis that $\mu = 0$ is to be either accepted or rejected (contrary to the philosophy of R. A. Fisher, who maintained that the hypothesis can be rejected but cannot be accepted), then there are two types of error to consider, the Type-I and Type-II errors. (Recall that the Type-II error is accepting the hypothesis when it is actually false.) This framework provides a basis for deciding on the critical region. At this point we could show how the error rate for Type-II errors can be computed, and discuss the balancing of Type-I and II error rates according to the Neyman–Pearson theory of hypothesis testing. However, here we take the point of view that this approach is difficult and not natural to the scientist. We feel that the scientist is better served by the less rigorous reporting style characterized by giving estimates, standard deviations of the estimates, and significance test P values where these will be helpful to his reader.

3.   Fisher was highly critical of Neyman and Pearson for their "commercial" view of the scientist as a producer of decisions. There are some deep issues involved here that we cannot go into, except to say that there are two strong

sides to the question and the question is still very much alive among applied statisticians today. Here it must only be added that if the scientist is to be regarded as a decision maker, then the Neyman–Pearson theory does not go far enough. If the scientist is to go further than reporting his set of data, discussing it as fully as possible, and relating it to the body of current knowledge—if he is to regard himself as a decision maker—then he must consider more than error rates. He must weigh past evidence in with the data he has accumulated, and he must weigh the penalties or costs involved in making the various types of error. Neyman and Pearson attempted only to deal with the error rates on a mathematical basis. Later, A. Wald attempted to deal with the whole problem of present data, past data, the whole body of current knowledge, error rates, and costs of possible errors. His decision theory approach does lead to an answer as to how small the P value should be in order to reject the hypothesis, but the decision theory approach has not been adopted in the everyday use of statistics in medicine and biology. The reasons are partly inertia, partly the complexity of the approach, and partly the resistance to formalize the more subjective of the inference processes of the scientist.

PROBLEMS

5. Return to Problem 1, and test the hypothesis that the two measurement methods are equivalent.

6. Return to Problem 2, and test the hypothesis that operant conditioning has no effect on blood pressure reduction. (Save your solution for later comparisons with the nonparametric approach of Chapter 12.)

7. Return to Problem 3, and test the hypothesis that tensile strengths for tape-closed wounds are equivalent to tensile strengths for sutured wounds. (Save your results for comparison with results to be obtained in Problem 1, Chapter 12).

3.   DETERMINING THE SAMPLE SIZE

One of the most important uses of statistical theory is in the planning of studies. Knowing that the reliability of the sample mean, as measured by its estimated standard deviation, depends on the sample size, a study might be planned that generates enough observations to assure a desired reliability.

Recall that the standard deviation of the sample mean $\bar{d}$ is equal to $\sigma/\sqrt{n}$, where n is the sample size and σ is the population standard deviation. Suppose that an investigator would like to estimate a population mean from a sample large enough to assure (with probability 95%) that the sample mean is within a distance D of the population mean. In symbols, this requirement can be written as

(5)
$$Pr(|\bar{d} - \mu| \le D) = .95.$$

The event

$$|\bar{d} - \mu| \le D$$

is the same as the event

$$\left|\frac{\bar{d} - \mu}{\sigma/\sqrt{n}}\right| \le \frac{D}{\sigma/\sqrt{n}},$$

since we have simply divided both sides of the inequality by $\sigma/\sqrt{n}$. Thus we can rewrite equation (5) as

(5')
$$Pr\left(\left|\frac{\bar{d} - \mu}{\sigma/\sqrt{n}}\right| \le \frac{D}{\sigma/\sqrt{n}}\right) = .95.$$

Recall (Chapter 3) that if the distribution of the d's is exactly normal, then the distribution of $\bar{d}$ is exactly normal and, in fact, $(\bar{d} - \mu)/(\sigma/\sqrt{n})$ has a standard normal distribution. Thus we have that

(6)
$$Pr\left(\left|\frac{\bar{d} - \mu}{\sigma/\sqrt{n}}\right| \le 1.96\right) = .95.$$

Even if the distribution of the d's is not normal, the central limit theorem assures us that $(\bar{d} - \mu)/(\sigma/\sqrt{n})$ has an approximate standard normal distribution and the $=$ symbol in (6) can be replaced by $\approx$ (approximately equals).

Comparing equation (5') and (6) yields the result

$$\frac{D}{\sigma/\sqrt{n}} = 1.96.$$

Solving for n yields

(7)
$$n = \frac{(1.96)^2 \sigma^2}{D^2}.$$

The investigator has specified D and, through the degree of assurance desired, he has specified the normal percentile, 1.96. The only remaining value needed for calculation of the sample size is the magnitude of the population standard deviation, σ. This standard deviation can often be approximated from personal knowledge, past data, or information in the scientific literature. Note that if the approximation is too large the only damage done is to call for a sample size larger than required. The result is an estimate more reliable than necessary.

More generally, if we want our estimate $\bar{d}$ to be within D of the population mean with probability $1 - \alpha$, equation (7) becomes

(7')
$$n = \frac{(z_{\alpha/2})^2 \cdot \sigma^2}{D^2},$$

where $z_{\alpha/2}$ is the upper $\alpha/2$ percentile point of a standard normal distribution. These percentiles are given in Table C2.

Note that, rather than specifying individual values of σ and D, it is sufficient to specify the value of the ratio σ/D in order to use equation (7) or (the more general) equation (7').

Comments

13.   Consider equation (7') which gives a method of choosing the sample size n. Note that:
(i)   As σ^2 increases, n increases. Thus populations with high variability require more observations (than populations with low variability) to achieve a desired precision.
(ii)   As α decreases, n increases. If you want a higher probability of being within D of the true value of μ, you must increase the sample size.
(iii)   As D increases, n decreases. If you are willing to accept being further away from the true value of μ, you can get away with smaller samples.

Summary of Statistical Concepts

In this chapter the reader should have learned that:
(i)   Confidence intervals can be used to provide (interval) estimates of unknown population parameters.
(ii)   Hypothesis tests are procedures that are directly related to confidence intervals and are useful for making inferences about the values of population parameters.
(iii)   The t distribution arises, as a derived distribution, in inference problems concerning the mean of a normal population.

(iv)   It is possible to determine the sample size, when planning studies, to assure that (with high probability) the sample mean will be within a specified distance of the population mean.

PROBLEMS

8.   Use the triglyceride measurements of fresh blood samples, the second column of Table 1, to estimate the standard deviation of triglyceride levels in the study population represented. How large a sample of individual subjects would be necessary to estimate the population mean triglyceride level to within 10 mg/ 100 ml, with approximate probability 99%? [Hint: Use the sample standard deviation s, in place of the unknown value of σ, in (7′). With this substitution, the actual probability will only approximate the desired 99%.]

9.   Explain why formula (7′), for determining the sample size, only depends on σ and D through the ratio σ/D. [For example, the configuration $\alpha = .05$, $\sigma = 1$, $D = .8$ yields the same n as the configuration $\alpha = .05$, $\sigma = 3$, $D = 2.4$ because in both cases $\sigma/D = 1.25$.]

10.   Define sample standard deviation, standard error, confidence interval, confidence coefficient, t distribution with $n - 1$ degrees of freedom, nonparametric, outlier, null hypothesis, test statistic, significance probability, critical region, Type-I error, Type-II error, power, two-tailed test, one-tailed test, two-sided alternative, and one-sided alternative.

Supplementary Problems

1.   Hayes, Carey, and Schmidt (1975), in a study of retinal degeneration associated with taurine deficiency in the cat, analyzed amino acid concentrations for nine animals on a casein diet. Their finding for the amino acid isoleucine was expressed as the "mean ± standard deviation: 60 ± 14". Determine a 95% confidence interval for μ, the true mean concentration of isoleucine in plasma for the (conceptual) population of cats fed a casein diet for the prescribed period of time. [Note that, as abstracted here, this is a one-sample (as opposed to a paired-sample) problem, and Comment 9, is appropriate. Note also that this problem does not give the basic data, but only certain summary statistics. The reader who peruses the scientific literature will find this is often the case when statistical analyses are given. That is, sometimes the authors of articles choose not to include the basic data, and sometimes the absence of the data is a consequence of a journal's editorial policy. In any event, the omission of the basic data is undesirable because it is usually not possible to check published statistical results without the basic data. Furthermore, the brief phrase "mean ± standard deviation: 60 ± 14" is ambiguous. Is the value 14, which Hayes *et al.* call the standard deviation, the sample standard deviation (s, in our notation), or the estimate of the standard deviation of the sample mean ($\widehat{SD}$, in our notation)? Obviously, we need to make the proper identification before we can compute the confidence interval for μ. As we have pointed out in the text, $\widehat{SD}$ is often called the standard error of the (sample) mean. With this in mind, and by noting that the value 14 is, in the amino acid concentration context, too large to be $\widehat{SD}$, we can be reasonably sure that the authors simply abbreviated sample standard deviation to standard deviation, and thus we take $s = 14$.]

2. Questions 2 and 3 refer to the data in Table 4. These data are from a study of White, Brin, and Janicki (1975) on the effect of long-term smoking of marijuana on the functional status of blood lymphocytes. Table 4 gives blastogenic responses *in vitro* to pokeweed mitogen in microcultures of peripheral blood lymphocytes from a group of 12 healthy, long-term marijuana smokers and a group of matched control subjects.

TABLE 4.   Blastogenic Responses of Lymphocytes to Pokeweed Mitogen

	Radioactivity (dpm) per culture	
Experiment	Smokers	Controls
1	141,448	100,540
2	163,225	153,372
3	99,984	110,029
4	94,467	120,627
5	167,983	150,436
6	107,180	173,707
7	75,893	99,772
8	126,051	90,498
9	76,932	86,072
10	86,691	107,214
11	90,587	101,932
12	115,015	106,852

Source: S. C. White, S. C. Brin, and B. W. Janicki (1975).

Are there significant differences between the smokers and nonsmokers? Use an appropriate hypothesis test.

3. Compute 99% confidence limits for the long-run average difference between smokers and controls.

4. Questions 4–6 refer to the data in Table 5. The data are a subset of the data obtained by Kaneto, Kosaka and Nakao (1967) and are cited by Hollander and Wolfe (1973). The experiment investigated the effect of vagal nerve stimulation on insulin secretion. The subjects were mongrel dogs with varying body weights. Table 5 gives the amount of

TABLE 5. Blood Levels of Immunoreactive Insulin(μU/ml)

Dog	Before	After
1	350	480
2	200	130
3	240	250
4	290	310
5	90	280
6	370	1,450
7	240	280

Source: A. Kaneto, K. Kosaka, and K. Nakao (1967).

immunoreactive insulin in pancreatic venous plasma just before stimulation of the left vagus nerve and the amount measured 5 minutes after stimulation for seven dogs.

Test the null hypothesis that stimulation of the vagus nerve has no effect on the blood level of immunoreactive insulin against the one-sided alternative hypothesis that stimulation increases the blood level.

5. Determine 90% confidence limits for the mean of the "after minus before" population.

6. The 1450 "after" value for dog 6 might be considered an outlier. Explain. Change 1450 to 145 and determine the effect this change has on the value of $\bar{d}$.

7. Return to the data of Table 3, Chapter 1 and test the hypothesis of no difference between lags 1 and 2.

8. For the data of Table 3, Chapter 1, determine 95% confidence limits for the mean of the "lag 2 minus lag 1" population.

9. Questions 9 and 10 refer to the data in Table 6. The data are a subset of the data obtained by Cole and Katz (1966) and are also cited by Hollander and Wolfe (1973). Cole and Katz were investigating the relation between ozone concentrations and weather fleck damage to tobacco crops in southern Ontario, Canada. One of the objective measurements reported was oxidant content of dew water in parts per million (ppm) ozone. Table 6 gives the oxidant contents of twelve samples of dew that were collected during the period of August 25–30, 1960 at Port Burwell, Ontario.

Test the hypothesis that the mean oxidant content μ of dew water was .25 against the alternative that it was not equal to .25. [Hint: This is one-sample, as opposed to paired-sample, data. Comment 9 is relevant.]

10. Give a 95% confidence interval for the mean oxidant content μ.

TABLE 6.   Oxidant Content of Dew Water

Sample	Oxidant Content (ppm ozone)
1	0.32
2	0.21
3	0.28
4	0.15
5	0.08
6	0.22
7	0.17
8	0.35
9	0.20
10	0.31
11	0.17
12	0.11

Source: A. F. W. Cole and M. Katz (1966).

REFERENCES

Cole, A. F. W. and Katz, M. (1966). Summer ozone concentrations in southern Ontario in relation to photochemical aspects and vegetation damage. *J. Air Poll. Control Ass.* **16**, 201–206.

Hayes, K. C., Carey, R. E., and Schmidt, S. Y. (1975). Retinal degeneration associated with taurine deficiency in the cat. *Science* **188**, 949–951.

Hoel, P. G. (1971a). *Elementary Statistics*, 3rd edition. Wiley, New York.

*Hoel, P. G. (1971b). *Introduction to Mathematical Statistics*, 4th edition. Wiley, New York.

Hollander, M. and Wolfe, D. A. (1973). *Nonparametric Statistical Methods*. Wiley, New York.

Kaneto, A., Kosaka, K., and Nakao, K. (1967). Effects of stimulation of the vagus nerve on insulin secretion. *Endocrinology* **80**, 530–536.

*Lin, P. E. and Stivers, L. E. (1974). On difference of means with incomplete data. *Biometrika* **61**, 325–334.

White, S. C., Brin, S. S., and Janicki, B. W. (1975). Mitogen-induced blastogenic responses of lymphocytes from marihuana. *Science* **188**, 71–72.

Wood, P. D. (1973). Personal communication.

CHAPTER 5

Analysis of the Two-Sample Location Problem Using Sample Means

In this chapter we analyze data consisting of two independent random samples, sample 1 from population 1 and sample 2 from population 2. It is convenient to think of sample 1 as a control sample and sample 2 as a treatment sample. We want to know if there is a treatment effect; that is, does the mean of population 2 differ from the mean of population 1? The null hypothesis asserts that there is no treatment effect, that the population means are equal.

In Section 1 we estimate the unknown population means by their respective sample means, and a confidence interval is obtained for the difference in population means. Section 2 describes how to test the hypothesis that the population means are equal. The test is directly related to the confidence interval of Section 1.

Section 3 considers the determination of sample sizes. Section 4 presents a test of the equality of population variances.

1.	A CONFIDENCE INTERVAL FOR THE DIFFERENCE BETWEEN THE POPULATION MEANS

Two Independent Samples versus Matched Pairs

In Chapter 4 we saw how to test for and estimate a treatment effect when the basic data occur in pairs. For example, in Table 5 of Chapter 4 the basic data are seven pairs of measurements, one pair for each dog, where the first observation in a pair is the blood level of immunoreactive insulin just before stimulation of the left vagus nerve, and the second observation in a pair the blood level 5 minutes after stimulation. However, in many experiments, it is not possible for the subjects (in the insulin experiment, the dogs) to serve as their own controls. For example, consider the "study on walking in infants" described in Chapter 1. Suppose our goal was to compare "active-exercise" with "no-exercise." Clearly, we could not subject an infant to both active exercise and no exercise. One way to circumvent this difficulty and still apply the techniques of Chapter 4 is to form matched pairs. Thus one can form pairs of infants in such a way that within each pair, the infants are as homogeneous as possible with respect to factors that might affect the basic variable of interest—time to walking alone. Then within each pair one infant would be assigned the "active-exercise" treatment, the other infant the "no-exercise" treatment, and we could apply the techniques of Chapter 4. However, the success of matched pairs depends on the degree to which the subjects within pairs are homogeneous.

In the "walking in infants" study, and in many other studies designed to compare two (or more) treatments, it is often more convenient to divide the subjects into separate groups. It is usually best to keep the sizes of the groups equal or nearly equal (we say more about this in Section 3), and it is important that the subjects be randomly assigned to the treatments (this is discussed in detail in Section 2 of Chapter 6). The methods presented in this chapter are for this "unpaired" case, in which there is no pairing; instead, we have two independent samples, one sample corresponding to treatment 1 (perhaps a control), the other sample corresponding to treatment 2. As we emphasize in Chapter 6, random assignment of the subjects to the treatments eliminates bias and tends to make the treatment groups comparable on *all* relevant characteristics, including relevant characteristics that are unknown to the investigator.

Effects of Massive Ingestion of Vitamin C

Since ascorbic acid is a dietary essential for humans, many investigators have considered the question of what is a reasonable daily intake. The National Academy of Sciences, National Research Council, in their 1973 recommended dietary allowances specified a recommended daily intake of 45 mg of ascorbic acid for human males and females. (See Johnson, 1974.)

Claims (Pauling, 1971) have been made that large daily doses of ascorbic acid can prevent or cure the common cold. Partially as a result of such claims, consumption of ascorbic acid by the public has increased. Although there is evidence that vitamin C is relatively innocuous even in relatively large amounts, there has been some concern about massive ingestion. Keith (1974) has performed a study designed to investigate the effects of massive ingestion of L-ascorbic acid upon food consumption, growth, kidney weight, and morphology and proximate body composition of young male guinea pigs. Male guinea pigs were force-fed ascorbic acid supplements of 250, 500, and 1000 mg daily for nine weeks; a control group was included. In Table 1 we present the proximate body compositions (in terms of percent of body weight) of crude lipid in the carcasses of the control and 1000-mg groups.

TABLE 1.   Crude Lipid Compositions (%)

Control Group	1000-mg Group
23.8	13.8
15.4	9.3
21.7	17.2
18.0	15.1

Source: R. E. Keith (1974).

One way to attempt to answer the question of whether the population means are equal is to compare the means of the two samples. Let us designate the control group observations (sample 1) as X's and the 1000-mg group observations (sample 2) as Y's. Then the mean of sample 1 is

$$\bar{X} = \frac{23.8 + 15.4 + 21.7 + 18.0}{4} = 19.73,$$

and the mean of sample 2 is

$$\bar{Y} = \frac{13.8 + 9.3 + 17.2 + 15.1}{4} = 13.85.$$

We see that

$$\bar{Y} - \bar{X} = -5.88.$$

Now, $\bar{Y} - \bar{X}$ is an estimate of $\mu_2 - \mu_1$, where μ_1 is the mean of population 1 and μ_2 is the mean of population 2. How reliable is $\bar{Y} - \bar{X}$ as an estimate of $\mu_2 - \mu_1$? We know, for example, that with four different control guinea pigs and four different 1000-mg guinea pigs, the corresponding value of $\bar{Y} - \bar{X}$ would not be the same as the value of -5.88 that was obtained. If the population means were equal, we would expect $\bar{Y} - \bar{X}$ to be close to zero. Is the value -5.88 that was obtained far enough away from zero to conclude that, in fact, the population means are not equal?

Reliability of the Difference in Sample Means

Consider the problem of making an inference about the difference between two population means, based on independent, random samples from the two populations. Let the sample size of the sample from population 1 be denoted as m and the sample size of the sample from population 2 be denoted as n. Set

$$\Delta = \mu_2 - \mu_1,$$

the difference between the two population means.

A natural estimate of Δ is the statistic

(1)
$$\hat{\Delta} = \bar{Y} - \bar{X},$$

where $\bar{X}$ is the mean of sample 1 and $\bar{Y}$ is the mean of sample 2. (A different estimate of Δ is presented in Comment 4 of Chapter 13.) To assign a reliability to the estimate $\hat{\Delta}$ of the true difference Δ, we need to consider the standard deviation of $\hat{\Delta}$. It can be shown that the standard deviation of $\hat{\Delta}$ is

(2)
$$SD(\hat{\Delta}) = \sqrt{\frac{\sigma_1^2}{m} + \frac{\sigma_2^2}{n}},$$

where σ_1 is the standard deviation of population 1 and σ_2 is the standard deviation of population 2. If we do not know σ_1 and σ_2, as is the usual case, we can estimate σ_1 and σ_2 by the respective sample standard deviations, s_1 and s_2.

To compute s_1:
1. Sum the X values from sample 1 and call this sum of m observations $\sum X$.
2. Square each X value, sum the m squared values, and call this sum $\sum X^2$.
3. Compute

(3)
$$s_1 = \sqrt{\frac{\sum X^2 - \dfrac{(\sum X)^2}{m}}{m-1}}.$$

To compute s_2:
1. Sum the Y values from sample 2 and call this sum of n observations $\sum Y$.
2. Square each Y value, sum the n squared values, and call this sum $\sum Y^2$.
3. Compute

(4)
$$s_2 = \sqrt{\frac{\sum Y^2 - \dfrac{(\sum Y)^2}{n}}{n-1}}.$$

We can use s_1 and s_2 to obtain an estimator of the standard deviation of $\hat{\Delta}$. In the right-hand side of equation (2) replace σ_1 by s_1 and σ_2 by s_2 to obtain the estimator

(5)
$$\widehat{SD} = \sqrt{\frac{s_1^2}{m} + \frac{s_2^2}{n}}.$$

An *approximate* $1-\alpha$ confidence interval for Δ is then obtained as

(6)
$$\hat{\Delta} \pm (z_{\alpha/2}) \cdot (\widehat{SD})$$

where $\widehat{SD}$ is given by (5) and where $z_{\alpha/2}$ is the upper $\alpha/2$ percentile point of a standard normal distribution (see Table C2). By "approximate" we mean here that the coverage probability is approximately $1-\alpha$ and tends to $1-\alpha$ as m and n get large.

Let us now obtain the interval, given by (6), for the crude lipid data of Table 1. We use a confidence coefficient of 95% so that, in our notation, $\alpha = .05$, $(\alpha/2) = .025$, and (from Table C2) we find $z_{.025} = 1.96$. The estimator $\hat{\Delta}$ has already been computed,

$$\hat{\Delta} = \bar{Y} - \bar{X} = -5.88,$$

but we need to obtain s_1^2 and s_2^2 for use in equation (5).
From Table 1, we find

$$\sum X = 23.8 + 15.4 + 21.7 + 18.0 = 78.9,$$
$$\sum X^2 = (23.8)^2 + (15.4)^2 + (21.7)^2 + (18.0)^2 = 1598.49,$$

$$\sum Y = 13.8 + 9.3 + 17.2 + 15.1 = 55.4,$$

$$\sum Y^2 = (13.8)^2 + (9.3)^2 + (17.2)^2 + (15.1)^2 = 800.78.$$

Substitution into equations (3) and (4), with $m = 4$ and $n = 4$, yields

$$s_1^2 = \frac{1598.49 - \dfrac{(78.9)^2}{4}}{3} = 14.06,$$

$$s_2^2 = \frac{800.78 - \dfrac{(55.4)^2}{4}}{3} = 11.16.$$

From (5), we obtain

$$\widehat{SD} = \sqrt{\frac{14.06}{4} + \frac{11.16}{4}} = 2.51.$$

The desired approximate 95% confidence interval is found from (6) to be

$$-5.88 \pm (1.96) \cdot (2.51) = -5.88 \pm 4.92$$

$$= -10.80 \text{ and } -.96.$$

The confidence interval given by (6) is only approximate. The approximation is based on the fact that the distribution of $\hat{\Delta}/\widehat{SD}$ is approximately normal (subject to the very mild mathematical assumption that σ_1^2 and σ_2^2 are finite). The approximation improves as m and n get large, but cannot be trusted for sample sizes as small as the $m = 4$, $n = 4$ case of our example. When the sample sizes are small we must rely on further information about the parent distributions. If we know, or are willing to assume, that the parent populations are normal and the variances are the same in the two populations (i.e., $\sigma_1^2 = \sigma_2^2$), an exact $1 - \alpha$ confidence interval can be given. The common variance is estimated by pooling the two sample variances, s_1^2 and s_2^2, to obtain a **pooled estimate, s_p^2**:

(7)
$$s_p^2 = \frac{(m-1)s_1^2 + (n-1)s_2^2}{(m-1) + (n-1)}.$$

An alternative way of writing s_p^2, and easier for computational purposes than equation (7), is

(7')
$$s_p^2 = \frac{\sum X^2 - \dfrac{(\sum X)^2}{m} + \sum Y^2 - \dfrac{(\sum Y)^2}{n}}{m + n - 2}.$$

The *exact* $1-\alpha$ interval for Δ is

(8)
$$\hat{\Delta} \pm (t_{\alpha/2,m+n-2}) \cdot s_p \cdot \sqrt{\frac{1}{m}+\frac{1}{n}},$$

where $t_{\alpha/2,m+n-2}$ is the upper $\alpha/2$ percentile point of the t distribution with $m+n-2$ degrees of freedom. Values of these percentile points are given in Table C5.

Let us obtain the interval given by (8) for the crude lipid data of Table 1. We find, using (7'),

$$s_p = \sqrt{\frac{1598.49-\dfrac{(78.9)^2}{4}+800.78-\dfrac{(55.4)^2}{4}}{6}} = \sqrt{12.61} = 3.55.$$

If we seek a 95% confidence interval, $(\alpha/2) = .025$, and from Table C5, $t_{.025,6} = 2.447$. Thus the desired interval is

$$-5.88 \pm (2.447) \cdot (3.55) \cdot \sqrt{\frac{1}{4}+\frac{1}{4}} = -5.88 \pm 6.14$$

$$= -12.02 \text{ and } .26.$$

Comments

1. The confidence interval defined by (8) is exact (i.e., the true coverage probability is exactly $1-\alpha$) when the two parent populations are normal with equal population variances. A test for the equality of population variances is presented in Section 4. A nonparametric confidence interval, which does not require the assumption of normal parent populations, is described in Chapter 13.

2. The estimator $\hat{\Delta}$ and the confidence intervals based on $\hat{\Delta}$ are very sensitive to outlying observations. For example, in Table 1, change the Y value 9.3 to 33 and note how this changes the value of $\hat{\Delta} = \bar{Y} - \bar{X}$. In Chapter 13 we present another interval estimator of $\mu_2 - \mu_1$ that is less sensitive to outliers than the interval estimators based on $\hat{\Delta}$.

PROBLEMS

1. Consider the kidney data of Table 1, Chapter 13. Determine a 90% confidence interval for Δ using interval (8) of this section. (Save this result for comparison with the 88.8% nonparametric interval to be obtained in Section 2 of Chapter 13.)

2. Consider the ozone concentration data of Table 3, Chapter 13. Determine a 95% confidence interval for Δ using interval (8) of this section. (Save this result for comparison with the 95.2% interval to be obtained in Problem 4 of Chapter 13.)

3. Construct an example that illustrates the extreme sensitivity of the estimator $\hat{\Delta}$ to outlying observations. Why is this sensitivity undesirable?

2.	**A HYPOTHESIS TEST FOR THE DIFFERENCE BETWEEN THE POPULATION MEANS**

Does the Mean of the Control Population Equal that of the 1000-mg Population?

Suppose attention were focused on some hypothesis concerning Δ. We show how a significance test can be made, comparing the data with what would be expected under the hypothesis. The most common null hypothesis H_0 is that the populations have the same mean; that is, $\mu_1 = \mu_2$, or equivalently,

$$H_0: \Delta = 0.$$

A test of H_0 can be based on

(9)
$$T' = \frac{\bar{Y} - \bar{X}}{s_p \cdot \sqrt{\dfrac{1}{m} + \dfrac{1}{n}}},$$

where s_p^2 is the pooled estimate given by (7'). If the parent populations are normal with the same standard deviation, then when H_0 is true the distribution of T' is the t distribution with $m + n - 2$ degrees of freedom. Tests for significance are based on how far T' is from zero.

To test H_0, versus the alternative $\Delta \neq 0$, at the α level, compute T', and if T' is larger than $t_{\alpha/2, m+n-2}$ or smaller than $-t_{\alpha/2, m+n-2}$, the null hypothesis is rejected. If not, the null hypothesis is accepted.

For the data of Table 1 we have

$$T' = \frac{-5.88}{(3.55) \cdot \sqrt{\dfrac{1}{4} + \dfrac{1}{4}}} = -2.34.$$

At $\alpha = .05$, $(\alpha/2) = .025$, and from Table C5 we find $t_{.025,6} = 2.447$. Since $-2.447 \leq T' \leq 2.447$, the null hypothesis H_0 is accepted at the $\alpha = .05$ level. This could have been noted in the previous section, since there we saw that the 95% confidence interval, obtained with equation (8), included the point 0. Note, however, that the significance test P value is very close to 5%, and the investigator might well take the position that his data suggest that ascorbic acid decreases the lipid fraction.

Discussion Question 1. Could you study the effect of ascorbic acid on the lipid fraction with a *paired replicates* experiment of the type analyzed in Chapter 4?

Discussion Question 2. Do the results of the study described in this Chapter enable the experimenter to conclude that ascorbic acid affects lipid metabolism in male guinea pigs?

Comments

3. In this section we illustrated the two-sided test based on T'. The one-sided α level test of H_0, versus the alternative $\Delta > 0$, rejects when T' exceeds $t_{\alpha, m+n-2}$, and accepts otherwise. Similarly, the one-sided α level test of H_0, versus the alternative $\Delta < 0$, rejects when T' is less than $-t_{\alpha, m+n-2}$, and accepts otherwise.

4. In this section, our hypothesis tests were formulated for the hypothesis $\Delta = 0$. To test $\Delta = \Delta_0$ (say), where Δ_0 is any specified value, simply redefine T' to be

$$T' = \frac{(\bar{Y} - \bar{X}) - \Delta_0}{s_p \cdot \sqrt{\dfrac{1}{m} + \dfrac{1}{n}}},$$

and refer it to percentiles of the t distribution with $m + n - 2$ degrees of freedom (Table C5). Significantly large values indicate $\Delta > \Delta_0$; significantly small values indicate $\Delta < \Delta_0$.

When the population standard deviations σ_1 and σ_2 are known (this is rarely the case in practice), the test statistic to be used is

$$T_{\sigma_1, \sigma_2} = \frac{(\bar{Y} - \bar{X}) - \Delta_0}{\sqrt{\dfrac{\sigma_1^2}{m} + \dfrac{\sigma_2^2}{n}}},$$

and T_{σ_1, σ_2} should be referred to the standard normal distribution (Table C2). More specifically, the two-sided α level test of H_0: $\Delta = \Delta_0$ versus $\Delta \neq \Delta_0$ rejects H_0 when T_{σ_1, σ_2} is larger than $z_{\alpha/2}$ or smaller than $-z_{\alpha/2}$; if not H_0 is accepted. The one-sided α level test of H_0 versus $\Delta > \Delta_0$ rejects if $T_{\sigma_1,\sigma_2} > z_\alpha$; if not H_0 is accepted. The one-sided α level test of H_0 versus $\Delta < \Delta_0$ rejects H_0 if $T_{\sigma_1,\sigma_2} < -z_\alpha$ and accepts H_0 otherwise.

In the case where σ_1 and σ_2 are known, the $1 - \alpha$ confidence interval for Δ is

$$\bar{Y} - \bar{X} \pm (z_{\alpha/2}) \cdot \sqrt{\frac{\sigma_1^2}{m} + \frac{\sigma_2^2}{n}}.$$

5. The tests of this section are exact when the two parent populations are normal with equal population variances. Competing nonparametric tests, which do not require assumption of normal populations, are presented in Chapter 13. See also Problem 6.

6. The test based on T' (and the test based on the statistic T'' introduced in Problem 6) is discussed by Armitage (1973, pp. 118–126), who gives references to related procedures based on the t distribution.

PROBLEMS

4. Consider the kidney data of Table 1, Chapter 13. Use T' to test the hypothesis $\Delta = 0$ versus the alternative $\Delta < 0$. (Later you get a chance to compare your results with those obtained via the nonparametric test of Chapter 13.)

5. Consider the ozone concentration data of Table 3, Chapter 13. Use T' to test the null hypothesis that the population means are equal against the alternative hypothesis that the mean of the acute population is less than that of the chronic population. State what assumptions you are making. (Later, compare the solution to this problem with the solution to Problem 1 of Chapter 13.)

6. Corresponding to the approximate confidence interval given by equation (6) of Section 1, an approximate α level test of $H_0: \Delta = \Delta_0$ (where Δ_0 is a specified constant) can be based on the statistic

$$T'' = \frac{(\bar{Y} - \bar{X}) - \Delta_0}{\sqrt{\frac{s_1^2}{m} + \frac{s_2^2}{n}}}.$$

Assuming σ_1^2 and σ_2^2 are finite, the distribution of T'' can be approximated by a standardized normal, and the significance of T'' can be obtained by referring T'' to Table C2. The test, although only approximate, does not assume normality of the populations, nor does it assume that the populations have equal standard deviations. The approximation improves as m and n increase. Apply the test of $H_0: \Delta = 0$ based on T'' to the crude lipid compositions data, and compare your results with the results of the test of this section based on T'. (Note that, since $m = n$, we have $T'' = T'$. However, T' is to be referred to student's t distribution and T'' is to be referred to the standard normal distribution.)

7. Apply the test based on T'', introduced in Problem 6, to the kidney data of Table 1, Chapter 13. Compare your results with those obtained in Problem 4.

8. Apply the test based on T'', introduced in Problem 6, to the ozone concentration data of Table 3, Chapter 13. Compare your results with those obtained in Problem 5.

3.	DETERMINING SAMPLE SIZES

The following discussion of required sample sizes is patterned closely after the corresponding discussion in Section 3 of Chapter 4. Here the focus is on the parameter $\Delta = \mu_2 - \mu_1$. We suppose that we have information concerning the population standard deviations. Often it is sufficient to take the two standard deviations to be equal, say, $\sigma_1 = \sigma_2 = \sigma$, and we consider only this case. Furthermore, we assume that σ is known (or, at least that the experimenter can specify a reasonable value of σ).

Now the estimate for the difference in population means is the difference in sample means, $\hat{\Delta} = \bar{Y} - \bar{X}$. The derivation of required sample size utilizes the following result.

> Let $\bar{X}$ denote the sample mean of a sample of size m from a normal population with mean μ_1 and variance σ_1^2. Let $\bar{Y}$ denote the sample mean of an independent sample of size n from a normal population with mean μ_2 and variance σ_2^2. Then $\hat{\Delta} = \bar{Y} - \bar{X}$ has a normal distribution with mean $\Delta = \mu_2 - \mu_1$ and variance $\dfrac{\sigma_1^2}{m} + \dfrac{\sigma_2^2}{n}$.

In particular, under the assumption that $\sigma_1 = \sigma_2 = \sigma$, the standard deviation of $\hat{\Delta}$ is

$$SD(\hat{\Delta}) = \sqrt{\frac{\sigma^2}{m} + \frac{\sigma^2}{n}} = \sigma \cdot \sqrt{\frac{1}{m} + \frac{1}{n}}.$$

Suppose that an investigator would like to estimate Δ from samples large enough to assure (with probability .95) that $\hat{\Delta}$ is within a distance D of Δ. In symbols, this requirement can be written as

(10)
$$Pr(|\hat{\Delta} - \Delta| \le D) = .95.$$

The event

$$|\hat{\Delta} - \Delta| \leq D$$

is the same as the event

$$\left| \frac{\hat{\Delta} - \Delta}{\sigma \cdot \sqrt{\dfrac{1}{m} + \dfrac{1}{n}}} \right| \leq \frac{D}{\sigma \cdot \sqrt{\dfrac{1}{m} + \dfrac{1}{n}}},$$

since we have simply divided both sides of the inequality by $\sigma \cdot \sqrt{\dfrac{1}{m} + \dfrac{1}{n}}$. Thus we can rewrite equation (10) as

(10′)
$$Pr\left\{ \left| \frac{\hat{\Delta} - \Delta}{\sigma \cdot \sqrt{\dfrac{1}{m} + \dfrac{1}{n}}} \right| \leq \frac{D}{\sigma \cdot \sqrt{\dfrac{1}{m} + \dfrac{1}{n}}} \right\} = .95.$$

But $(\hat{\Delta} - \Delta) / \left\{ \sigma \cdot \sqrt{\dfrac{1}{m} + \dfrac{1}{n}} \right\}$ has a standard normal distribution so we also know that

(11)
$$Pr\left\{ \left| \frac{\hat{\Delta} - \Delta}{\sigma \cdot \sqrt{\dfrac{1}{m} + \dfrac{1}{n}}} \right| \leq 1.96 \right\} = .95.$$

Comparing equations (10′) and (11) yields the result

(12)
$$\frac{D}{\sigma \cdot \sqrt{\dfrac{1}{m} + \dfrac{1}{n}}} = 1.96.$$

Equation (12) is a single equation in the two unknowns, m and n. There are many combinations of m and n that satisfy equation (12), but there are a number of reasons for choosing equal sample sizes, $m = n$. In this case, the required number of observations in each sample is

(13)
$$n = \frac{2(1.96)^2 \cdot \sigma^2}{D^2}.$$

Two reasons for choosing equal sample sizes are that this choice minimizes the total number of observations required and it also reduces the errors resulting from using the procedure based on T', with pooled sample standard deviations, for populations whose true standard deviations are not actually equal.

More generally, if we want our estimate to be within D of the true value with probability $1-\alpha$, equation (13) becomes

(13′)
$$n = \frac{2(z_{\alpha/2})^2 \sigma^2}{D^2},$$

where $z_{\alpha/2}$ is the upper $\alpha/2$ percentile point of a standard normal distribution. These percentiles are given in Table C2.

For the crude lipid composition problem, suppose we want our estimator $\hat{\Delta}$ to be within 5 of the true value of $\Delta = \mu_2 - \mu_1$, with probability .95. Suppose further we are willing to assume $\sigma_1 = \sigma_2 = 4$. Then equation (13′), with $D = 5$, $\sigma^2 = 16$, and $z_{\alpha/2} = z_{.025} = 1.96$, gives

$$n = \frac{2(1.96)^2(16)}{25} \approx 5.$$

Note that rather than specifying individual values of σ and D, it is sufficient to specify the value of the ratio σ/D in order to use equation (13) or (the more general) equation (13′). By doing this, we are saying we want our confidence interval to represent so many standard deviation units.

Comments

7. Consider equation (13′) which gives a method of choosing the sample sizes. Note that:

(i) As σ^2 increases, n increases. Thus populations with high variability require more observations (than populations with low variability) to achieve a desired precision.

(ii) As α decreases, n increases. If you want a higher probability of being within D of the true value of $\mu_2 - \mu_1$, you must increase the sample size.

(iii) As D increases, n decreases. If you are willing to accept a larger deviation from the true value of $\mu_2 - \mu_1$, you can get by with smaller samples.

PROBLEMS

9. Compare the sample size formula (13′) of this section with the analogous formula (7′) of Chapter 4. Explain why the factor 2 appears in (13′) but not in (7′) of Chapter 4.

4.	TESTING THE ASSUMPTION OF EQUAL POPULATION VARIANCES

The confidence interval given by (8), and the test based on T' [see equation (9)], require that the underlying populations are normal with *equal* variances. The assumption of equal variances can be tested using the statistic $\mathscr{F}$:

(14)
$$\mathscr{F} = \frac{s_2^2}{s_1^2}.$$

Suppose population 1 is normal with mean μ_1 and variance σ_1^2, and population 2 is normal with mean μ_2 and variance σ_2^2. Let $H_{0'}$, denote the hypothesis of equality of variances:

$$H_{0'}: \sigma_1^2 = \sigma_2^2.$$

When $H_{0'}$ is true, the statistic $\mathscr{F}$ has an **F distribution** with $n-1$ degrees of freedom in the numerator and $m-1$ degrees of freedom in the denominator. The F distributions (see Table C9) are identified by two positive integers, rather than one as with the t distribution. The first integer, ν_1, is called the degrees of freedom of the numerator and the second integer, ν_2, is called the degrees of freedom of the denominator. As with the t distribution, the F distribution usually occurs as a derived distribution, rather than a distribution that is assumed for the basic data.

To test $H_{0'}$ versus the alternative $\sigma_1^2 \neq \sigma_2^2$ at the α level, reject $H_{0'}$ if $\mathscr{F}$ is larger than $F_{\alpha/2, n-1, m-1}$ or if $\mathscr{F}$ is smaller than $F_{1-(\alpha/2), n-1, m-1}$. If not, $H_{0'}$ is accepted. Here $F_{\alpha/2, n-1, m-1}$ is the upper $\alpha/2$ percentile point of an F distribution with $n-1$ degrees of freedom in the numerator and $m-1$ degrees of freedom in the denominator, and $F_{1-(\alpha/2), n-1, m-1}$ is the lower $\alpha/2$ percentile point (i.e., the upper $1-\alpha/2$ percentile point) of an F distribution with $n-1$ degrees of freedom in the numerator and $m-1$ degrees of freedom in the denominator. (We have, for convenience, chosen the two tails to have equal areas, although this is not quite the best test.)

For the data of Table 1, where $m = n = 4$, we have, in Section 1, determined that

$$s_1^2 = 14.06, \, s_2^2 = 11.16.$$

From (14) we then obtain

$$\mathscr{F} = \frac{11.16}{14.06} = .79.$$

With $\alpha = .05$, we find from Table C9, $F_{.975,3,3} = .065$ and $F_{.025,3,3} = 15.4$. Since $\mathcal{F} = .79$ is between .065 and 15.4, we accept $H_{0'}$.

Comments

8.   In this section we illustrated the two-sided test based on $\mathcal{F}$. The one-sided α level test of $H_{0'}$, versus the alternative $\sigma_2^2 > \sigma_1^2$, rejects when $\mathcal{F}$ exceeds $F_{\alpha,n-1,m-1}$, and accepts otherwise. Similarly, the one-sided α level test of $H_{0'}$ versus the alternative $\sigma_2^2 < \sigma_1^2$, rejects when $\mathcal{F}$ is less than $F_{1-\alpha,n-1,m-1}$, and accepts otherwise.

9.   In this section, the $\mathcal{F}$ test is formulated for the hypothesis $\sigma_2^2 = \sigma_1^2$, or equivalently, $(\sigma_2^2/\sigma_1^2) = 1$. To test $(\sigma_2^2/\sigma_1^2) = \gamma_0$ (say), where γ_0 is any specified positive value, simply redefine $\mathcal{F}$ to be

$$\mathcal{F} = \frac{s_2^2}{\gamma_0 \cdot s_1^2},$$

and refer it to percentiles of the F distribution with $n - 1$ degrees of freedom in the numerator and $m - 1$ degrees of freedom in the denominator. Significantly large values indicate $(\sigma_2^2/\sigma_1^2) > \gamma_0$; significantly small values indicate $(\sigma_2^2/\sigma_1^2) < \gamma_0$.

10.   The statistic $\mathcal{F}$ can also be used to furnish confidence intervals for σ_2^2/σ_1^2. Assuming the two underlying populations are normal, a $100(1-\alpha)\%$ confidence interval for σ_2^2/σ_1^2 is

$$\left(\frac{s_2^2/s_1^2}{F_{\alpha/2,n-1,m-1}}, \frac{s_2^2/s_1^2}{F_{1-(\alpha/2),n-1,m-1}} \right).$$

For the data of Table 1, a 95% confidence interval for σ_2^2/σ_1^2 is

$$\left(\frac{.79}{15.4}, \frac{.79}{.065} \right) = (.05, 12.2).$$

11.   The F percentiles have the property that

$$(15) \qquad F_{\alpha,\nu_1,\nu_2} = \frac{1}{F_{1-\alpha,\nu_2,\nu_1}}.$$

For example, find in Table C9 that $F_{.975,3,8} = .069$ and $F_{.025,8,3} = 14.5$ and note that $.069 = (1/14.5)$, agreeing with (15).

To determine critical points for tests based on $\mathcal{F}$, the user will find equation (15) very helpful. As an example, suppose $n = 6$, $m = 5$, and we desire a two-sided

$\alpha = .002$ test of $H_{0'}$: $\sigma_1^2 = \sigma_2^2$. Then $(\alpha/2) = .001$, and our procedure is

$$\text{reject } H_{0'} \quad \text{if } \mathscr{F} > F_{.001,5,4} \text{ or } \quad \text{if } \mathscr{F} < F_{.999,5,4},$$

$$\text{accept } H_{0'} \quad \text{if } F_{.999,5,4} \leq \mathscr{F} \leq F_{.001,5,4}.$$

We can determine $F_{.001,5,4}$ directly from Table C9. We find $F_{.001,5,4} = 51.7$. To find $F_{.999,5,4}$ we use (15) and Table C9. From (15),

$$F_{.999,5,4} = \frac{1}{F_{.001,4,5}} = \frac{1}{31.1} = .032,$$

where the value $F_{.001,4,5} = 31.1$ is obtained from Table C9. Thus the $\alpha = .002$ two-sided test is

$$\text{reject } H_{0'} \quad \text{if } \mathscr{F} > 51.7 \quad \text{or if } \mathscr{F} < .032,$$

$$\text{accept } H_{0'} \quad \text{if } .032 \leq \mathscr{F} \leq 51.7.$$

12. The test for equality of variances described in this section, based on the statistic $\mathscr{F}$, is not robust with respect to the assumption of normality in the sense that when the underlying populations are not normal, the true probability of a Type-I error when using the $\mathscr{F}$ test (that is supposed to have a Type-I error probability equal to α) may be quite far from α. Similarly, the true confidence coefficient of the confidence interval defined in Comment 10 may be quite far from $1 - \alpha$. Nonparametric tests for equality of variances, and confidence intervals for the ratio of variances, which do not require the assumption of underlying normal populations, are discussed in Chapter 5 of Hollander and Wolfe (1973).

Summary of Statistical Concepts

In this chapter the reader should have learned that:
(i)   Dividing the experimental subjects into two groups is a very useful alternative to matched pairs for the problem of comparing two treatments. The assignment of subjects to treatment groups should be randomized—this is emphasized in Chapter 6.
(ii)   The t distribution (earlier encountered in Chapter 4) again arises as a derived distribution, here in inference problems concerning $\Delta = \mu_2 - \mu_1$, the difference between two population means. Specifically, if the two populations are normal with the same standard deviation, then under the null hypothesis H_0: $\mu_2 = \mu_1$, the basic test statistic T' has a t distribution.

(iii) It is possible to determine the sample sizes when planning two sample studies to assure that (with high probability) $\bar{Y} - \bar{X}$ will be within a specified distance of $\mu_2 - \mu_1$.

(iv) The ratio of sample variances, $\mathcal{F} = s_2^2 / s_1^2$, can be used to test the hypothesis of equal population variances, and to provide a confidence interval for σ_2^2 / σ_1^2.

PROBLEMS

10. Consider the ozone concentration data of Table 3, Chapter 13. Use the $\mathcal{F}$ statistic to test the hypothesis of equal variability in the acute and chronic populations.

11. Consider the nuclear volume data of Table 4, Chapter 13. Use the $\mathcal{F}$ statistic to test the hypothesis of equal variability in the two underlying populations.

12. Define pooled estimate.

Supplementary Problems

1. Young, Leary, Zimmerman, and Strobel (1973) considered the question of whether dietary deficiency is related to the development of myopia. The data in Table 2, a subset of the data obtained by Young *et al.*, give vertical spectacle refractions for the right eye from 20 monkeys that had been on a low protein diet (2.5–3%) for an average of 32 months and for 17 monkeys that had been on a high protein diet (22–23%) for an average of 28 months.

Use T'' (see Problem 6) to test for a possible effect of the different levels of protein on refractive characteristics of the right eye.

2. Raisman and Field (1971) considered a quantitative evaluation of the relative distribution of synapses on dendritic shafts and spines in samples from the preoptic area and the ventromedial nucleus of male and female rats. Here we consider their data from the synapses in the neuropil of the preoptic area, and we investigate the possibility that the neuropil of the preoptic area is sexually dimorphic. We view the 10 male rat observations in Table 3 as a random sample of size ten from a male rat population (population 1) and the 8 female rat observations as a random sample of size eight from a female rat population (population 2). Use T' to test the hypothesis of equal means for the male and female populations. Give a 95% confidence interval for $\mu_2 - \mu_1$.

3. In the study discussed in Supplementary Problem 2, Raisman and Field investigated the relative distribution of synapses on dendritic shafts and spines in samples taken from the ventromedial nucleus of male and female rats. Their data, based on six male rats and six

TABLE 2. Vertical Spectacle Refractions for the Right Eye for Low and High Protein Groups of Rhesus Monkeys

Low Protein Group	High Protein Group
1.27	−6.00
−4.98	0.25
−0.50	1.25
1.25	−2.00
−0.25	3.14
0.75	2.00
−2.75	0.75
0.75	1.75
1.00	0.00
3.00	0.75
2.25	0.75
0.53	0.25
1.25	1.25
−1.50	1.25
−5.00	1.00
0.75	0.50
1.50	−2.25
0.50	
1.75	
1.50	

Source: F. A. Young, G. A. Leary, R. R. Zimmerman, and D. A. Strobel (1973).

female rats, are given in Table 4. Use T' to test for differences between population means. Give a 99% confidence interval for $\mu_2 - \mu_1$.

4.   The data in Table 5 are from a study by Dick, Deodhar, Provan, Nuki, and Buchanan (1971). Table 5 gives values of uptake of intravenously administered radioactive technetium (^{99m}Tc), measured over the knee joints in nine (apparently) healthy volunteers and in 11 (unoperated) patients with rheumatoid arthritis. Is there evidence of different uptake rates for healthy volunteers and patients with rheumatoid arthritis?

5.   Consider the "study on walking in infants" of Chapter 1. Do the data of Table 2, Chapter 1, indicate that the mean time to walking alone for the active-exercise group is significantly less than that of the no-exercise group?

*6.   Suppose s_1^2 is very close to s_2^2. In such cases, how does the length of interval (6) compare with the length of interval (8)?

TABLE 3. Ratio (R) of the Number of Nonamygdaloid Synapses on Dendritic Shafts to the Number on Dendritic Spines for the Preoptic Area

Male	Female
32.3	13.9
27.2	13.5
23.8	11.7
23.4	10.7
22.6	10.5
20.9	9.7
18.2	9.1
14.6	6.2
13.4	
12.9	

Source: G. Raisman and P. M. Field (1971).

TABLE 4. Ratio (R) of the Number of Nonamygdaloid Synapses on Dendritic Shafts to the Number on Dendritic Spines for the Ventromedial Nucleus

Female	Male
6.2	6.1
5.1	4.8
4.5	4.7
4.0	3.9
3.8	3.7
3.3	3.4

Source: G. Raisman and P. M. Field (1971).

TABLE 5. ^{99m}Tc Uptakes (c.p.m.) in Healthy Subjects and Rheumatoid Arthritis Patients

Volunteers	Rheumatoid Arthritis Patients
2397	3592
3002	3820
2662	4689
2205	2870
2811	3522
1981	2932
2755	4135
2862	3868
2917	4218
	3961
	4875

Source: W. C. Dick, S. D. Deodhar, C. J. Provan, G. Nuki, and W. W. Buchanan (1971).

7. Generalize the formula s_p^2, given by (7) for the two sample case, to obtain a pooled estimate of σ^2 in the k sample case. That is, suppose we have k independent samples, sample 1 from population 1, sample 2 from population 2, . . . , sample k from population k. Assume that the k populations have equal variances σ^2. Define a pooled estimate of σ^2.

8. Suppose $\bar{X}$ is the sample mean of a sample of size 12 from a normal population with mean 10 and variance 1, and that $\bar{Y}$ is the sample mean of an independent sample of size 15 from a normal population with mean 12 and variance 2. What is the distribution of $\hat{\Delta} = \bar{Y} - \bar{X}$?

9. Recall the study of myopia and dietary deficiency referenced in Supplementary Problem 1. Could you design a study to investigate the relationship between myopia and dietary deficiency using a paired replicates experiment of the type described in Chapter 4? Explain.

10. An investigator wishes to compare two treatments for a disease. Suppose the investigator *does not* assign the patients to the treatments at random. How would the absence of random assignments open the door for various biases to enter the study?

REFERENCES

Armitage, P. (1973). *Statistical Methods in Medical Research.* Wiley, New York.

Dick, W. C., Deodhar, S. D., Provan, C. J., Nuki, G., and Buchanan, W. W. (1971). Isotope studies in normal and diseased knee joints: ^{99m}Tc uptake related to clinical assessment and to synovial perfusion measured by the ^{133}Xe clearance technique. *Clin. Sci.* **40,** 327–336.

Hollander, M. and Wolfe, D. A. (1973). *Nonparametric Statistical Methods.* Wiley, New York.

Johnson, P. E. (1974). Recommended dietary allowances, revised 1973. *Food Nutr. News* **45(2),** 1.

Keith, R. E. (1974). The effects of massive ingestion of L-ascorbic acid upon food consumption, growth, kidney weight and morphology and proximate body composition of young male guinea pigs. M.Sc. thesis, Florida State University.

Pauling, L. (1971). *Vitamin C and the Common Cold.* Freeman, San Francisco.

Raisman, G. and Field, P. M. (1971). Sexual dimorphism in the preoptic area of the rat. *Science* **173,** 731–733.

Young, F. A., Leary, G. A., Zimmerman, R. R., and Strobel, D. A. (1973). Diet and refractive characteristics. *Amer. J. Optom. Arch. Amer. Acad. Optom.* **50,** 226–233.

CHAPTER 6

Surveys and Experiments in Medical Research

In this chapter the application of statistical techniques to clinical research is discussed in some detail. In the first section it is pointed out that the patients seen by a physician and included in a clinical study technically cannot be considered a random sample. Yet statistical procedures can be applied validly in the analysis of clinical data under certain circumstances, particularly if treatments are randomly assigned to patients.

In Section 2 we discuss methods of executing a valid **clinical trial**, that is, a medical experiment on human beings. Random allocation of treatments to patients presents some difficult problems in medical ethics; these are discussed in Section 3. Finally, in Section 4 some other research approaches, alternative to the randomized clinical trial, are discussed.

1. APPLICATION OF STATISTICS IN LABORATORY AND CLINICAL STUDIES

To this point we have emphasized applications in which there is an identifiable, existing parent population of objects. We have acted as if the sample were obtained from the parent population according to a carefully defined technical procedure called random sampling. In the two sample situation, we again considered that there are two existing populations, each sampled randomly and independently. Such problems and procedures fall under the statistical classification of sample surveys.

In biomedical research, the sample survey situation is not the most common form of study. The laboratory investigator may be studying intact animals, treated in several ways. He wants to compare results for the several treatments but the animals are not randomly selected from some large population. The clinician, who is attempting to describe the results he has obtained with a particular therapy, cannot say that he has obtained his patients as a random sample from a parent population. How can statistics help in measuring the credibility of scientific inferences in such studies?

First, we consider the one-sample case. Here the data are to be compared with other results outside the study or are presented for descriptive purposes. For example, a surgeon may do a particular vein graft procedure (coronary bypass) to supplement coronary blood flow in patients with angina. Suppose he decides to summarize his results on his first 100 patients. He may measure improvement in blood flow (or improvement on some measured response to an exercise test) for each patient. He can certainly compute the mean for his hundred patients, and regard his report as pure description. But this is not very satisfactory. The clinical scientist usually attempts to generalize from his results. The surgeon attempts to convince himself and his readers that these results on 100 patients *typify* the results to be expected for this procedure in his hands and in the hands of others, at least for a particular type (population) of patient. Therefore, he goes to some lengths to provide assurance that these patients are like a random sample from his potential, conceptual supply of patients. He says they are "consecutive" patients (or "unselected" patients) of a given type. Further, he describes how the patients were diagnosed and what their characteristics were at the time of operation so that his readers can see what type of "population" he is operating on. Thus there is some concept of an underlying parent population, of which the 100 are a sample. There is a desire to generalize, to make an inference from sample mean to "population" mean, and the standard error of the sample mean can help in assessing the reliability of the sample mean for this purpose. However, it should be underlined here that the assumption that the collection of 100 simulates a random sample of this surgeon's patients, and this surgeon's surgical skill, may be a tenuous one. The broader inference to patients in *other* institutions operated on by *other* surgeons is still more dangerous. And the standard error of the mean cannot be trusted to measure all of these uncertainties, when random sampling has not been done.

There are some things the clinical investigator can do to aid in generalizing, or to guard against errors in generalizing, from a single series of patients. First, he can describe the group of patients in some detail so that his readers can see the nature of the patients he has treated, their age range, the severity of their disease, and so forth. Second, in measuring the effects of treatment, he can report measurements both before and after surgery, so that each patient serves as his own control, so to speak. But the single sample study remains difficult to evaluate, with or without statistical aids.

In many clinical studies and in most laboratory studies a control group is included, so that comparisons can be made between experimental effect and control effect within the study itself. Such studies are called **comparative experiments**. In the simplest comparative experiment there are two treatment groups, for example, one group of mice receiving treatment A and another group receiving treatment B. Although the mice might be assumed to simulate samples of mice from two conceptual populations (imagine treating all mice of strain X with A and, alternatively, treating all mice of strain X with B), there is nothing in the standard statistical procedures that aids in justifying this broad generalization to all mice of strain X, since the mice in the experiment are not a random sample from strain X. Confidence intervals and tests of significance are useful here, but they rest on a model that is strictly conceptual and usually beyond any direct verification. Note the difference between this situation and the sample survey situation, in which the populations are well defined and samples can be considered random samples because they were drawn according to prescribed random sampling procedures.

In 1935 Fisher (1966) proposed a method by which statistical analysis of comparative experiments could be put on a much firmer footing. He proposed that the experimental units (say, mice) be randomly allocated to the several treatments under study, using randomizing procedures similar to the random selection procedures we have described for surveys. Randomization serves two purposes. First, the groups of units receiving the several treatments tend to be comparable on all variables, known and unknown. Second, such randomization provides a secure foundation on which statistical measures of credibility (confidence coefficients, P values) can be justified.

If an experiment is randomized, (i.e., if the experimental units are randomly allocated to the several treatments) then the statistical inference can speak validly to the question of whether the difference observed between treatment groups can be attributed to the treatments themselves or is explainable solely in terms of differences that might arise due to chance imbalance in randomized groups. The standard errors measure the variability due to the random variation *purposely* built into the experiment by randomization.

Of course, Fisher randomization provides a basis only for statistical inference concerning the experimental units at hand. If these units have not been randomly selected from a larger population, and usually they have not, then inference to a larger sphere must be done extra-statistically, or at least without the benefit of an explicitly defined random sample and parent population. Standard errors and other statistical aids are useful in making these broader inferences, but they do not provide full measures of uncertainty.

PROBLEMS

1. Gunther and Bauman (1969) reported a clinical study in which women in labor were randomly assigned to one of two spinal anesthetics, lidocaine plus epinephrine or mepivicaine without epinephrine. One of the observations then recorded was the length of time from full dilation to delivery. The study included 67% of all women delivering babies at the Stanford Medical Center and 88% of all caudal anesthetics during a period of $11\frac{1}{2}$ months. To what population of women can the statistical inference be applied most directly? Can the inference be generalized to women arriving at the Stanford Medical Center in future years? Can the statistical logic support a generalization to other women, treated by other obstetricians, in other medical centers?

2.　RANDOMIZED CLINICAL EXPERIMENTS

Fisher's idea of random allocation of experimental units to treatments was firmly established in agricultural research by 1940. The idea of randomization was developed for clinical experiments on human beings in the 1940s and is now rather widely accepted, especially in the comparison of drug therapies. There is a large literature, including several books, specifically directed to the problems of conducting valid, randomized experiments (called **randomized clinical trials**) on human subjects. The books by Hill (1962, 1971), a pioneer in this area, and the book edited by Witts (1964) contain much of the early history, specific case studies of trials, together with discussion of the principles and techniques of implementation. Here we touch on several of the important general aspects of clinical trials.

It is important to state again the two objects or advantages of randomly allocating the treatments to the patients in a comparative study. First, the random allocation eliminates bias in the assignment of treatment to patients and will tend to make the treatment groups comparable on all relevant factors but the treatment itself. Note that the random allocation tends to balance the groups on *all* relevant characteristics of the patients in the trial, whether or not the investigator has measured the characteristics or is even aware of them! Second, the random allocation furnishes a logical foundation on which probability calculations can be made for the purpose of statistical inference. The probability calculations help to separate apparent treatment effects from the chance variation that might be associated with the random allocation itself.

After the treatments have been randomly assigned to the patients, it is necessary to take steps to preserve the unbiasedness of the experiment as it progresses through the actual treatment of the patients to the evaluation of the outcome of therapy. It is well known and documented that the physician himself and his teammates can influence the results of therapy, both through their own psychological encouragement and optimism and through more thorough attention to the details of administering and monitoring one treatment or the other. Of course, much of this comes about subconsciously. The effects of mock (placebo) treatments and other effects that come about through patient attitudes and expectations are also well known and documented. To obviate these physician and patient biases, clinical trials are run in a **double blind** manner where possible. Double blind means that neither physician nor patient are aware of which treatment the patient has received. This can often be accomplished easily in drug trials, where a tablet identical to the one under test can be produced that lacks only the active ingredient. Ingenious ways of double blinding have been devised, and it is wise to consult people with experience and broad knowledge of these techniques before concluding that double blinding is not feasible for a given situation. Even in situations in which true double blinding is not possible, certain compromises can be made that are far better than giving up altogether.

Something should be said about the actual procedures for randomization. Suppose, for simplicity, two treatments (one of which may be a placebo) are to be compared. If all patient candidates for the trial are known at the start of the trial, it is a trivial matter to list them by name, number them, and go to a table of random numbers to select at random half of them for assignment to one of the treatments, leaving the remaining half for the other treatment. However, in the usual clinical trial the candidates are not known in advance. They come to the clinic or hospital, willy-nilly, and must be treated with one or another treatment as they present themselves. In this case a sequential list of patient numbers can be made up in advance, with treatments randomly assigned to the numbers ahead of time. As the patients are admitted to the study, they are assigned a number in sequence, and thus they receive a treatment assignment. Since the decision as to which patient is a proper candidate for the trial is usually somewhat subjective, depending on various diagnostic criteria and matters of judgment, it is very important that the decision maker be ignorant of the treatment assignments that have been allocated to the various patient numbers as he makes his decision for admission. Of course, it is all the better if the physician can be blindfolded even after the patient is admitted to the study and the therapy assigned and begun.

The random assignment is often arranged by inserting treatment assignment cards into sequentially numbered envelopes, to be opened one by one as the patients are admitted to the study and assigned an envelope study number. To assure balance in the number of patients assigned to the several treatments, the random assignments are balanced to achieve equal numbers in the two treatment groups in successive blocks of six (or eight, say) assignments. The method of balancing should not be revealed to the investigator choosing the patients for treatment, since this would give him information about the next treatment assignment toward the end of each block.

Of course, the envelope method will not leave the physician blindfold as to the ultimate treatment assignment for the patient. In certain drug trials it is possible to make up the two drugs in a numbered series of packages without treatment identification, keeping code for the series separate for decoding at analysis of the data. Later we discuss methods of random allocation other than the so-called completely randomized experiment outlined thus far. Randomization can be done within certain types of subject, for example, males and females separately, to assure better balance in the number of each sex assigned to each treatment. We mention the more complicated randomization procedures here only to emphasize that *some* sort of randomization is highly desirable in clinical trials, but it need not be complete randomization. This is analogous to the requirement in sample survey work that the sampling be random sampling but not necessarily simple random sampling. For more discussion of the types of randomization and the appropriate statistical analyses, see Chapter 10.

PROBLEMS

2. In Problem 1, how would you set up the actual procedure for randomization of women to caudal anesthetic? Do you think that blindfold precautions would be important in this experiment? How would you set them up?

*3. How would you set up a double-blind randomized clinical trial to test the theory that aspirin with an antiacid additive provides better relief, due to ease in digestion, than aspirin alone?

*4. A clinical trial that provoked immense interest and comment from both clinicians and biostatisticians was a randomized trial of oral hypoglycemic drugs for mildly ill adult diabetics. To the surprise and consternation of the investigators and the medical community in general, the drugs under study seemed to *cause*, rather than *prevent*, heart disease in patients. The trial seemed well designed and carefully executed, but it has come under careful scrutiny and much criticism. The report of the results by the University Group Diabetes Program (Klimt *et al*. 1970) together with a detailed criticism by Feinstein (1971) and rejoinder by Cornfield (1971) provide an excellent lesson in the design and evaluation of clinical trials. Read the papers in order; criticize the first paper and compare your critique with Feinstein's; then attempt to answer Feinstein's criticism and compare your efforts with Cornfield's (1971).

3.	**ETHICS AND STATISTICAL ASPECTS OF THE CLINICAL TRIAL**

There is general agreement that medical science is obligated to work for better treatments for disease. But there is some hazard in recommending newly developed medical treatments for general practice. At first blush, a treatment may appear to be effective, even more effective than the current treatment of choice, and then later the new treatment may be found to be disappointing in efficacy, or it may prove to have harmful side effects. Thus careful experimentation, first in the laboratory and then in the clinic, is necessary to assure the goal of better treatments. If there is any doubt that more careful and thorough clinical experiments are needed in medicine, the doubt will vanish with consideration of the many therapies that have been accepted into medical practice and then found to be ineffective or toxic, such as Thalidomide, MER 29, cold treatment for ulcers, 5-Fluoruracil for bladder cancer, estrogen treatment for heart disease, and many more. Currently, millions of tonsillectomies and appendectomies are done each year, with measurable risk of operative mortality, and there is continuing doubt concerning the benefits derived from this surgery by the overwhelming majority of patients. Each year hundreds of radical mastectomies are done, without convincing evidence that this treatment is more effective than the physically and psychologically less traumatic simple mastectomy. In the light of these experiences in modern medicine, it is no wonder that many have argued for the wider use of the randomized clinical trial as a test of new therapies before their wide acceptance into clinical practice.

The randomized clinical trial does present a dilemma to the physician-scientist, however. At the point where a new therapy is thought to be effective, more effective than the currently accepted therapies, the physician would seem obligated to prescribe the therapy that seems best for the patient, namely, the new therapy. It would seem unethical to experiment with the patient by putting him in a trial where he has some chance of receiving a second-best treatment. Some will argue that the physician cannot know for certain which is the better treatment at this point, so that it is quite all right to choose the therapy at random. We feel that it is more forthright to admit that the physician who puts a patient in a clinical experiment is surely subjecting the patient to the chance of getting a therapy that, in the physician's best judgment at the moment, is not the best treatment for that patient. The purpose of the trial is to confirm the physician's theory. We will argue subsequently that there are a number of other factors that usually counterbalance any disadvantage to the patient and accrue to his net benefit. Nevertheless, the point should be clearly stated that clinical experiments are carried out with certain risks or costs to patients in the experiment for the purpose of improving the treatment of patients in the future.

The ethical dilemma of the clinical experiment can most easily be seen by visualizing yourself as one patient in a series of 101 patients to be treated for a particular disease. If you are first in the sequence, you would want the physician to choose the treatment that he feels, in the light of all knowledge and experience to date, is the very best treatment for you. On the other hand, if you are the 101st patient in the sequence, you would be best treated if the physician were to experiment on the first 100 in such a way that he has gained maximum information about current therapies by the time he gets to you. From this example it is clear that the physician-scientist must make some difficult judgments as to when a clinical experiment is appropriate for the maximum benefit for patients, current and future.

One of the great ethical supports for the modern physician-scientist is the current practice of requiring the informed consent of the patient before he is admitted to any clinical experiment. This practice had its beginnings in the Nuremburg Code and was given further force by the World Health Organization in the Declaration of Helsinki, adopted by the 18th World Medical Assembly (1964). In the United States, the National Institutes of Health require the informed consent of the patient in every clinical experiment carried out with NIH financial assistance. The NIH has further required (1971) that in every medical center in which clinical experimentation is carried out under NIH financing, there must be a committee that examines each experimental plan to see that the rights and welfare of the patients are protected. The committee is to be made up of persons knowledgeable in "institutional commitments and regulations, applicable law, standards of professional conduct and practice, and community attitudes." The approval of this committee, together with the informed consent of the patient, can help to reassure the physician by allowing him to share the ethical burden of judging whether the experiment is justified. It must be admitted that this committee system is far from perfect in practice. A common deficiency is the lack of a nonscientist, patient advocate on most committees.

Although there can be no denial of the fact that the patient in the clinical trial is in an experiment and is not in the same situation as he is as an ordinary patient, there may be some compensating factors. These are not all present in all clinical trials, but they are commonly found and can be strong inducements for the patient and sound factors for persuading the physician that the experiment does not compromise the net welfare and care of the patients in the experiment:

1. Often the patient cannot get the new treatment *unless* he participates in the experiment. For example, the new treatment may be a new drug, approved by the FDA only for experimental use. This is a strong argument for experimenting with new treatments when they are new, rather than waiting till they are well accepted and widely available before attempting to give them a careful evaluation.

2. It should be emphasized that all patients with the disease being studied need not be included in the trial. There will be some patients for whom one of the treatments is mildly contraindicated or strongly indicated, because of con-comitant disease or other conditions. Such patients should be given the treatment indicated and should *not* be included in the study. The study will still be valid, provided that the decision for inclusion or exclusion is made *prior* to random assignment of treatment, rather than after the patient has been included in the study and assigned the contraindicated treatment. Since the group of patients can be carefully selected, this selection can be done in such a way that the decision as to which treatment is the treatment of choice is a matter of some doubt, and the ethical dilemma of withholding one or the other is not quite so acute, though it will still be real.

3. It is a matter of record that in many clinical experiments, the patients do better than the investigators anticipated, no matter which treatment they receive, as a result of closer nursing attention, more frequent laboratory tests, and more frequent visits by the study physicians. For example, in a study of the use of anticoagulants in the treatment of patients following myocardial infarction and reported by Achor *et al.* (1964), past experience in the general hospitals involved indicated that the current treatment showed a mortality rate of more than 20% in the first month following the heart attack. In the actual experiment comparing the current treatment with a new treatment, the experience with *both* treatments was substantially better than this. Thus even the patients receiving the inferior of the two treatments were probably better off for being in the experiment.

Incidentally, the new treatment turned out to be inferior to the established treatment in this trial.

4.   In many clinical experiments it is possible to promise the patients that they will receive the new treatment later, if they happen to be assigned the old treatment and the new treatment ultimately turns out to be more effective. This can be done, for example, in a chronic disease such as angina, or hayfever. It may not be possible where the treatments are varieties of surgical procedure or the trial involves treatment of an acute and fatal disease.

5.   Usually there are other inducements to the patient. Foremost is the scientific contribution he is making by helping in the generation of new scientific knowledge that will be of direct help to others. Other benefits may be free medical care during the experiment, and sometimes following the experiment. This point must be handled very delicately, however, because the patient should be protected against undue influence of short-term financial benefits. A patient with little money may be influenced to undergo a treatment of uncertain benefit and risk simply because the treatment is free. This point is especially important as our society tries to recognize and correct some of the more glaring inequities in the treatment of the disadvantaged.

PROBLEMS

5.   Give arguments pro and con for a randomized clinical trial to determine the efficacy or benefits of tonsillectomy. Do your arguments change if the operation is appendectomy? How do they change if the operation is heart transplant? Note that in the last instance, the operation is not an accepted and common mode of therapy; it is done only experimentally, for patients on the brink of cardiac death, and at only a few medical centers in the world.

•6.   Gehan and Freireich (1974) have written a paper that strongly defends the nonrandomized clinical study, at least in a variety of particular circumstances. Study their paper. Try to answer their arguments against the randomized study and in favor of the alternatives that they describe.

4.   RETROSPECTIVE STUDIES, PROSPECTIVE STUDIES, AND RANDOMIZED CLINICAL EXPERIMENTS

In these times the medical scientist is more and more frequently faced with the problem of determining whether some specific factor is a low-level public health hazard. For example, we list the following problems of interest at the time of this writing:

1. Do birth control pills cause thromboembolic death?
2. Do atomic energy plants cause birth defects among babies born of women living in areas proximate to the plants?
3. Does air pollution cause lung cancer?
4. Do oral hypoglycemic drugs cause heart attacks?
5. Does Halothane anesthesia cause liver damage?
6. Does smoking cause heart attacks?
7. Does diethylstilbestrol in our meat cause cancer?
8. Does estrogen therapy for cancer or for hypercholesterolemia cause heart attacks?
9. Do anesthetic gases in the operating room environment cause birth defects in babies born of operating room personnel?

There are reasons for believing that the answer to each of these questions is yes, but all are controversial. If yes, the added risk, although small for the individual involved, may sum across the United States population to a significant public health problem. The purpose of this section is to outline some strategies for studying this kind of question, pointing out the advantages and disadvantages of each strategy.

All the strategies are directed toward discovering if a person, exposed to factor (X), for example, estrogen, has a higher risk of some untoward event (E), for example, heart attack, than a person unexposed $(C,$ for control).

Figure 1 is useful for showing the four possibilities for the individual person:

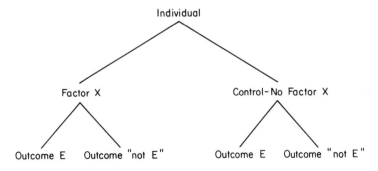

FIGURE 1. Possible exposure and outcome combinations.

Four commonly used strategies for attempting to discover whether persons exposed to factor X have a higher risk of outcome E are

I   Retrospective study (of past events).
II   Prospective study (of past events).
III Prospective study (of on-going events).
IV   Randomized clinical trial (on-going events).

We now proceed to describe these four strategies and their relative advantages and disadvantages.

Strategy I: The Retrospective Study of Past Data

The retrospective strategy can be described by referring to Figure 1. By some sort of case-finding method, persons are identified who have experienced the event (E) in question. Then the history of these cases is traced to determine whether they have been exposed to the factor X. The strategy is called retrospective because the direction of the study goes backward from effect to the hypothesized cause. In most studies it is also important to study some comparable control subjects who have *not* experienced the event, in order to demonstrate that the exposure to X in the study group is unusually high, relative to the rate of exposure in the control group.

A good example of the retrospective method is the early British study of thromboembolic death and birth control drugs reported by Inman and Vessey (1968). Thromboembolic deaths were identified from death certificates, and the exposure to the pill was traced by interview of the woman's physician and a check of her various medical records. Control women were women in the same age range, under the care of the same physicians.

Another large set of retrospective studies is the first few dozen studies of lung cancer and smoking. These studies are reviewed in a report to the U.S. Surgeon General (1964). Lung cancer cases or deaths were identified in a variety of ways, together with controls selected in a multitude of ways (e.g., stomach cancer cases in the same hospitals). Then smoking histories were obtained by interview of the cases, the controls or their families. The results consistently showed higher smoking rates among the lung cancer cases.

The advantage of the retrospective method is that it can be extremely fast and inexpensive, compared to the other strategies.

The disadvantages are several. Often there is difficulty in choosing a control group that is comparable to the event group on all variables except the event experience itself. Also, there is great opportunity for biases to affect the measurements, particularly the measurements of exposure to X. This is especially worrisome when the observations are subjective (e.g. smoking history) and the interviewers or subjects know the nature of the hypothesis under study. Another disadvantage of the retrospective study is that the risk of E for persons exposed to X cannot be evaluated directly. The retrospective approach does not follow a group of persons exposed to factor X in order to count events; consequently, no ratio of events per exposure can be obtained directly. Instead, one obtains a measure of exposure per event, a ratio with no practical interpretation.

Strategy II: Prospective Study of Past Data

This strategy depends on the existence of records of a peculiar form. Samples of persons exposed to X and persons not exposed to X are identified in the records. Then the records of the selected persons are traced to determine if they have ever experienced event E to the present time. The method is called prospective because it proceeds from exposure forward to the event.

A good example of this type of study is one of the later smoking studies, reported by Dorn (1958). There it was found that in the United States veteran insurance records, questions on smoking habits were asked at the initiation of the policies. Of course, the deaths of policy holders were also a matter of record, so that the existing data could be used to trace the experience of the smokers and the nonsmokers prospectively in time, using past data.

The prospective study of past data has several advantages. It may be quite fast, depending on the accessibility and quality of the data. It does furnish numerators and denominators for direct calculation of risks in the exposed and unexposed groups. Since it depends on existing data, it may not be subject to quite as many sources of bias as the usual retrospective study.

The disadvantage of the prospective study of past data is that it depends on the existence of the right data, collected reliably and completely, and readily accessible. The investigator who finds such data available is fortunate, indeed.

Strategy III: Prospective Study (of On-going Events)

This approach calls for the selection of samples for current study. Persons exposed to X and others unexposed are followed, and the incidence of the event E is observed in the two groups. Several of these studies have been carried out now with tens of thousands of smokers and nonsmokers, followed over periods of years.

The prospective study of current events has several 'obvious advantages. The methods of observation and recording can be designed and executed by the investigators so that the right data can be collected and the biases can be minimized.

The disadvantages are the relatively large cost of such studies, and, more important, the time necessary to carry them out. For low-hazard factors, in which the incidence of the event may be in the order of only a few events per thousand persons per year (even less in the case of thromboembolic death and the pill), the number of person-years of observation required in the exposed and unexposed groups may be prohibitively high.

Strategy IV: The Randomized Experiment

All three strategies just discussed are subject to the well-known criticism that association or correlation does not imply causation. Even though the incidence of E is higher among persons exposed to X, it can always be claimed that both X and E are caused by a common factor, Y, and changing X without changing Y would leave the risk of E unaffected. Fisher claimed that some people have a genetic predisposition to smoke and to contract lung cancer. If this is so and if this is the sole reason for the correlation we see, then a smoker with this predisposition cannot alter his chances for lung cancer by quitting his habit; he just makes himself miserable, for want of a cigarette. One might think of getting around this argument by checking the incidence of lung cancer among persons who have quit smoking, and indeed these people do have an incidence intermediate between smokers and persons who have never smoked. However, although this does seem to add weight to the theory of causality, it can also be explained in terms of weaker genetic propensities for both smoking and for lung cancer among people who find it possible to give up the habit once started.

We do have one strategy that can demonstrate causality directly. That strategy is the randomized clinical trial. Here the subjects are assigned to the exposure and nonexposure groups by some random mechanism, and the groups tend to be balanced on all factors, known and unknown, except for the factor X. Then, if the two groups differ in E by a statistically significant margin, this *can* be laid to X as the causal factor.

The randomized clinical study has the advantage of being a controlled experiment as definitive as the scientific method can be in revealing causal relationships.

However, it is not practical for many studies of side effects or low level environmental hazards, and it has the further disadvantage of using human subjects for experimentation. As discussed in Section 3, experiments with human subjects are difficult enough to justify when the treatments are therapies for patients needing medical treatment. Experimental exposure to hazards that are randomly assigned and otherwise avoidable is even harder to justify. Yet there are situations in which the clinical trial to assess hazard is appropriate. Exposure to very low levels of toxicity that are almost certainly transient, randomly protecting persons who would otherwise remain exposed (to normal cholesterol diet, say), substituting placebo or a known safe drug for a commonly used drug suspected of a fairly rare side effect—these are all examples of what may be ethical studies in some circumstances.

Discussion Question 1. In asking whether exposure to a factor X increases the risk of an untoward event E among persons in a population, the medical investigator might find it helpful to develop his study plan, in part, by considering the four classical strategies: the retrospective study, the prospective study of an existing data bank, the current prospective study, and the randomized clinical trial. What are the advantages and disadvantages of these four approaches?

Summary of Statistical Concepts

In this chapter the reader should have learned that:
(i)   Randomized clinical trials have come to be a mark of careful scientific investigation in clinical medicine. The main purposes of randomly assigning experimental units to treatments are to eliminate bias in making the assignments and to provide a valid foundation for statistical inference.
(ii)   The random, double-blind assignment of treatments to patients in a properly run clinical trial can be well justified on ethical grounds.
(iii)   In addition to the randomized clinical trial, there are other approaches to the study of clinical questions concerning therapy, drug side effects, public health hazards, and other population problems. The retrospective study and the prospective study are methods of controlled study that are not randomized but offer the possibility of valid inference, sometimes at a great gain in economy of time and cost.

PROBLEMS

•7. Suppose that there were to arise some reason to speculate that aspirin causes stomach cancer. Sketch designs of studies of this question along the four lines outlined in this section. Estimate the cost of each study, based on rough estimates of the cost per observation, using 100 cases per 100,000 persons per year as the incidence of the disease and assuming that 50% of all people make use of aspirin or a similar pain killer. Suppose that the study is to be completed in three years, and that a doubling of the risk of stomach cancer would be considered important enough to deserve detection.

8. Which of the questions listed at the beginning of this section could be studied ethically by the method of randomized experiment on human subjects?

9. Define clinical trial, comparative experiment, randomized clinical trial, double blind.

Supplementary Problems

1. In a television advertisement for a particular gasoline, 10 black cars and 10 white cars each start with a small amount of fuel, the black cars burning brand X and the white, brand A. Of course, the 10 white cars all go further than any of the black cars. How would you use the concepts of randomization and blinding to improve the credibility of this experiment?

2. In many cases the randomization of treatments to individual subjects is not feasible, even though the individual subjects are to be treated. For example, in studies involving grade school students it may be far more convenient to treat all students alike in any given classroom. This is quite acceptable if there are a number of classrooms and the treatments are randomly assigned to the classrooms. The experimental units are classrooms, then; the statistical analysis is somewhat more complicated, and confidence intervals usually will be much longer than if students had been randomized (though this is not necessarily so). Try to think of some examples in medicine or biology in which convenience or feasibility might dictate random assignment to groups of subjects, rather than individual subjects. Hint: plots of ground in agriculture, cages in a laboratory, clinics, surgeons.

3. Someone has said that there is never a right and ethical time to do a clinical trial comparing a new therapy with the current standard therapy—it is always too early or too late. Can you elaborate on this?

4. In some clinical trials it is possible to randomize within a patient, giving each patient two drugs simultaneously. For example, patients with psoriasis can be found who have similar skin lesions on each side of their body, for example, both hands might be affected by the disease. Then, a standard ointment might be applied to one side and an experimental ointment to the other. If the assignment of side to treatment is random, the result will be a matched pair study, readily analyzed by one-sample methods as in Chapter 4. But suppose an investigator, not versed in statistical principles but eager to objectively demonstrate that the test ointment is better, assigns the test ointment to the patient's side that appears *worse*. His reasoning is that if the test ointment appears better than the standard, with this handicap the results will be persuasive. How do you feel about this approach compared with random assignments?

5. Some opponents of the randomized, blinded, clinical trial, especially in the field of psychiatry, have argued that a necessary or at least important component in the efficacy of a psychoactive drug, for example, a tranquilizer, is the confidence the physician and the patient have in this efficacy. This factor is lost, of course, if one of the drugs in the trial is a placebo, and the active drug and placebo cannot be identified. Hence an active drug that *would* be efficacious if identified appears no better than placebo and is lost to medical practice. Do you agree with this position? Explain.

6. Skip ahead, for the moment, to Section 1 of Chapter 8, where we discuss a study entitled "Anesthetic Gases in the Operating Room Air—Harmful or Not?" Read the description of the study and state whether the study was a retrospective study of past events, a prospective study of past events, a prospective study of on-going events, or a randomized clinical trial.

7. Recall the Medi-Cal study described in Chapter 1. Was that study a retrospective study of past events, a prospective study of past events, a prospective study of on-going events, or a randomized clinical trial?

8. Skip ahead, for the moment, to Problem 3 of Chapter 8, which deals with a study on tonsillectomy rates. Is the study a retrospective study of past events, a prospective study of past events, a prospective study of on-going events, or a randomized clinical trial?

9. Cite some sources of error that are more likely to arise in retrospective studies than in prospective studies.

10. Explain why, especially for rare diseases, the retrospective study may be the only feasible approach.

REFERENCES

Achor, R. W. *et al.* (Cooperative Study). (1964). Sodium heparin vs. sodium warfarin in acute myocardial infarction. *J. Amer. Med. Assoc.* **189,** 555–562.

Advisory Committee to the Surgeon General of the Public Health Service (1964). *Smoking and Health*, Public Health Service Publication No. 1103. U.S. Government Printing Office, Washington, D.C.

Cornfield, J. (1971). The University Group Diabetes Program: A further statistical analysis of the mortality findings. *J. Amer. Med. Assoc.* **217,** 1676–1687.

Declaration of Helsinki: Recommendations guiding doctors in clinical research (1964). *World Med. J.* **11,** 281.

Dorn, H. F. (1958). The mortality of smokers and non-smokers. *American Statistical Association, Social Statistics Section, Proceedings* **1,** 34–71.

Feinstein, A. R. (1971). Clinical Biostatistics VIII: An analytical appraisal of the University Group Diabetes Program (UGDP) study. *Clin. Pharmacol. Ther.* **12,** 167–191.

Fisher, R. A. (1966). *The Design of Experiments*, 8th edition. Hafner, New York.

Gehan, E. A. and Freireich, E. J. (1974). Non-randomized controls in cancer clinical trials. *N. Engl. J. Med.* **290,** 198–203.

Gunther, R. E. and Bauman, J. (1969). Obstetrical caudal anesthesia; I. A randomized study comparing 1% mepivicaine with 1% lidocaine plus epinephrine. *Anesthesiology* **31,** 5–19.

Hill, A. B. (1971). *Principles of Medical Statistics*, 9th edition. Oxford University Press, London.

Hill, A. B. (1962). *Statistical Methods in Clinical and Preventive Medicine*. Oxford University Press, London.

Inman, W. H. W. and Vessey, M. P. (1968). Investigation of deaths from pulmonary, coronary, and cerebral thrombosis and embolism in women of child-bearing age. *Brit. Med. J.* **27,** 193–199.

Klimt, C. R., Knatterud, G. L., Meinert, C. L., and Prout, T. E. (1970). The University Group Diabetes Program: A study of the effects of hypoglycemic agents on vascular complications in patients with adult-onset diabetes. *Diabetes* **19** (Supplement 2), 747–830.

*Turnbull, B. W., Brown, B. Wm. Jr., and Hu, M. (1974). Survivorship analysis of heart transplant data. *J. Amer. Statist. Assoc.* **69,** 74–80.

U.S. Department of Health, Education, and Welfare (1971). *The Institutional Guide to DHEW Policy on Protection of Human Subjects*, DHEW Publication No. (NIH) 72–102. U.S. Government Printing Office, Washington, D.C.

Witts, L. J. (Ed.) (1964). *Medical Surveys and Clinical Trials*, 2nd edition. Oxford University Press, London.

CHAPTER | 7

Statistical Inference for Dichotomous Variables

In Chapters 3, 4, and 5 we analyzed data in which the observation for an individual unit is a quantitative measurement (e.g., weight, cholesterol level, or I.Q.). In many studies the observation consists merely of recording whether or not a particular response or attribute is present. Examples are dead or alive, male or female, cured or not, pregnant or not, with a specified disease or without. For convenience we speak of the presence of the attribute as a success, the absence as a failure. Such characteristics or variables are called dichotomous or binary variables.

Suppose our experiment consists of n independent trials. On each trial we observe either a success or a failure. We assume that the probability of success is the same for each trial, and we denote this probability of success by the symbol p. The probability of a failure on a given trial is thus $1-p$.

Let the total number of successes in the n trials be denoted by B. The sample rate of success is $\hat{p} = B/n$. The statistic $\hat{p}$ is a natural estimator of the corresponding unknown success probability, p. However, we would also like to know how to obtain or approximate the standard deviation of $\hat{p}$. Furthermore, it is informative to compute a confidence interval for p and to test hypotheses about p. The standard deviation of $\hat{p}$ is given in Section 1, where we also show how to obtain a confidence interval for p. Hypothesis tests for the parameter p are presented in Section 2.

---

1.   A CONFIDENCE INTERVAL FOR A SUCCESS PROBABILITY p

Independent Bernoulli Trials

We use the term "n independent repeated Bernoulli trials" to describe an experiment consisting of n trials where

    I   Each individual trial has two outcomes which we may conveniently label as success and failure.

    II The probability p of success stays the same throughout the trials.

    III The n trials are independent.

In each particular example you encounter in this chapter, and in your own experience, you will have to carefully consider whether the given experiment under consideration does indeed satisfy assumptions I, II, and III.

In this chapter we make inferences about the parameter p, using the statistic B, which is defined to be the number of successes in n independent Bernoulli trials. The distribution of B is the binomial distribution; we defer discussion of this distribution to Section 2.

Taste Testing

A firm that manufactures a particular best-selling children's medicine wants to change the formula to increase therapeutic efficacy. To determine whether the taste is also changed, a taste test is arranged. Samples are submitted to an expert taster; the samples are submitted to the taster in triplets, each triplet having two samples of the old formula and one of the new. The taster is asked to select the sample that is different in each triplet. Suppose the taster succeeds in correctly identifying the new formula in 48 of 100 triplets. Does this constitute strong evidence that the change in formula changed the taste as perceived by the expert?

Let us say a given trial is a success if the expert taster correctly identifies the new formula. Let p denote the probability of success. Does this taste-testing problem fit into the framework of independent, repeated Bernoulli trials? Clearly, assumption I is satisfied. But consider the possibility that a taste carry-over effect, from trial to trial, might destroy the validity of assumptions II and III.

We assume that sufficient time is allowed between trials to remove the possibility of a carry-over effect. If assumptions I, II, and III are indeed satisfied, how should we estimate p? A natural estimator for p is $\hat{p}$, the observed frequency of success:

$$\hat{p} = \frac{B}{n}.$$

For our taste-testing experiment,

$$\hat{p} = \frac{48}{100} = .48.$$

To obtain an interval estimate of the true success probability, p, based on the sample success rate, $\hat{p}$, we need to have an estimate of the standard deviation of $\hat{p}$. The standard deviation of $\hat{p}$ can be shown to be equal to

(1) $$SD(\hat{p}) = \sqrt{\frac{p(1-p)}{n}}.$$

Since the value of p is not known, $\hat{p}$ is substituted for p to obtain an estimate of the standard deviation of $\hat{p}$:

(2) $$\widehat{SD}(\hat{p}) = \sqrt{\frac{\hat{p}(1-\hat{p})}{n}}.$$

The right-hand side of (2) is sometimes called the standard error of $\hat{p}$.
    Thus for the taste-testing data,

$$\widehat{SD}(\hat{p}) = \sqrt{\frac{(.48)(.52)}{100}} = .050.$$

The central limit theorem allows us to conclude that the distribution of $\hat{p}$ is approximately normal (the approximation improves as n increases, see Comment 4), and a confidence interval can be based on this large sample approximation. For a $1-\alpha$ confidence coefficient, the confidence limits are

(3) $$\hat{p} \pm z_{\alpha/2}\widehat{SD}(\hat{p})$$

where $z_{\alpha/2}$ is the upper $\alpha/2$ percentile of the normal distribution (Table C2). For the taste-test data, 95% ($\alpha=.05$) confidence limits for p are

$$.48 \pm 1.96(.050) = .48 \pm .098$$

$$= .38 \text{ and } .58.$$

Suppose the old formula and new formula have the same taste. Then the true value of p would be 1/3. Since the 95% confidence limits do not include the value 1/3 and the lower limit .38 exceeds 1/3, this can be taken as strong evidence that $p>1/3$ and that the change in formula did, in fact, change the taste of the medicine. We return to this question, in a more formal hypothesis-testing framework, in Section 2.

Risk of Death Among Men Who Have Suffered a Heart Attack

The Coronary Drug Project Research Group (1974) reported that among 2789 men, aged 30–64 who had experienced one or more heart attacks (myocardial infarctions) and had apparently recovered and were in good health, 358 men (12.8%) died within the ensuing 3 years of study. Let us say that there is a success in a given trial if the corresponding postinfarction man does not die within the ensuing 3 years of study. We have $n = 2789$ trials. We find

$$\hat{p} = \frac{2431}{2789} = .872,$$

and

$$1 - \hat{p} = \frac{358}{2789} = .128.$$

Note that whereas $\hat{p}$ is the estimated probability of not dying in the ensuing 3 years of study, $1-\hat{p}$ is the estimated probability of dying within the ensuing 3 years. The estimated standard deviation of $\hat{p}$ is

$$\widehat{SD}(\hat{p}) = \sqrt{\frac{(.128)(.872)}{2789}} = .006.$$

Similarly,

$$\widehat{SD}(1-\hat{p}) = .006.$$

[Explain why $\widehat{SD}(1-\hat{p}) = \widehat{SD}(\hat{p})$.]
  Confidence limits for p, with $\alpha = .05$, are found from (3):

$$.872 \pm 1.96(.006) = .872 \pm .012$$

$$= .860 \text{ and } .884.$$

Similarly, 95% confidence limits for $1-p$ are

$$.128 \pm 1.96(.006) = .128 \pm .012$$

$$= .116 \text{ and } .140.$$

On the basis of these calculations, and the assumption that the 2789 men can be viewed as a random sample from the vast population of postinfarction men in the country, our interval estimate of the 3-year risk of death is 11.6 to 14.0%. In fact, the 2789 were not selected by random sampling techniques at all. They were recruited into a clinical trial comparing several cholesterol-lowering agents against placebo as prophylactic agents for heart disease. The 2789 are the men assigned to placebo among the 53 clinics across the United States which partici-pated in the study. There may be reason to believe that these men were more aware and more attentive of their problems and received more careful medical surveillance, with a resultant bias relative to the randomly selected postinfarct United States male. This is one of the reasons for including this group as a control group in this randomized clinical drug trial.

Discussion Question 1. For the death rate data the limits for the 3-year risk of death (for 95% confidence) were found to be 11.6% and 14.0%. Do you think that, on the basis of these limits, it is reasonable to estimate the *annual* mortality rate for such males to be in the range of 3.9 to 4.7%?

Discussion Question 2. The following article by C. Nolte-Watts appeared in the September 2, 1975 issue of the *St. Petersburg Times*, a newspaper published in St. Petersburg, Florida. Read the article and consider these questions. Is the indepen-dent Bernoulli trials model appropriate? How would you define trial and success? What is your estimate of the probability of success? Can you give a confidence interval for the probability of success? Does the article provide sufficient informa-tion to compare Freedent with "regular" gum, in terms of sticking?

Denture wearers confirm 'no-stick' gum ad claim

By CAROLYN NOLTE-WATTS
St. Petersburg Times Staff Writer
(c) 1975 Times Publishing Co.

Wrigley's is advertising a new gum, Freedent, on nationwide television commercials and in newspaper ads.

A woman holding a pack of gum says, "New Freedent gum won't stick to my dental work. I had to give up chewing ordinary gum because of the sticking problem. But Freedent, the new gum from Wrigley's, is specially formulated not to stick to most dental work. So I'm free to enjoy chewing again . . ."

WE ASKED EIGHT people who wear dentures to try Freedent. They said they stopped chewing regular gum because it stuck to their dental work.

Seven of the eight said Freedent did not stick to their dentures. One woman said it did. Most of the testers also commented however that the flavor was gone in a short time — less than 15 minutes.

A Wrigley's company spokesman explained that most dentures are made of acrylic resin, a material that tends to remain dry, causing most gum to stick to dental work.

HE SAID FREEDENT was made of the same ingredients used in other gums, but the ingredients were compounded differently to make the final product "wetter," eliminating the sticking problem.

Five sticks of Freedent cost 15 cents.

Watch This Space is a column in which The St. Petersburg Times puts advertisers' claims to the test.

"New Freedent® gum won't stick to my dental work."

"I had to give up chewing ordinary gum because of the sticking problem. But Freedent, the new gum from Wrigley's, is specially formulated not to stick to most dental work. So I'm free to enjoy chewing again. And I like the fresh mint flavor of Freedent. Everyone does."

Try some yourself. New Freedent in the bright blue pack. You'll like it.

Comments

1. The confidence limits given by (3) are based on a large sample approximation. See Comment 4 for guidance on when the large sample approximation is satisfactory. Table C6 contains for the small sample sizes $n = 1, 2, \ldots, 10$, confidence intervals for p of the form (p_L, p_U). These intervals have corresponding confidence coefficients that are greater than or equal to $1 - \alpha$. A graphical procedure for finding the confidence limits given in Table C6 is presented by Clopper and Pearson (1934).

To illustrate the use of Table C6, we consider the following example. Suppose that a surgeon obtains five cures in eight operations using a new procedure. What are 95% confidence limits for the probability p of a cure? Here $n = 8$, and the number of successes B is 5. Entering Table C6 at $n = 8$, $B = 5$, and $\alpha = .05$, the limits (with confidence coefficient at least 95%) are found to be $p_L = .2449$ and $p_U = .9148$.

2. In designing a study to estimate a success probability p, we can try to choose the sample size n so that the estimate $\hat{p}$ is within a distance D of p, with probability $1 - \alpha$. To do this we equate D to $z_{\alpha/2} SD(\hat{p})$:

$$D = z_{\alpha/2} \sqrt{\frac{p(1-p)}{n}}.$$

Solving for n, we have:

(4)
$$n = \frac{(z_{\alpha/2})^2 p(1-p)}{D^2}.$$

Expression (4) cannot be used without substituting a guess or estimate for p itself, since p is not known. It is useful to note that $p(1-p)$ attains its maximum value at $p = 1/2$ and decreases to zero as p approaches zero or one. Thus in cases in which p is not known at all, the most conservative sample size would be that obtained by substituting $1/2$ for p in the sample size equation. In other situations n should be calculated using the value of p closest to $1/2$ among the values of p that are regarded as possibilities.

3. Bernoulli trials are so named in honor of the French mathematician Jacques Bernoulli. His book *Ars Conjectandi* (1713) contains a profound study of such trials and is viewed as a milestone in the history of probability theory.

PROBLEMS

1. In Table 6, Chapter 2, some data of Yerushalmy *et al.* are given, showing that only 22 out of 30 X-rays on people with tuberculosis were actually designated as positive by the

X-ray reader. Determine an approximate 95% confidence interval for the sensitivity of the X-ray test, that is, for the probability that the test will make the correct indication (positive) on a person with the disease.

2.   Recall the Medi-Cal study discussed in Chapter 1. From Table 1 of Chapter 1 we see that out of 56 Medi-Cal children, 30 received intramuscular injections of antibiotics. Determine an approximate 90% confidence interval for the probability that a Medi-Cal child will receive an intramuscular injection (rather than an oral administration) of antibiotics.

3.   Neuringer and Neuringer (1974) performed an experiment on pigeons designed to study the nature and extent of learning by following a food source. In the researchers' "condition 2" type training, four out of five birds did not learn to "peck the disk." (See the Neuringer and Neuringer paper for details on the training involved in condition 2 and on the criterion for learning to peck the disk.) Use Table C6 to obtain a 90% confidence interval for p, the probability that a bird who undergoes condition-2 training, will fail to learn to peck the disk.

4.   Kalman (1974) performed a study designed to compare measurement techniques for digoxin in the blood serum. Blood samples were taken from 41 patients taking digoxin for cardiovascular disease. Each blood sample was split and the halves sent to two separate laboratories for measurement of digoxin in the blood. He found that laboratory A obtained the same result as laboratory B on 8 of the 41 samples, laboratory A obtained a higher measurement than laboratory B on 20 of the samples, and laboratory B obtained a higher measurement than laboratory A on the remaining 13 samples. Let p denote the probability that laboratories A and B will obtain the same result on a given sample. Determine a 95% confidence interval for p.

5.   Consider an experiment in which $n = 10$, $B = 8$, and hence $\hat{p} = (B/n) = .8$. Compare the 95% confidence limits for p obtained via the large sample approximation [expression (3)] with those obtained from Table C6.

2.	HYPOTHESIS TESTS FOR A SUCCESS PROBABILITY p

Tests for the Probability of Correctly Identifying the New Formula

Returning to the taste-testing example of Section 1, the hypothesis that the new formula tastes the same as the old formula is equivalent to the hypothesis that $p = 1/3$. (Recall that p is the probability that, on any given trial, the expert taste tester correctly identifies the new formula.) A statistical test of this null hypothesis

$$H_0: p = 1/3$$

can be based on the calculation of the distance (in standard deviation units) between the observed frequency of success ($\hat{p} = .48$) and the hypothesized rate $p = 1/3$. We consider the deviate

$$\frac{\hat{p} - 1/3}{SD(\hat{p})},$$

realizing that we must estimate $SD(\hat{p})$. In Section 1 the standard deviation of $\hat{p}$ is estimated by substituting the sample proportion $\hat{p}$, for p, in the expression $SD(\hat{p}) = \sqrt{p(1-p)/n}$. To compute the test statistic for testing the hypothesis that $p = 1/3$, it is proper to substitute the *hypothesized* value of p in the expression for $SD(\hat{p})$.

More generally, a formal test of the null hypothesis

$$H_0: p = p_0,$$

where p_0 is the hypothesized success probability, is based on the test statistic

(5) $$B^* = \frac{\hat{p} - p_0}{\sqrt{\dfrac{p_0(1-p_0)}{n}}}.$$

Equivalently, multiplying the numerator and denominator of the right-hand side of (5) by n, we find

(5') $$B^* = \frac{B - np_0}{\sqrt{np_0(1-p_0)}},$$

where B is the number of successes in the sample of size n.

The two-sided α level test of $H_0: p = p_0$, versus the alternative $p \neq p_0$, rejects H_0 if B^* is larger than $z_{\alpha/2}$ or smaller than $-z_{\alpha/2}$, otherwise H_0 is accepted.

The one-sided α level test of H_0, versus the alternative $p > p_0$, rejects when B^* exceeds z_α and accepts otherwise. Similarly, the one-sided α level test of H_0, versus the alternative $p < p_0$, rejects when B^* is less than $-z_\alpha$ and accepts otherwise.

For the taste-testing problem, if the new formula did taste different, the expert tester should have a success probability p that exceeds $1/3$. Thus in this case the one-sided upper tail test of $H_0: p = 1/3$, versus the alternative $p > 1/3$, is appropriate.

We find, using (5'),

$$B^* = \frac{48 - 100(1/3)}{\sqrt{100(1/3)(2/3)}} = 3.11.$$

At $\alpha = .05$, $z_\alpha = 1.645$, and since B^* is greater than 1.645, the null hypothesis

$p = 1/3$ is at $\alpha = .05$ rejected in favor of the alternative $p > 1/3$. Furthermore, referring $B^* = 3.11$ to Table C2, we find the one-sided P value to be .0009. This can be taken as strong evidence that the new formula does taste different from the old formula.

The normal approximation based on B^* yields a close approximation to the actual P value, provided the sample size is large. In the case of binary variables, the rule of thumb is that both types of events (i.e., successes and failures) must have expected values at least five *under the hypothesis*. That is, $np_0 \geq 5$ *and* $n(1 - p_0) \geq 5$. Since the hypothesis here is that $p = 1/3$, the expected numbers of successes and failures are $(1/3)(100) \approx 33$ and $(2/3)(100) \approx 67$, so that the normal approximation is quite good.

The Binomial Distribution

The hypothesis test of $H_0 \colon p = p_0$, based on B^* [see equation (5')], uses the fact that as n gets large, the distribution of $\hat{p}$ tends to the normal distribution. If the sample size is too small to justify the normal approximation to the distribution of $\hat{p}$, then exact probabilities can be computed. Consider the following example. Suppose that a surgeon obtains five cures in eight operations using a new procedure, where the probability p of cure (for a given operation) heretofore has been established to be only .30. Assuming all other aspects of the before-after comparison are similar (an assumption that may be difficult to establish), is the apparently increased cure rate on this sample of eight statistically significant? Here $n = 8$, $B = 5$, $\hat{p} = 5/8$, and the null hypothesis is $H_0 \colon p = .30$. The one-sided alternative, corresponding to an increased cure rate, is $H_a \colon p > .30$. Since $np = 2.4$ under the hypothesis, the normal approximation is not recommended in this case. We compute it anyway, for comparison with the exact calculations that follow. The test statistic is, from (5),

$$B^* = \frac{(5/8) - .30}{\sqrt{\dfrac{(.3)(.7)}{8}}} = 2.01.$$

Referring B^* to Table C2, the one-tailed P value for this normal deviate is found to be approximately .02.

Under the null hypothesis, $p = .30$, the distribution of B can be computed exactly, using the formula for the **binomial distribution**. This formula, for the probability of x successes in n independent Bernoulli trials when the probability of success for any one trial is p, can be found in most college algebra books, for example, Hart (1967). We have, in general,

(6)
$$Pr(B = x) = \binom{n}{x} p^x (1-p)^{n-x},$$

where $\binom{n}{x}$ (called a **binomial coefficient**) denotes the number of patterns with x successes and $n - x$ failures (see Comment 7) and where equation (6) is valid for each possible value $x = 0, 1, 2, \ldots, n$. The probabilities given by (6) are called binomial probabilities. These probabilities can be found in a table of the binomial distribution, such as the table published by the National Bureau of Standards (1950). Table C7 contains a small set of binomial probabilities. [Recall also that we encountered the binomial coefficients in Chapter 3. In the sampling terminology of Chapter 3, $\binom{N}{n}$ is the number of different samples of size n that can be formed from a population of N objects.]

By direct calculation, using (6), or from Table C7, we obtain the following values for the case $n = 8$, $p = .30$.

$$Pr(B = 0 | p = .30) = \binom{8}{0}(.3)^0(.7)^8 = (.7)^8 = .0576$$

$$Pr(B = 1 | p = .30) = \binom{8}{1}(.3)^1(.7)^7 = 8(.3)^1(.7)^7 = .1977$$

$$Pr(B = 2 | p = .30) = \binom{8}{2}(.3)^2(.7)^6 = 28(.3)^2(.7)^6 = .2965$$

$$Pr(B = 3 | p = .30) = \binom{8}{3}(.3)^3(.7)^5 = 56(.3)^3(.7)^5 = .2541$$

$$Pr(B = 4 | p = .30) = \binom{8}{4}(.3)^4(.7)^4 = 70(.3)^4(.7)^4 = .1361$$

$$Pr(B = 5 | p = .30) = \binom{8}{5}(.3)^5(.7)^3 = 56(.3)^5(.7)^3 = .0467$$

$$Pr(B = 6 | p = .30) = \binom{8}{6}(.3)^6(.7)^2 = 28(.3)^6(.7)^2 = .0100$$

$$Pr(B = 7 | p = .30) = \binom{8}{7}(.3)^7(.7)^1 = 8(.3)^7(.7)^1 = .0012$$

$$Pr(B = 8 | p = .30) = \binom{8}{8}(.3)^8(.7)^0 = (.3)^8 = .0001.$$

Note that the sum of the probabilities, $Pr(B = 0 | p = .30) + Pr(B = 1 | p = .30) + \cdots + Pr(B = 8 | p = .30)$, is equal to 1.

The probability, under the null hypothesis $p = .30$, that the n independent trials (operations) yield a value of B greater than or equal to the observed value of 5 (equivalently, the probability of obtaining a value of $\hat{p}$ equal to or greater than the observed value of 5/8) is thus found to be

$$Pr(B \geq 5 | p = .30) = .0467 + .0100 + .0012 + .0001 = .0580.$$

Note that this probability is the P value.

The analogous one-tail normal probability value obtained earlier is .02, which is a fairly good approximation for such a small sample. Furthermore, the normal approximation can be improved by making a so-called **correction for continuity** (see Comment 8). Suppose B has the binomial distribution with parameters n and p. To approximate $Pr(B \geq x)$, using the normal approximation to the distribution of B, we find the area under a standard normal curve to the right of the "corrected" deviate

$$z = \frac{x - \frac{1}{2} - np}{\sqrt{np(1-p)}}.$$

Similarly, to approximate $Pr(B \leq x)$, we find the area under a standard normal curve to the left of the "corrected" deviate

$$z = \frac{x + \frac{1}{2} - np}{\sqrt{np(1-p)}}.$$

Thus to approximate $Pr(B \geq 5 | p = .30)$, we determine the area under a standard normal curve to the right of

$$z = \frac{5 - \frac{1}{2} - 8(.30)}{\sqrt{8(.30)(.70)}} = 1.62.$$

From Table C2 the area under the standard normal curve to the right of $z = 1.62$ is .0526. Thus this one-tailed P value of .0526 based on the normal approximation agrees very well indeed with the exact binomial value, even in this case in which our rule of thumb would rule out the approximation as not good enough.

Calculating the Probability of a Type-I Error and the Probability of a Type-II Error for Tests Based on the Binomial Distribution

Consider testing the hypothesis

$$H_0: p = .30$$

based on $n = 8$ independent Bernoulli trials. If the alternative hypothesis under consideration is $p > .30$, a reasonable test procedure is to reject H_0 if B, the number of successes, is too large. For example, the following tests are both of the form "reject H_0 if B is too large."

$$\text{Test 1:} \quad \text{Reject } H_0 \text{ if } B \geq 6,$$
$$\text{Accept } H_0 \text{ if } B \leq 5$$

$$\text{Test 2:} \quad \text{Reject } H_0 \text{ if } B \geq 5,$$
$$\text{Accept } H_0 \text{ if } B \leq 4.$$

One way to compare Test 1 and Test 2 is to compute for each test the probability α of Type-I error and the probability β of Type-II error. First we consider the problem of computing α. Recall (see, e.g., Comment 6 of Chapter 4) that α is the probability of rejecting H_0 when H_0 is true; in symbols

$$\alpha = Pr(\text{rejecting } H_0 | H_0 \text{ is true}).$$

Thus for Test 1

$$\alpha = Pr(B \geq 6 | p = .30)$$
$$= Pr(B = 6 | p = .30) + Pr(B = 7 | p = .30) + Pr(B = 8 | p = .30)$$
$$= .0100 + .0012 + .0001$$
$$= .0113,$$

where we have obtained the values .0100, .0012, .0001 by entering Table C7 at $n = 8$, $p = .30$ and respective x values 6, 7, and 8.

Similarly, for Test 2,

$$\alpha = Pr(B \geq 5 | p = .30)$$
$$= Pr(B = 5 | p = .30) + Pr(B = 6 | p = .30) + Pr(B = 7 | p = .30)$$
$$+ Pr(B = 8 | p = .30)$$
$$= .0467 + .0100 + .0012 + .0001$$
$$= .0580.$$

Test 1 has a smaller probability of a Type-I error ($\alpha = .0113$) than Test 2 ($\alpha = .0580$). This is an advantage of Test 1 compared to Test 2, but we now see that when the probabilities of Type-II errors are computed, the comparison favors Test 2.

Recall that β is the probability of accepting H_0 when H_0 is false. The value of β depends on which particular alternative to H_0 is true. We compute β for Tests 1 and 2 under the alternative hypothesis that $p = .50$. For Test 1,

$$\beta = Pr(B \le 5 | p = .50)$$

$$= Pr(B = 0 | p = .50) + Pr(B = 1 | p = .50) + Pr(B = 2 | p = .50)$$

$$+ Pr(B = 3 | p = .50) + Pr(B = 4 | p = .50) + Pr(B = 5 | p = .50)$$

$$= .0039 + .0312 + .1094 + .2187 + .2734 + .2187$$

$$= .8553,$$

where the values .0039, .0312, .1094, .2187, .2734, and .2187 are obtained by entering Table C7 at $n = 8$, $p = .50$ and respective x values 0, 1, 2, 3, 4, and 5. Similarly, for Test 2,

$$\beta = Pr(B \le 4 | p = .50)$$

$$= Pr(B = 0 | p = .50) + Pr(B = 1 | p = .50) + Pr(B = 2 | p = .50)$$

$$+ Pr(B = 3 | p = .50) + Pr(B = 4 | p = .50)$$

$$= .0039 + .0312 + .1094 + .2187 + .2734$$

$$= .6366.$$

Summarizing these calculations, we have:

	α	β	$1 - \beta$
Test 1:	.0113	.8553	.1447
Test 2:	.0580	.6366	.3634

In the context of the experiment in which a surgeon is using a new procedure to improve the probability of a cure, a Type-I error corresponds to concluding the new procedure has a cure rate that is greater than .30, when, in fact, its cure rate is only .30. Similarly, a Type-II error, when the alternative $p = .50$ is true, corresponds to mistakenly concluding the new operation is no better or worse than the old ($p = .30$), when, in fact, it does increase the cure rate to $p = .50$.

Discussion Question 3.   Comparing Test 1 and Test 2, we found β values, for the alternative $p = .50$ to be .8553 and .6366. Can you give some reasons why these probabilities of Type-II errors are so high?

Discussion Question 4.   Can you think of some experiments in which the variables are dichotomous and the nature of the scientific problem being considered leads directly to the hypothesis H_0: $p = 1/2$? Can you furnish some examples where the hypothesis H_0: $p = 1/5$ arises in a natural way?

Comments

4.   It can be shown, using the central limit theorem, that the distribution of $\hat{p}$ is approximately normal with mean p and standard deviation $\sqrt{p(1-p)/n}$. Figure 1 gives the exact distribution of $\hat{p}$ when $p = .30$ for three different sample sizes, $n = 10, 20$, and 30. For example, from Table C7 entered at $n = 10, p = .30$, we find

$$Pr(\hat{p} = 0 | p = .30) = Pr(B = 0 | p = .30) = .0282$$

$$Pr(\hat{p} = .1 | p = .30) = Pr(B = 1 | p = .30) = .1211$$

$$Pr(\hat{p} = .2 | p = .30) = Pr(B = 2 | p = .30) = .2335$$

$$Pr(\hat{p} = .3 | p = .30) = Pr(B = 3 | p = .30) = .2668$$

$$Pr(\hat{p} = .4 | p = .30) = Pr(B = 4 | p = .30) = .2001$$

$$Pr(\hat{p} = .5 | p = .30) = Pr(B = 5 | p = .30) = .1029$$

$$Pr(\hat{p} = .6 | p = .30) = Pr(B = 6 | p = .30) = .0368$$

$$Pr(\hat{p} = .7 | p = .30) = Pr(B = 7 | p = .30) = .0090$$

$$Pr(\hat{p} = .8 | p = .30) = Pr(B = 8 | p = .30) = .0014$$

$$Pr(\hat{p} = .9 | p = .30) = Pr(B = 9 | p = .30) = .0001$$

$$Pr(\hat{p} = 1 | p = .30) = Pr(B = 10 | p = .30) = .0000.$$

We note that $Pr(\hat{p} = 1 | p = .30)$ is not really zero, but is zero to four decimal places. [The actual value is $(.30)^{10} = .0000059$.] Furthermore, for obvious reasons, in Figure 1 we did not attempt to draw bars of heights .0014, .0001, and .000059 at the $\hat{p}$ values .8, .9, and 1, respectively. Nevertheless, the reader should be aware that those values have positive probabilities of occurring. In drawing Figure 1 we treated $Pr(\hat{p})$ as zero whenever it was actually less than .005. (The values for $n = 20$ and $n = 30$ were obtained from the already-cited Tables of the Binomial Probability Distribution published by the National Bureau of Standards.)

For $n = 30$, the shape of the distribution seems quite symmetrical and bell shaped. If p is very close to either zero or one, the distribution of $\hat{p}$ cannot be very symmetrical unless n is very large. A rule of thumb for application of the normal approximation is that np and $n(1-p)$ be five or more.

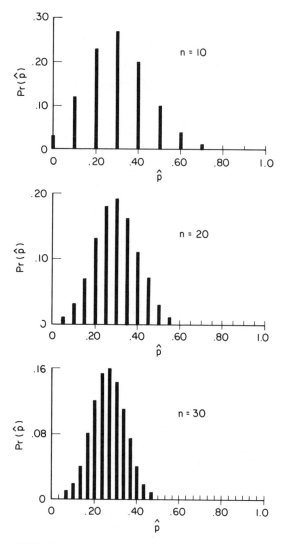

FIGURE 1. Distribution of $\hat{p}$ when $p = .30$, for sample sizes $n = 10, 20,$ and 30.

5.   Table C7 does not contain entries for cases in which p is greater than $1/2$. This is because a binomial probability for $p > 1/2$ can be reduced to a corresponding binomial probability with $p < 1/2$. Suppose, for example, a surgeon wishes to calculate $Pr(B = 6)$ where B is the number of cures in nine independent operations and where it is believed the chance of cure for each operation is .8. If we call a cure a success, at first glance it appears that we cannot use Table C7 because $p > 1/2$. However, if we call a noncure a "success", we are then seeking the probability of three successes in nine independent trials, where the success probability is .2. Entering Table C7 at $n = 9$, $p = .2$, and $x = 3$, yields the value .1762 for the desired probability.

6.   In the special case in which $p = 1/2$, $Pr(B = x) = Pr(B = n - x)$. For example, with $n = 10$ and $p = .5$, note from Table C7 that $Pr(B = 3) = .1172$ and $Pr(B = 10 - 3) = Pr(B = 7) = .1172$. [Thus in Table C7 for the $p = .5$ entries $Pr(B = x)$ need only have been given for x values less than or equal to $n/2$.]

7.   It can be shown that $\binom{n}{x}$, the number of patterns with x successes and $n - x$ failures is given by the following expression:

(7)
$$\binom{n}{x} = \frac{n!}{x! \cdot (n - x)!},$$

where the quantity $n!$ (read n *factorial*) is just the product of the first n positive integers. That is,

$$n! = n(n - 1)(n - 2) \cdots (2)(1).$$

For equation (7) to remain valid when $x = 0$ and when $x = n$, the value $0!$ is defined to be equal to 1.

Recall that we have already encountered binomial coefficients. In Chapter 3 we were concerned with the number of different samples of size n that could be formed from a population of N objects. This number is equal to the binomial coefficient $\binom{N}{n}$.

8.   In the text we have applied a continuity correction to approximate $Pr(B \geq 5)$ for the case $p = .30$, $n = 8$. The correction entailed finding the area under a standard normal curve to the right of the corrected deviate

$$z = \frac{5 - \frac{1}{2} - 8(.30)}{\sqrt{8(.30)(.70)}}.$$

The rationale for this correction is as follows. The histogram for the distribution of B can be represented by rectangles, of width one unit, centered at the possible values $0, 1, 2, \ldots, 8$ of B. The heights of the rectangles at the possible values of B are the corresponding probabilities of these possible values. Thus the height of the rectangle at 4 is .1361, the height at 5 is .0467, and so forth. (You can obtain the heights from Table C7 entered at $n = 8$, $p = .30$.) Figure 2 shows a portion of the histogram of B along with the continuous normal curve that is to approximate the distribution.

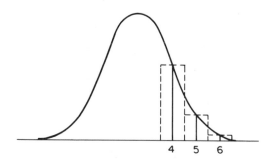

FIGURE 2.   Graphical justification for the continuity correction.

Note that $Pr(B \geq 5)$ is given by the sum of the areas of all rectangles to the right of $x = 4\frac{1}{2}$ in Figure 2. If one only summed areas to the right of $x = 5$, $\frac{1}{2}$ of the contribution from $Pr(B = 5)$ would be lost. Thus the normal approximation should approximate the sum of the rectangular areas to the right of $4\frac{1}{2}$ rather than 5.

Summary of Statistical Concepts

In this chapter the analysis of binary (i.e., dichotomous or "go, no-go") data is discussed for the case of a single sample. Using the techniques described, the reader should have learned to:
(i)   Estimate the true success probability p from the observed number of successes B in n independent Bernoulli trials.
(ii)   Estimate the standard deviation of $\hat{p}$.
(iii)   Calculate confidence limits for the unknown value of p.
(iv)   Test hypotheses about the unknown success probability p.

PROBLEMS

6. In clinical trials designed to compare the efficacies of two antiarthritic drugs, it is the practice to select patients who have symmetrical joint problems, say, similar severity of involvement of the finger joints of both hands. Drug A is then randomly assigned to one side and drug B to the other, and the assigned treatments are then applied in a double blind manner if possible. Suppose that 10 patients are treated and are then asked to express a preference in terms of the side seemingly afforded the most relief from arthritic pain. When the treatment identifications are decoded it is found that drug A won in eight cases and drug B won in two cases. What is the P value? Obtain this value by the normal approximation, with and without the correction for continuity, and also by using Table C7 (the table of binomial probabilities). [Hint: Let p denote the probability that a patient will prefer drug A to drug B.]

7. A biology examination consists of 25 multiple choice questions. Each question has five possible answers, only one of which is correct. Student A obtains a score of eight correct on the exam. Test the hypothesis that student A was simply guessing the answers.

8. Return to the measurement study of Problem 4. Test the hypothesis that the laboratories are not biased relative to each other in their measurement techniques for digoxin. [Hint: If they are not biased, then the proportion of untied measurements in which laboratory A has a greater value than laboratory B should be $1/2$.]

9. Use formula (6) to obtain the distribution of the number of successes in five independent Bernoulli trials when the probability of success for each trial is $p = 1/7$.

10. Define binomial distribution, binomial coefficient, correction for continuity.

Supplementary Problems

1. In a study of patients who were given exhaustive diagnostic tests for prostate cancer, including surgical removal of the lymph nodes around the aorta in the pelvic region, Dr. Gordon Ray (1975) found that 14 of 25 patients found to have nodal involvement at surgery showed an elevated level of acid phosphatase in their blood serum prior to surgery, whereas only 4 showed an elevated serum acid phosphatase among 25 patients with no nodal involvement revealed at surgery. Taking each group, that is, positive nodal involvement and negative nodal involvement, to be random samples of prostate cancer patients of these two categories, estimate the sensitivity rate and the specificity rate for serum acid phosphatase as a predictor of nodal involvement. Compute an estimated standard deviation for each estimate. Compute a 95% confidence interval for the sensitivity and a 95% confidence interval for the specificity, using the normal approximation.

2. Fisher (1959), in his book on the foundations of statistical inference, discussed a genetics problem in which an animal is bred and produces a litter of seven animals, all showing a recessive trait. Under the circumstances of the experiment, the parent is either heterozygous, in which case, 1/2 of its progeny would show the recessive trait, or the parent is homozygous recessive, in which case all its progeny would show the recessive trait. Test the hypothesis that the parent is heterozygous, based on the seven progeny produced.

3. In clinical trials for comparison of topical ointments for treatment of the swollen joints of arthritis, patients are chosen who have roughly similar involvement of the joints in both hands. Then the ointments to be compared (call them A and B) are randomly allocated to the two hands of each patient, neither patient nor physician being told which hand is being treated with which ointment. At the end of a period of several weeks the patient is asked which treatment he prefers. The treatments are then identified, and the number of preferences for each drug is counted. Suppose that in such a randomized, double blind trial, the new drug B is preferred to the standard drug A by 20 of the 25 patients. Test the hypothesis that the two drugs are equally effective.

Now suppose that in a similar clinical trial involving 50 patients each patient is asked for a preference but allowed to say that he can't distinguish between the two ointments. Suppose that 25 patients say that they have no preference, 20 prefer B, and 5 prefer A. Test the hypothesis that the two drugs are equally effective.

4. In a retrospective study at the Mayo Clinic, reported by O'Fallon, Labarthe, and Kurland (1975), that was performed to investigate the hypothesis that rauwolfia, a drug commonly given for hypertension, causes breast cancer, investigators identified 450 cases of breast cancer and 475 women with cholelithiasis (gallstones) as a control group. Twenty-nine of the breast cancer cases had taken rauwolfia prior to the diagnosis of breast cancer, and 38 of the control women had taken the drug. Calculate the percentage of rauwolfia exposure in each group and the estimated standard deviation for each percentage. (In the next chapter, you will learn how to estimate the standard deviation of the difference between the two percentages, and to test the hypothesis of no difference in true rates for the two groups.) The investigators, finding *less* rauwolfia use among the breast cancer cases, proceeded to explore the reasons. You might wish to read their paper for a careful analysis and discussion of the question of rauwolfia and breast cancer, including their critical review of previous papers that claimed that rauwolfia caused breast cancer.

*5. In studies of drugs and other factors that cause damage or death to the fetus, the proportion of births that are male is of interest since this proportion may be altered if one sex (usually the male) is more sensitive to the toxic factor. Suppose the rate of male births is to be studied for babies born of mothers living close to atomic power plants, to test the theory that the environment around such plants is hazardous to the fetus. How large a sample size would you advise if a decrease from 51% males to 48% males is considered important enough to want to know about and follow up.

6. The data in Table 1 are abstracted by Wiorkowski (1975) from an earlier paper by Dern and Wiorkowski (1969) on some genetic aspects of biological activity of adenosine triphosphate (ATP) in red blood cells. We have data on the ATP activity level for the father and mother in 14 "randomly selected families."

TABLE 1. Adenosine Triphosphate Activity Level
(μ moles ATP per gram of hemoglobin)

Family	Father	Mother
1	3.72	4.43
2	4.54	3.79
3	5.05	4.66
4	4.10	5.42
5	4.26	4.39
6	4.09	5.29
7	4.83	4.99
8	4.24	4.38
9	5.43	4.73
10	5.23	5.34
11	4.56	5.29
12	5.16	4.71
13	3.77	5.13
14	4.15	4.18

Source: J. Wiorkowski (1975).

The familial matching provides something of a control for current environment, for example, diet. Note that in 10 of the 14 families the mother has a higher level than the father. Use the large sample test based on B^* [equation (5′)] to test the hypothesis of no sex difference. [Note that the statistic B in the numerator of (5′) is simply the number of positive differences obtained when computing the "mother" minus "father" values. That is, B is the number of positive "signs," and consequently this test is called the sign test. See also Comment 5 of Chapter 12.]

Note that these data could have been handled as measured data, by calculating differences, computing the test statistic T (equation (4) of Chapter 4) based on the mean difference, and referring the test statistic T to a t-table, 13 degrees of freedom, to obtain a P value. This latter method uses the quantitative data rather than the signs of the differences alone, and usually but not always provides a smaller P value if the null hypothesis is false and the underlying data emanate from a normal population. Alternative methods for analyzing matched pairs data are presented in Chapter 12. The Chapter 12 test procedure is based on signs and ranks.

*7.   In reading Pap smears (that is, judging whether cells scraped from the cervix indicate the beginnings of cancer or not), pathologists have difficulty in agreeing on whether a certain minor atypical cellular appearance is abnormal or not. Suppose two pathologists classify the same set of slides independently, grading them from zero, indicating normal, to

4, indicating invasive carcinoma (that is, frank cancer). Suppose the results are as follows (a hypothetical but not unrealistic set of data):

		Pathologist A					
		0	1	2	3	4	
	0	21	3	1	0	0	25
	1	1	4	2	1	0	8
Pathologist B	2	0	2	6	4	0	12
	3	0	0	2	2	0	4
	4	0	0	0	0	3	3
		22	9	11	7	3	52

Thus 52 slides were read. The pathologists agreed on $21+4+6+2+3 = 36$ of them, but gave differing readings on the others. One interesting question is whether one pathologist tends to see more pathology than the other. Is there evidence for this in the table? Let the null hypothesis assert that neither sees more disease than the other. Compute the P value.

8. Suppose that a particular neurosurgeon determines that in his long experience with certain brain cancer cases, only 10% survive as long as 4 months following diagnosis. He decides to treat his patients, following surgery, with various modes of chemotherapy, searching for a drug or drug combination that will prolong the lives of his patients. He decides that he will not do a randomized comparison of each drug with control, but use his previous experience of 10% survival at 4 months as the benchmark. He will treat 10 patients with each new drug therapy, and declare that he has found an effective drug if he treats 10 patients with the drug and at least 5 patients of the 10 survive to 4 months or longer. Use Table C7, the table of exact binomial probabilities, to calculate the probability that an ineffective drug will be erroneously declared a winner. Use the table to calculate the probability that a drug that increases the 4-month survival rate to 70% will be declared a loser.

9. Suppose that a clinical laboratory in a hospital does about 300 white blood cell counts a day using an electronic counting device. About 10% of these are abnormally high on the average, simply because the hospital serves a clientele with a spectrum of diseases, some of which cause elevated white blood cell counts. Suppose that on one particular day the laboratory does white blood cell counts on 260 blood samples and 35 of them are abnormal. Is this number statistically significant when compared with the overall average of 10%? Would you regard this as persuasive evidence that the automatic counting equipment needs adjustment? This statistical method has been advocated by some as a method for statistical quality control in the hospital laboratory.

10. Consider the following test of H_0: $p = .4$, versus the alternative hypothesis $p > .4$, based on $n = 9$ independent Bernoulli trials.

$$\text{Reject } H_0 \quad \text{if } B \geq 7,$$

$$\text{Accept } H_0 \quad \text{if } B \leq 6.$$

What is the value of α, the probability of a Type-I error, for this test? What is the value of β, the probability of a Type-II error, when the alternative $p = .6$ is true? What is the value of β when the alternative $p = .8$ is true? [Hint: To use Table C7 for cases where p is greater than .5, see Comment 5.]

REFERENCES

*Bernoulli, J. (1713). *Ars Conjectandi.*

*Clopper, C. J. and Pearson, E. S. (1934). The use of confidence or fiducial limits illustrated in the case of the binomial. *Biometrika* **26,** 404–413.

Coronary Drug Project Research Group (1974). Factors influencing long-term prognosis after recovery from myocardial infarction—three year findings of the Coronary Drug Project. *J. Chron. Dis.* **27,** 267–285.

Dern, R. and Wiorkowski, J. (1969). The hereditary component of pre- and post-storage erythrocyte adenosine triphosphate levels. *J. Lab. Clin. Med.* **73,** 1019–1029.

Fisher, R. A. (1959). *Statistical Methods and Scientific Inference*, 2nd edition. Hafner, New York.

Hart, W. L. (1967). *Contemporary College Algebra and Trigonometry.* Heath, Boston.

Kalman, S. (1974). Personal communication.

National Bureau of Standards (1950). *Tables of the Binomial Probability Distribution.* Applied Mathematics Series 6, U.S. Government Printing Office, Washington, D.C.

Neuringer, A. and Neuringer, M. (1974). Learning by following a food source. *Science* **184,** 1005–1008.

Nolte-Watts, C. (1975). Denture wearers confirm 'no-stick' gum ad claim. St. Petersburg Times. September 2.

O'Fallon, W. M., Labarthe, D. R. and Kurland, L. T. (1975). Rauwolfia derivatives and breast cancer. *Lancet*, August 16. 292–296.

Ray, G. (1975). Personal communication.

*Wiorkowski, J. (1975). Unbalanced regression analysis with residuals having a covariance structure of intra-class form. *Biometrics* **31,** 611–618.

CHAPTER 8

Comparing Two Success Probabilities

In Chapter 7 we described inferential procedures for a single success probability. These procedures are based on the proportion of successes observed in n independent Bernoulli trials. Recall that each observation could be classified a success or failure (depending on whether or not a specified attribute was present). In this chapter the object is to compare two unknown success probabilities on the basis of the corresponding rates of success in independent samples.

Section 1 describes a confidence interval for the difference between the two unknown probabilities. Section 2 presents a test of the hypothesis that the two probabilities are equal. This hypothesis test can be cast in a category of statistical tests known as chi-squared tests, and the introduction of "chi-squared" in this section is a forerunner to the more general coverage of chi-squared tests to be found in Chapter 9.

1.	A CONFIDENCE INTERVAL FOR THE DIFFERENCE BETWEEN THE SUCCESS PROBABILITIES

Anesthetic Gases in the Operating Room Air—Harmful or Not?

A few years ago a Russian medical scientist at an international meeting of anesthesiologists made the observation that pregnant nurses working in the operating room seemed to have an unusually large number of miscarriages. He suggested the possibility that anesthetic gases in the operating room atmosphere, breathed in and getting into the blood of the mother, could be harming the fetus. This speculation stimulated a number of investigators to go back to their home countries and study the question.

At Stanford University, Cohen, Bellville, and Brown (1971) organized a survey of operating room nurses at the Stanford Medical Center and two hospitals in the surrounding area. The operating room nurses were each privately interviewed by one of two trained interviewers. For control or comparison, the floor nurses at the same hospitals were also interviewed.

Only married nurses between the ages of 25 and 50 were interviewed. They were asked about any pregnancies they had had within the last 5 years. For each pregnancy, they were asked whether the pregnancy ended in a miscarriage or proceeded successfully to delivery.

Sixty-seven operating room nurses and ninety-two general duty nurses were interviewed. The operating room nurses reported 36 pregnancies within the preceding 5 years, and the general duty nurses reported 34 pregnancies. The numbers of miscarriages are reported in Table 1.

TABLE 1. Miscarriage Rates for Operating Room Nurses and General Duty Nurses

	Operating Room Nurses	General Duty Nurses
Number interviewed	67	92
Number of pregnancies	36	34
Number of miscarriages	10	3
Rate of miscarriage	27.8%	8.8%

Source: E. N. Cohen, J. W. Bellville, and B. Wm. Brown, Jr. (1971).

In Table 1 note that the miscarriage rate for operating room nurses is 27.8% and the rate for general duty nurses is only 8.8%. On the surface, this difference would seem to corroborate the suspicion that operating room nurses are exposing their unborn babies to unusual hazard. However, the numbers of miscarriages are small, 10 in 36 pregnancies for the operating room nurses and 3 in 34 pregnancies for the general duty nurses. Could this large a difference come about through chance variation, even though the actual hazard is the same for the two types of nurse? In other words, how reliable are these miscarriage rates, and, in particular, how safe are we in concluding from these data that the true rate for operating room nurses exceeds that for general duty nurses?

Interval Estimates

The rates 27.8% and 8.8% displayed in Table 1 are estimates of true unknown probabilities. For example, let p_1 denote the unknown probability that a pregnant operating room nurse will suffer a miscarriage. We estimate this probability by the observed proportion $\hat{p}_1 = (10/36) = .278$. Similarly, let p_2 denote the unknown probability that a pregnant general duty nurse will suffer a miscarriage. Then our estimate of p_2 is $\hat{p}_2 = (3/34) = .088$.

Following the methods of Chapter 7, each of the two samples can be analyzed separately. In particular, we can obtain confidence intervals for p_1 and p_2. From formula (3), Chapter 7, an approximate 95% confidence interval for p_1 is

$$.278 \pm (1.96)(.075) = .278 \pm .147$$

$$= .131 \text{ and } .425.$$

Similarly, an approximate 95% confidence interval for p_2 is

$$.088 \pm 1.96(.049) = .088 \pm .096$$

$$= 0 \text{ and } .184.$$

(Note that the lower limit as calculated above actually is $.088 - .096 = -.008$. However, we know that p_2 must be nonnegative, thus we automatically increase the lower limit to zero.)

Thus it seems that there is a higher rate for operating room nurses than for general duty nurses. It is useful to attach a standard deviation to the difference in the two rates. An estimate of the standard deviation for the difference is computed as follows:

$$(1) \qquad \widehat{SD}(\hat{p}_1 - \hat{p}_2) = \sqrt{[\widehat{SD}(\hat{p}_1)]^2 + [\widehat{SD}(\hat{p}_2)]^2},$$

where, from equation (2) of Chapter 7,

$$\widehat{SD}(\hat{p}_1) = \sqrt{\frac{\hat{p}_1(1-\hat{p}_1)}{m}}, \qquad \widehat{SD}(\hat{p}_2) = \sqrt{\frac{\hat{p}_2(1-\hat{p}_2)}{n}},$$

and m is the sample size of sample 1, n is the sample size of sample 2. We find

$$\widehat{SD}(\hat{p}_1) = \sqrt{\frac{(10/36)(26/36)}{36}} = .075,$$

$$\widehat{SD}(\hat{p}_2) = \sqrt{\frac{(3/34)(31/34)}{34}} = .049.$$

Thus from equation (1)

$$\widehat{SD}(\hat{p}_1 - \hat{p}_2) = \sqrt{[.075]^2 + [.049]^2}$$

$$= \sqrt{.008}$$

$$= .090.$$

A confidence interval for the true difference $p_1 - p_2$, with (approximate) confidence coefficient $1 - \alpha$, is

(2) $$(\hat{p}_1 - \hat{p}_2) \pm (z_{\alpha/2}) \cdot \widehat{SD}(\hat{p}_1 - \hat{p}_2).$$

For the data of Table 1, with $\alpha = .05$, we find the interval to be

$$(.278 - .088) \pm 1.96(.090) = .190 \pm .176$$

$$= .014 \text{ and } .366.$$

The confidence interval does not include zero, so that we can exclude (at $\alpha = .05$) a claim of equal hazard for the two groups.

Comments

1. Suppose we want to determine sample sizes so that our estimate $\hat{p}_1 - \hat{p}_2$ of the true difference $p_1 - p_2$ will be within D of the true value, with probability equal to $1 - \alpha$. Equating the desired precision D to the actual precision $(z_{\alpha/2}) \cdot SD(\hat{p}_1 - \hat{p}_2)$ yields the equation

$$D = z_{\alpha/2} \cdot \sqrt{\frac{p_1(1-p_1)}{m} + \frac{p_2(1-p_2)}{n}},$$

where m is the sample size of sample 1 and n is the sample size of sample 2. Taking

$m = n$ and solving for n yields

(3)
$$n = \frac{(z_{\alpha/2})^2 \cdot [p_1(1-p_1) + p_2(1-p_2)]}{D^2}.$$

We cannot use equation (3) as it stands, because p_1 and p_2 are not known. (The purpose of the experiment is to obtain information about the unknown values of p_1, p_2, and $p_1 - p_2$.) However, the term in square brackets in the numerator of the right-hand side of (3) is largest when $p_1 = p_2 = \frac{1}{2}$. Thus a sufficiently large sample would be

$$n = \frac{(z_{\alpha/2})^2 \cdot [(\frac{1}{2})(\frac{1}{2}) + (\frac{1}{2})(\frac{1}{2})]}{D^2} = \frac{(z_{\alpha/2})^2}{2(D^2)}.$$

This sample size assures the desired reliability regardless of the values of p_1 and p_2. In situations in which it is known that p_1 and p_2 are definitely less than some maximum value p^* (say) which is less than $\frac{1}{2}$, p^* can be substituted for p_1 and p_2 in (3) yielding

(3′)
$$n = \frac{(z_{\alpha/2})^2 [2(p^*)(1-p^*)]}{D^2}.$$

The same is true if p_1 and p_2 are known to be greater than some value p^* which is greater than $\frac{1}{2}$.

PROBLEMS

1. Cohen, Bellville, and Brown (1971) reported a second study of miscarriages among operating room personnel. They mailed questionnaires to female M.D. anesthetists in the state of California and to a comparable group of female medical specialists who did not work in the operating room. The reported results are shown in Table 2. Compute estimates of the risks of miscarriage for the two groups, with estimated standard deviations. Compute 95% confidence limits for the difference in rates.

2. Recall the Medi-Cal study described in Chapter 1. Compute 90% confidence limits for the difference between rates of intramuscular injections for the Medi-Cal and non-Medi-Cal groups.

3. Bunker and Brown (1973) conducted a pilot questionnaire survey of Stanford Medical School students designed to see if medical doctors allow as much surgery on their children as do parents who are medical laymen. Medical school students are a good group to study because a large proportion of them come from medical families. Of the 363 students in the study, 295 responded to the questionnaire. Table 3 summarizes the rates of tonsillectomy for students with M.D. parents and students without M.D. parents. Compute 95% confidence limits for the difference in true rates.

TABLE 2. Miscarriage Rates for M.D. Anesthetists and Other Medical Specialists

	Anesthetists	Other Specialists
Number of respondents	50	81
Number of pregnancies	37	58
Number of miscarriages	14	6
Rate of miscarriage	37.8%	10.3%

Source: E. N. Cohen, J. W. Bellville, and B. Wm. Brown, Jr. (1971).

TABLE 3. Tonsillectomy Rates for Stanford Medical Students with M.D. Parents and Stanford Medical Students without M.D. Parents

	Students with M.D.parents	Students without M.D. parents
Number reporting tonsillectomy	21	134
Number reporting no tonsillectomy	33	107
Rate of tonsillectomy	38.9%	55.6%

Source: J. P. Bunker and B. Wm. Brown, Jr. (1973).

---

2.   A HYPOTHESIS TEST FOR THE DIFFERENCE BETWEEN THE SUCCESS PROBABILITIES

Does the Probability that an Operating Room Nurse will Miscarry Equal the Probability that a General Duty Nurse will Miscarry?

The hypothesis that the two groups of nurses are exposed to equal risks of miscarriage, that is, that their true rates are equal, is of central interest. It is convenient to focus on the question of whether the data are consistent with this hypothesis. This can be done by computing the 95% confidence interval for the true difference, as we did in the preceding section, and seeing if the interval contains the value zero. However, a significance test of the null hypothesis, as next described, is the more common approach.

We start by assuming that the risk of miscarriage is the same for the two groups. This is called the null hypothesis. Formally, we can write the null hypothesis as

$$H_0: p_1 = p_2.$$

Then we note that the observed rates in the samples are not equal (27.8% vs. 8.8%) and we ask whether a discrepancy this large, or larger, could occur simply through chance variation when H_0 is true. Even if H_0 were true, the rates in the two samples would usually not be *identical*, and the question comes down to determining the probability of a difference as large or larger than the observed difference under the null hypothesis. If this probability proves to be very small, the null hypothesis H_0 would be untenable, and if the probability is substantial, H_0 would be regarded as quite reasonable in the light of the data.

The test procedure is as follows:

1.   Under the null hypothesis the best estimate of the common success rate is obtained by pooling the data in the two samples. Call this estimate $\hat{p}$. Letting B_1 denote the number of successes in sample 1, and B_2 the number of successes in sample 2, we have

(4)
$$\hat{p} = \frac{B_1 + B_2}{m + n}.$$

For the miscarriage data,

$$\hat{p} = \frac{10 + 3}{36 + 34} = .186.$$

2.   The standard deviation of the difference in rates is now estimated by

(5)
$$\widehat{SD}(\hat{p}_1 - \hat{p}_2) = \sqrt{\frac{\hat{p}(1-\hat{p})}{m} + \frac{\hat{p}(1-\hat{p})}{n}},$$

where m and n are the sample sizes (pregnancies) for operating room nurses and general duty nurses, respectively. We have

$$\widehat{SD}(\hat{p}_1 - \hat{p}_2) = \sqrt{\frac{(.186)(.814)}{36} + \frac{(.186)(.814)}{34}} = .093.$$

3. The difference $\hat{p}_1 - \hat{p}_2$ is then expressed in units of the estimated standard deviation by calculating the statistic A:

(6)
$$A = \frac{\hat{p}_1 - \hat{p}_2}{\widehat{SD}(\hat{p}_1 - \hat{p}_2)} = \frac{\left(\frac{10}{36}\right) - \left(\frac{3}{34}\right)}{.093} = 2.038.$$

To test H_0 versus the alternative $p_1 \neq p_2$, at the (approximate) α level, compute A, and if A is larger than $z_{\alpha/2}$ or smaller than $-z_{\alpha/2}$, the null hypothesis is rejected. If not, the null hypothesis is accepted as consistent with the data.

For the data of Table 1, we have found $A = 2.038$. At $\alpha = .05$, $(\alpha/2) = .025$, and from Table C2, we find $z_{.025} = 1.96$. Since $A = 2.038$ is greater than 1.96, the two-sided test at the $\alpha = .05$ level leads to rejection of H_0.

Actually, we can reject H_0 at values of α that are lower than .05. Referring $A = 2.038$ to Table C2, we note that 2.038 is the upper .02 percentile point of the standard normal distribution. It follows that the lowest value of α for which we can reject H_0 using the two-sided procedure is $\alpha = 2(.02) = .04$. Recall that this "lowest value" is called the two-sided P value. The one-sided P value, that is, the lowest value of α at which we can reject H_0 using a one-sided test (see Comment 2) is $P = .02$. Although statisticians differ in terms of a preference for reporting one-sided versus two-sided P values, the conservative approach is to report the two-sided P value.

We have found that the probability of a difference as large as or larger than $|\hat{p}_1 - \hat{p}_2| = .190$ would be only .04 if the risks of miscarriage were equal in the two groups. Since the P value is small, we conclude that the data are inconsistent with the null hypothesis.

The smaller the P value, the less faith we have in the null hypothesis. Generally, scientists do not regard one-sided P values above 10% as indications for rejecting the null hypothesis. One-sided P values in the 1 to 10% range may be taken seriously, but if other evidence strongly suggests that the null hypothesis is true, a P value in this range may not be taken to be definitive.

If the P value is not small enough to justify rejection of the null hypothesis, the conclusion is not necessarily acceptance of the null hypothesis. Instead, the indication may be that the data are not definitive—they are consistent with the null hypothesis and with other hypotheses. For example, in the study of miscarriages if the one-sided P value had been .20 instead of .02, the conclusion would have been that the observed difference was of a magnitude that might well occur by chance under the null hypothesis. In that case, although the data would still suggest an increased risk for women in the operating room, they would not *strongly* suggest that this is or is not so.

On the basis of the interval estimate and significance test, it can be concluded (pending further study) that nurses in the operating room do run a higher risk of miscarriage than general duty nurses. Ages of the two groups, family history, and other factors were examined, and no satisfactory alternative explanation was found. It was tentatively concluded that the cause lay in the operating room, and most likely in the anesthetic gases that are allowed to escape freely into the operating rooms from the anesthetising equipment.

The 2 × 2 Chi-Squared Test of Homogeneity

There is another way of developing the large sample analysis for comparing two proportions, and we exhibit it here for two reasons. It is extremely popular, and thus an essential part of a liberal education in statistics, but more importantly, it easily generalizes to more complex types of data (trichotomous, etc.). This approach is the chi-squared method for significance testing. The data can be exhibited in a so-called 2 × 2 or four-fold table, as follows:

	Sample 1	Sample 2	Totals
Successes	O_{11}	O_{12}	$n_{1.}$
Failures	O_{21}	O_{22}	$n_{2.}$
Totals:	$n_{.1}$	$n_{.2}$	$n_{..}$

(7)

The notation has been changed in anticipation of generalization in Chapter 9 to more complex sets of data. Now suppose that the hypothesis under test is H_0: $p_1 = p_2$. As before, the best estimate of the common success probability is $\hat{p}$ [given previously by (4) and rewritten here in the notation of display (7)]:

$$\hat{p} = \frac{O_{11} + O_{12}}{n_{.1} + n_{.2}} = \frac{n_{1.}}{n_{..}}.$$

Using this estimate, the expected numbers in the four cells, denoted by E_{11}, E_{12}, E_{21}, E_{22}, can be estimated as (note that we abuse notation slightly and use the symbol E for both expected frequency and an estimate of expected frequency):

$$E_{11} = \hat{p} \times n_{.1} = \frac{n_{1.} \times n_{.1}}{n_{..}}$$

$$E_{12} = \hat{p} \times n_{.2} = \frac{n_{1.} \times n_{.2}}{n_{..}},$$

(8)

$$E_{21} = (1 - \hat{p}) \times n_{.1} = \frac{n_{2.} \times n_{.1}}{n_{..}},$$

$$E_{22} = (1 - \hat{p}) \times n_{.2} = \frac{n_{2.} \times n_{.2}}{n_{..}}.$$

A measure of the discrepancy between the observed frequencies, the O's, and the expected frequencies under the hypothesis, the E's, can then be computed as follows:

(9) $$\chi^2 = \frac{(O_{11} - E_{11})^2}{E_{11}} + \frac{(O_{12} - E_{12})^2}{E_{12}} + \frac{(O_{21} - E_{21})^2}{E_{21}} + \frac{(O_{22} - E_{22})^2}{E_{22}}.$$

(The symbol χ is the Greek letter chi. Read χ^2 as chi-squared.) A shortened notation for this χ^2 statistic would be

(9') $$\chi^2 = \sum \frac{(O - E)^2}{E}$$

where we have omitted the subscripts on the O's and the E's, but it is to be understood that the summation $\sum$ is over all (in this case, four) cells. Note that the differences "observed minus expected," that is $O - E$, are squared, eliminating the balancing out of positive and negative discrepancies. Also, each squared difference is weighted by the inverse of the corresponding E, so that the differences involving small E's assume the greatest importance.

The statistic defined by equation (9) is simply the square of the test statistic defined by equation (6). Referring to the miscarriage rate data of Table 1, the data can be arranged as follows:

	Sample 1	Sample 2	Totals
Miscarriages	10	3	13
Births	26	31	57
Totals:	36	34	70

Thus, in the notation of display (7),

$$O_{11} = 10, O_{12} = 3, O_{21} = 26, O_{22} = 31,$$

$$n_{1.} = 13, n_{2.} = 57, n_{.1} = 36, n_{.2} = 34, n_{..} = 70.$$

From (8) we find

$$E_{11} = \frac{(13)(36)}{70} = 6.6857, \ E_{12} = \frac{(13)(34)}{70} = 6.3142,$$

$$E_{21} = \frac{(57)(36)}{70} = 29.3142, \ E_{22} = \frac{(57)(34)}{70} = 27.6857.$$

Thus, using (9),

$$\chi^2 = \frac{(10-6.6857)^2}{6.6857} + \frac{(3-6.3142)^2}{6.3142} + \frac{(26-29.3142)^2}{29.3142} + \frac{(31-27.6857)^2}{27.6857} = 4.15.$$

More directly,

$$\chi^2 = A^2 = (2.038)^2 = 4.15.$$

This test statistic would be referred to percentiles of the **chi-squared distribution** with one degree of freedom. It can be shown that if H_0 is true, the distribution of χ^2 has a form that is approximated by a specific mathematical function called the chi-squared distribution. Just as with the t distribution, there is a family of chi-squared distributions, one for each positive integer (called its degrees of freedom). In the case of the 2×2 table being analyzed in this chapter, the degrees of freedom integer is *one*, and we say that under H_0, χ^2 is distributed approximately as the chi-squared distribution with one degree of freedom.

More formally the χ^2 procedure, at the (approximate) α level, is

(10)
$$\text{reject } H_0 \quad \text{if } \chi^2 \geq \chi^2_{\alpha,1},$$
$$\text{accept } H_0 \quad \text{if } \chi^2 < \chi^2_{\alpha,1},$$

where we use the symbol $\chi^2_{\alpha,1}$ to denote the upper α percentile point of the chi-squared distribution with one degree of freedom. Values of these percentiles are given in row 1 of Table C8. It can be shown that the test procedure (10) is equivalent to the two-sided α level test, based on A, described in the preceding subsection. By equivalent we mean that if one procedure rejects H_0 for a given set of data at a given α, so will the other (at the same α level), and vice versa. In fact, as we have already noted, the chi-squared test statistic χ^2 is exactly the square of the standardized test statistic A, though we have left the algebraic verification of this fact to the reader. In general, it can be shown that the square of a standard normal variate is chi-squared distributed with one degree of freedom.

Note that only *large* values of χ^2 lead to rejection of H_0. A *small* value of χ^2 would indicate that the "observed minus expected" deviations are small, and thus consistent with the null hypothesis.

Entering Table C8 in row 1 (corresponding to one degree of freedom) we find $\chi^2_{.05,1} = 3.84$ and thus the $\alpha = .05$ test defined by (10) is

$$\text{reject } H_0 \quad \text{if } \chi^2 \geq 3.84,$$

$$\text{accept } H_0 \quad \text{if } \chi^2 < 3.84.$$

In the miscarriage rates example, the observed value of χ^2 is 4.15, and thus we reject H_0 at the .05 level. From Table C8, we can see that the P value is between .01 and .05. (From our earlier calculations, we know that the P value is approximately .04.)

There is a short cut formula for the calculation of the chi-squared test statistic, namely

(11)
$$\chi^2 = \frac{n_{..}(O_{11}O_{22} - O_{21}O_{12})^2}{n_{.1} \times n_{.2} \times n_{1.} \times n_{2.}}.$$

Note the square of the difference in the products of values on the two diagonals of display (7) is multiplied by the total number of observations, then divided by the product of the four marginal totals. For the miscarriage rate data of Table 1, we can use (11) to find (once again)

$$\chi^2 = \frac{70(10 \cdot 31 - 26 \cdot 3)^2}{(36)(34)(13)(57)} = 4.15.$$

The chi-squared test, as developed in this subsection, is called a chi-squared test of *homogeneity*. This is because, for display (7), we have considered the data to be based on a sample of size $n_{.1}$ from one population and a separate independent sample of size $n_{.2}$ from a second population. Thus for the Table 1 data $n_{.1} = 36$, $n_{.2} = 34$, and the null hypothesis specifies that $p_1 = p_2$, where p_1 denotes the probability that a pregnant operating room nurse will suffer a miscarriage and p_2 denotes the probability that a pregnant general duty nurse will suffer a miscarriage. The null hypothesis $p_1 = p_2$ is often called the homogeneity hypothesis since it specifies that the chance of success is the same for both populations.

In contrast to the homogeneity framework, data displays such as (7) can also arise when neither of $n_{.1}$, $n_{.2}$, $n_{1.}$, or $n_{2.}$ are fixed, but instead when each observation from a general population is cross-classified on the basis of two characteristics (having characteristic C, not having characteristic C; having characteristic D, not having characteristic D). The question is whether the occurrences of the characteristics are *independent*. In the next subsection, we see that the χ^2 statistic defined by (11) is also appropriate for a test of independence.

The 2×2 Chi-Squared Test of Independence

The data in Table 4, cited by Mosteller and Rourke (1973), are due to L. B. Ellis and D. E. Harken. Three hundred and twenty-two patients who are essentially cardiac invalids are cross-classified on two characteristics involving cardiac symptoms. Here the two characteristics are valvular calcification and mitral insufficiency.

TABLE 4. Classification of 322 Cardiac Invalids According to Valvular Calcification and Mitral Insufficiency

		Valvular calcification		
		Low	High	
Mitral Insufficiency	Low	142	75	217
	High	45	60	105
		187	135	322

Source: F. Mosteller and R. E. K. Rourke (1973).

Let C denote the event that a cardiac invalid has low valvular calcification and D denote the event that the invalid has low mitral insufficiency. Events C and D are independent if $Pr(C \text{ and } D) = Pr(C) \cdot Pr(D)$. Let us recast display (7) in the cross-classification notation.

(12)

	C	not C	Totals
D	O_{11}	O_{12}	$n_{1.}$
not D	O_{21}	O_{22}	$n_{2.}$
Totals:	$n_{.1}$	$n_{.2}$	$n_{..}$

Now, if E_{11} denotes the expected number of counts, under the null hypothesis of independence, falling in the northwest corner cell (that is, those patients having characteristic C and characteristic D), we have

$$E_{11} = n_{..} \times Pr(C \text{ and } D) = n_{..} \times Pr(C) \times Pr(D),$$

where we used the hypothesis of independence to rewrite $Pr(C \text{ and } D)$ as $Pr(C) \times Pr(D)$. It is natural to estimate $Pr(C)$ by $(O_{11} + O_{21})/n_{..}$ and $Pr(D)$ by $(O_{11} + O_{12})/n_{..}$. That is, $Pr(C)$ is estimated by the relative frequency of event C, and $Pr(D)$ is estimated by the relative frequency of event D. Thus under the hypothesis of independence, the E's are estimated as (note, we are again abusing notation and using the same symbol for the expected frequency and an estimate of the expected frequency)

$$E_{11} = n_{..} \times \left(\frac{O_{11} + O_{21}}{n_{..}}\right) \times \left(\frac{O_{11} + O_{12}}{n_{..}}\right) = \frac{n_{.1} \times n_{1.}}{n_{..}},$$

and similarly,

$$E_{12} = \frac{n_{.2} \times n_{1.}}{n_{..}},$$

$$E_{21} = \frac{n_{.1} \times n_{2.}}{n_{..}},$$

$$E_{22} = \frac{n_{.2} \times n_{2.}}{n_{..}}.$$

Since these E's agree with those given by equation (8) for the hypothesis of homogeneity, the formula for χ^2 [equation (9)-or the shortcut formula equation (11)] of the preceding subsection is also applicable for the test of independence. Again, the value of χ^2 is to be referred to percentiles of the chi-squared distribution with one degree of freedom (row 1 of Table C8). Large values of χ^2 indicate dependence between the occurrences of the characteristics; small values of χ^2 are consistent with the null hypothesis of independence.

For the Table 4 data we have, using equation (11),

$$\chi^2 = \frac{322(142 \cdot 60 - 45 \cdot 75)^2}{(187)(135)(217)(105)} = 14.8.$$

From row 1 of Table C8 we find $\chi^2_{.001,1} = 10.83$, and since $\chi^2 = 14.8$ is greater than 10.83 we can reject the hypothesis of independence at $\alpha = .001$. (From Table C8 we know the P value is less than .001, but Table C8 is not detailed enough to provide the exact P value.) Thus there is strong evidence that, in cardiac invalids, the degree of valvular calcification and the degree of mitral insufficiency are dependent.

Discussion Question 1.   The study of nurses described in this chapter was an observational study in that the study subjects were not randomly allocated to the two groups. They chose to be operating room or general duty nurses, and there may well be something about women electing to work in the operating room that is associated with proneness to miscarriage. Can you think of a feasible way of setting up a randomized experiment? You might think of reassigning randomly selected operating room nurses to general duty, or randomly selecting certain operating rooms throughout the country, venting the escaping gases in these rooms and studying personnel in the vented and unvented rooms.

Discussion Question 2.   The study of nurses described in this chapter has the disadvantage of depending on the memory of the nurses interviewed. It may be that operating room nurses have a better memory for miscarriages or are better at detecting early miscarriage than general duty nurses. How would you design a study that avoided these problems?

Comments

2.   In this section we illustrated the two-sided test based on A. The one-sided α level test of H_0, versus the alternative $p_1 > p_2$, rejects when A exceeds z_α and accepts otherwise. Similarly, the one-sided α level test of H_0, versus the alternative $p_1 < p_2$, rejects when A is less than $-z_\alpha$ and accepts otherwise.

3.   In this section, the test based on A was formulated for the null hypothesis $p_1 = p_2$, or equivalently, $p_1 - p_2 = 0$. To test $p_1 - p_2 = \delta_0$ (say), where δ_0 is any specified nonzero value between -1 and 1, use the statistic A' defined as

$$A' = \frac{(\hat{p}_1 - \hat{p}_2) - \delta_0}{\widehat{SD}(\hat{p}_1 - \hat{p}_2)}.$$

[Note that the denominator of A' uses $\widehat{SD}(\hat{p}_1 - \hat{p}_2)$, as given by (1), rather than the estimate (5) based on pooling the two sample proportions together.] The statistic A' should be referred to percentiles of the normal distribution. Significantly large values of A' indicate $p_1 - p_2 > \delta_0$; significantly small values of A' indicate $p_1 - p_2 < \delta_0$.

4.   The significance tests of Section 2 and the confidence interval of Section 1 depend on an approximation to exact probabilities. The approximation is close if

the samples are large. They should be large enough so that the E's, defined by (8), each be no smaller than five. In terms of the miscarriage rates example, the samples should be large enough so that the miscarriages and births expected in each sample number five or more. In fact, looking at Table 1, births and miscarriages number 10 and 26 in the sample of operating room nurses, but miscarriages in the sample of general duty nurses numbers 3, a little less than the 5 or more that assures a good approximation. Nevertheless the confidence coefficient of 95% is not seriously in error.

For the significance test, the null hypothesis leads to an estimate of risk of $\hat{p} = .186$ for both operating room nurses and general duty nurses. Hence the estimated expected numbers of miscarriages in the two samples, under the assumption that H_0 is true, are $(.186)36 = 6.7$ and $(.186)(34) = 6.3$, respectively—both greater than the suggested minimum level of five. Thus the approximations used yield P values close to the exact value.

The chi-squared test statistic can be corrected when the sample sizes are borderline in size, so as to guard against getting a P value that is off the correct value on the low side, possibly leading to erroneous rejection of the null hypothesis. The correction, called Yates' correction for continuity, is easily accomplished by modifying the formula for χ^2. The corrected statistic χ_c^2 is, in the miscarriage rates example,

$$\chi_c^2 = \frac{70 \cdot \left(|10 \cdot 31 - 26 \cdot 3| - \frac{70}{2} \right)^2}{(36)(34)(13)(57)} = 3.00.$$

The P value for this corrected chi-squared value, using Table C8 with one degree of freedom, is seen to be between .05 and .10. This is more conservative than the .04 value obtained without correction.

In the more general setting of display (7) and formula (11), the corrected χ^2 statistic is

(13)
$$\chi_c^2 = \frac{n_{..} \left(|O_{11}O_{22} - O_{21}O_{12}| - \frac{n_{..}}{2} \right)^2}{n_{.1} \times n_{.2} \times n_{1.} \times n_{2.}}.$$

This correction can be used in both the test for independence and the test for homogeneity.

5.   Display (12) is often called a 2×2 **contingency table**. Such tables, and larger contingency tables containing enumeration data, have been extensively studied by statisticians. In Chapter 9 we consider larger contingency tables.

Summary of Statistical Concepts

In this chapter the reader should have learned how to:
(i)   Obtain a confidence interval for the difference $p_1 - p_2$ between two success probabilities.
(ii)   Test hypotheses about the value of $p_1 - p_2$.
(iii)   View the test of the hypothesis $p_1 = p_2$ as a chi-squared test of homogeneity, a type of test that is covered in more generality in the next chapter.
(iv)   Test for independence in 2×2 tables, using the χ^2 statistic.

PROBLEMS

4.   Return to the data of Table 2, considered in Problem 1. Compute the significance test for the null hypothesis: (i) as a difference divided by an estimated standard deviation, (ii) as a chi-squared statistic by the shortcut formula, and (iii) as Yates' corrected chi-squared statistic. Are the data consistent with the null hypothesis? What is your conclusion concerning risk of miscarriage for the two groups?

5.   Return to the Medi-Cal data considered in Problem 2. Use Yates' corrected chi-squared statistic to determine if the data are consistent with the null hypothesis.

6.   Return to the tonsillectomy data of Table 3, considered in Problem 3. Use Yates' corrected chi-squared statistic to determine if the data are consistent with the null hypothesis.

7.   Consider the prevalence of insomnia data of Table 12, Chapter 9. Test the hypothesis that the prevalence rate of insomnia for women of ages 18–24 is the same as the rate for women of ages 25–34.

8.   Consider the prevalence of perspiring-hands data of Table 11, Chapter 9. Test the hypothesis that the prevalence rate of perspiring hands for men of ages 45–54 is the same as the rate for men of ages 55–64.

9.   Verify directly the equivalence of equations (9) and (11).

10.   Define chi-squared distribution, contingency table.

Supplementary Problems

1.   Jacobs (1975) reported that, in a study of the effect of the occurrence of the XYY chromosome constitution on male behavior, 7 of 197 males, institutionalized and requiring special security treatment on account of their dangerous, violent or criminal behavior, had the XYY constitution. None of 475 males randomly selected from the general population had the XYY chromosome constitution. Test the hypothesis of equal occurrence in both populations.

2. Mann, Vessey, Thorogood, and Doll (1975) did a retrospective study comparing women with myocardial infarction (M.I.) and a control group of women to determine if oral contraceptive practice was associated with a higher risk of myocardial infarction. They found that 17 of 58 M.I. patients had used oral contraceptives the month before admission to the hospital for myocardial infarction, whereas only 14 of 166 patients in the control group had used oral contraceptives in the month preceding their admission. Calculate the rates of usage for the two groups, the estimated standard deviation for each rate, the difference in rates, the estimated standard deviation of the difference, and the P value for the hypothesis of equal rates in the two groups.

It should be noted here that you are encouraged to compare the percentages in the two groups by standard techniques for two independent proportions, as done by the investigators. In fact, the data were obtained by matching three controls with each M.I. patient on marital status, age and year of admission to the hospital. The analysis should take such matching into account, by a method like taking differences within each quartet. The techniques would be expected to be more sensitive to a true difference, yielding a smaller P value if a true difference existed. The method of analysis is described by Fleiss (1973, Chapter 8) but it requires data for each quartet individually.

3. Mann *et al.* (1975), in the study described in Supplementary Problem 2, also checked to verify that the distribution of smokers confirmed the theory that smoking causes myocardial infarction. They reported the following results:

TABLE 5. Smoking Categories for Myocardial Infarction Patients and Control Patients

Smoking Category	Number of M.I. Patients	Number of Control Patients
Never smoked	12	60
Ex-smoker	2	14
1–14 cigarettes a day	12	50
15–19 cigarettes a day	7	11
20–24 cigarettes a day	17	18
25 or more cigarettes a day	9	4
Totals:	59	157

Source: J. I. Mann, M. P. Vessey, M. Thorogood, and R. Doll (1975).

Test the hypothesis that smoking history and myocardial infarction are independent, grouping the categories as nonsmokers and smokers, putting ex-smokers in the latter category.

In their 1975 paper Mann *et al.* used a test that is more sensitive (that is, more likely to reject the null hypothesis) when a trend exists. Here a possible trend could be the more smoking, the greater the risk of myocardial infarction. A test for trend is described in the next chapter (Section 9.3). Mann *et al.*, rejected the null hypothesis, and then went on to investigate the joint effect of oral contraception and smoking and the interaction between the two apparently causal factors.

4. In a randomized clinical study of two drugs for the treatment of acute asthma attack, Lewiston and Sloan (1975) found no difference between the two drugs, but they did note that among the 42 children treated, 28 showed complete relief within an hour of treatment and 14 did not. For those 38 for whom they had information on the onset of symptoms, 20 of 27 enjoying complete relief reported onset of symptoms within 24 hours of treatment, whereas only 3 of the 11 who were not relieved in the first hour reported such a short duration of symptoms. Test the hypothesis that duration of symtoms is independent of short-term relief.

5. Ray (1975) reported that 14 of 25 prostate cancer patients with elevated serum acid phosphatase proved to have involvement of the periaortic nodes, whereas only 4 of 25 patients with normal levels of serum acid phosphatase proved to have nodal involvement. Test the hypothesis that serum acid phosphatase is unrelated to nodal involvment, using the chi-squared statistic for the 2×2 table, with and without Yates' correction.

6. Adams and Mendenhall (1974), in a paper based on a survey of cardiologists to determine their desire for further medical training, reported the percent of cardiologists desiring classroom instruction in a variety of subjects. The cardiologists were classified by type of practice and by age. In particular the authors reported that, among cardiologists 40 years of age and older, 47 of 142 primary cardiologists practicing in institutions desired biostatistical instruction, whereas 29 of 103 institutional cardiologists practicing cardiology as a secondary specialist desired such training. Can this result be taken as strong evidence that the primary cardiologist is more interested in biostatistics than is the secondary cardiologist?

7. Goode and Coursey (1976), in a study of the theory that the tonsils serve as a reservoir harboring the virus that causes mononucleosis, obtained data on Stanford students seeking treatment for mononucleosis at the Stanford University Student Health Service. Among 46 students 21 years old diagnosed as having mononucleosis, they found that only 8 had had a tonsillectomy, whereas among 139 students of the same age, in the health center for other complaints, 48 had had a tonsillectomy. Test the null hypothesis of equal tonsillectomy rates in the two groups.

8. Vianna, Greenwald, and Davies (1971) considered a series of 101 Hodgkin's disease patients, with the purpose of testing the theory that the tonsils protect the body against invasion of the lymph nodes by a Hodgkin's disease virus. (The existence of such a virus has not been established.) Among the 101 Hodgkin's cases, they found 67 had had a tonsillectomy, whereas in a control group of 107 patients with other complaints, 43 had had a tonsillectomy. Postulate a suitable null hypothesis. Calculate the chi-squared statistic and test the null hypothesis.

•9. Johnson and Johnson (1972) studied the Hodgkin's-tonsillectomy theory described in Supplementary Problem 8, obtaining tonsillectomy data on 85 Hodgkin's cases and a sibling of each case. The data showed 41 tonsillectomies among the Hodgkin's cases and 33 tonsillectomies among the sibs. Note that the pairing of a case with sibling means that the rates for the two groups are not independent. The pairing should be taken into account in the analysis in order to achieve the best chance of detecting a departure from the null hypothesis. The null hypothesis asserts that Hodgkin's cases and their sibs have the same rates of tonsillectomy. A proper way to test the null hypothesis is to apply the one sample binomial test (Section 2 of Chapter 7) to Table 6, obtained by Johnson and Johnson.

TABLE 6.   Tonsillectomy Rates for Hodgkin's Disease Patients and Siblings

		Sibling		
		Tonsil-lectomy	No Tonsil-lectomy	Totals
Hodgkin's patients	Tonsillectomy	26	15	41
	No Tonsillectomy	7	37	44
	Totals:	33	52	85

Source: S. K. Johnson and R. E. Johnson (1972).

If there is no association between tonsillectomy and Hodgkin's disease, then the probability is 1/2 that a patient-sibling pair falls in the upper-right cell and 1/2 that it falls in the lower-left cell, given that the pair falls off the main diagonal. Since the pairs are independent, the ratio 15/22 can be compared with 1/2 by a binomial test as described in Chapter 7. Perform this test of the hypothesis that there is no association between tonsillectomy and Hodgkin's disease. What is the two-sided P value? [This test procedure is due to Q. McNemar (see, for example, Conover, 1971, pp. 127–130).]

•10. Consider a two-sample case in which the observed variable is trichotomous. For example, patients may be classified into three categories of response to treatment. The data could be displayed as follows:

Response Class	Treatment 1	Treatment 2	Totals
Better	O_{11}	O_{12}	$n_{1.}$
Same	O_{21}	O_{22}	$n_{2.}$
Worse	O_{31}	O_{32}	$n_{3.}$
Totals:	$n_{.1}$	$n_{.2}$	$n_{..}$

Give a χ^2 formula, analogous to equation (9), for this 3×2 table. State the expected numbers to be found in each of the six cells of the 3×2 table, under the homogeneity hypothesis that asserts that the response rates for the three types of response are the same for the two treatments. That is, let p_{11} denote the probability that a patient who receives treatment 1 will respond better, p_{21} the probability that a patient who receives treatment 1 will respond "same," p_{31} the probability that a patient who receives treatment 1 will respond "worse," p_{12} the probability that a patient who receives treatment 2 will respond "better," p_{22} the probability that a patient who receives treatment 2 will respond "same," and p_{32} the probability that a person who receives treatment 2 will respond "worse." The null hypothesis is

$$p_{11} = p_{12},$$

$$p_{21} = p_{22},$$

$$p_{31} = p_{32}.$$

REFERENCES

Adams, F. H. and Mendenhall, R. C. (Eds.) (1974). Profile of the cardiologist: Training and manpower requirements for the specialist in adult cardiovascular disease. *Amer. J. Cardiology* **34**, 389–456.

Bunker, J. P. and Brown, B. Wm., Jr. (1973). Personal communication.

Cannon, I. and Remen, N. (1972). Personal communication.

Cohen, E. N., Bellville, J. W., and Brown, B. Wm. Jr., (1971). Anesthesia, pregnancy and miscarriage: a study of operating room nurses and anesthetists. *Anesthesiology* **35**, 343–347.

*Conover, W. J. (1971). *Practical Nonparametric Statistics*. Wiley, New York.

Fleiss, J. L. (1973). *Statistical Methods for Rates and Proportions*. Wiley, New York.

Goode, R. L. and Coursey, D. L. (1976). Data submitted to R. G. Miller, Jr. for statistical evaluation. Data can be found in the article "Combining 2×2 contingency tables" by R. G. Miller, Jr., appearing in *Biostatistics Case Book, Volume One*, by Brown, B. Wm., Jr., Efron, B., Hyde, J., Leurgans, S., Miller, R. G., Jr., Moses, L. E., Wong, S., and Wolfe, R. Stanford University Technical Report No. 21, 1976.

Jacobs, P. A. (1975). XYY genotype (Letter to *Science*.) *Science* **189**, 1044.

Johnson, S. K. and Johnson, R. E. (1972). Tonsillectomy history in Hodgkin's disease. *N. Engl. J. Med.* **287**, 1122–1125.

Lewiston, N. and Sloan, E. (1975). Private communication of preliminary data.

Mann, J. I., Vessey, M. P., Thorogood, M., and Doll, R. (1975). Myocardial infarction in young women with special reference to oral contraceptive practice. *Brit. Med. J.* **2**, 241–245.

Mosteller, F. and Rourke, R. E. K. (1973). *Sturdy Statistics*. Addison-Wesley, Reading.

Ray, G. (1975). Personal communication.

Vianna, N. J., Greenwald, P., and Davies, J. M. P. (1971). Tonsillectomy and Hodgkin's disease: The lymphoid tissue barrier. *Lancet* **1**, 431–432.

*Yates, F. (1934). Contingency tables involving small numbers and the χ^2 test. *J. Roy. Statist. Soc. Suppl.* **1**, 217–235.

CHAPTER 9

Chi-Squared Tests

Chapter 9 is devoted to the analysis of categorical data. The variables consist of counts, the number of times events of certain types occur. The reader has been introduced to such data in Chapters 7 and 8.

Section 1 of Chapter 9 describes a chi-squared test for k specified probabilities. This test is a generalization of the binomial test, based on B^*, introduced in Chapter 7.

Recall that in Chapter 8 we presented 2×2 chi-squared tests of homogeneity and independence. Section 2 of this chapter generalizes those tests to $r \times k$ tables.

In Section 3 we specialize to $2 \times k$ tables, so that in essence we are analyzing a set of k proportions. Our aim is to detect a trend in the proportions, a trend suggested by the intrinsic nature of the scientific problem under consideration. The trend test of Section 3 can be viewed as a competitor of the $2 \times k$ chi-squared test of homogeneity covered in Section 2.

1.	THE CHI-SQUARED TEST FOR k PROBABILITIES (GOODNESS OF FIT)

Varieties of Sweet Peas

Beadle and Beadle (1966), in their introduction to the science of genetics, discuss a famous experiment by Bateson and Punnett (performed in 1906) concerning two varieties of sweet peas. One had purple flowers and a long pollen grain, the other had red flowers and a round pollen grain. If these four traits segregated independently (as the geneticist Gregor Mendel had reported for garden peas), the second hybrid generation should have yielded four distinct types in the theoretical ratio of $9:3:3:1$. Bateson and Punnett's observations for 256 plants are given in Table 1.

TABLE 1.   Outcomes Resulting from Sweet Pea Crossing Experiment

Types	Theoretical Probabilities	Observed Frequencies
Purple and long	9/16	177
Purple and round	3/16	15
Red and long	3/16	15
Red and round	1/16	49
		256

Source: G. W. Beadle and M. Beadle (1966).

According to Mendelian theory, the theoretical probabilities for the four types—purple and long, purple and round, red and long, red and round—are, respectively, $9/16, 3/16, 3/16$, and $1/16$. That is, letting $p_1 = Pr$(purple and long), $p_2 = Pr$(purple and round), $p_3 = Pr$(red and long), $p_4 = Pr$(red and round), the null hypothesis is

$$H_0: p_1 = \frac{9}{16}, p_2 = \frac{3}{16}, p_3 = \frac{3}{16}, p_4 = \frac{1}{16}.$$

To test H_0, we compare the observed frequencies $O_1 = 177$, $O_2 = 15$, $O_3 = 15$, $O_4 = 49$ with their corresponding expected frequencies under H_0.

Letting E_1 denote the expected frequency under H_0 of the type purple and long, we have

$$E_1 = p_1 \times 256 = \frac{9}{16} \times 256 = 144.$$

Similarly, the expected frequencies E_2 (of purple and round), E_3 (of red and long), and E_4 (of red and round) are

$$E_2 = p_2 \times 256 = \frac{3}{16} \times 256 = 48,$$

$$E_3 = p_3 \times 256 = \frac{3}{16} \times 256 = 48,$$

$$E_4 = p_4 \times 256 = \frac{1}{16} \times 256 = 16.$$

As a measure of closeness of the expected frequencies E_1, E_2, E_3, E_4 to the corresponding observed frequencies O_1, O_2, O_3, O_4, we utilize the chi-squared statistic

(1)
$$\begin{aligned}
\chi^2 &= \sum \frac{(O-E)^2}{E} \\
&= \frac{(O_1-E_1)^2}{E_1} + \frac{(O_2-E_2)^2}{E_2} + \frac{(O_3-E_3)^2}{E_3} + \frac{(O_4-E_4)^2}{E_4} \\
&= \frac{(177-144)^2}{144} + \frac{(15-48)^2}{48} + \frac{(15-48)^2}{48} + \frac{(49-16)^2}{16} \\
&= 121.
\end{aligned}$$

This test statistic is to be referred to percentiles of the chi-squared distribution with (in this case of four categories) three degrees of freedom, "large" values indicating a departure from H_0.

More generally, suppose we are testing a null hypothesis that specifies k theoretical probabilities, corresponding to k categories, are equal respectively to $p_1, p_2, \ldots, p_k$. Suppose a sample of size n yields the observed frequencies $O_1, O_2, \ldots, O_k$. Under the null hypothesis, the expected frequencies of the k categories are

$$E_1 = p_1 \times n, \quad E_2 = p_2 \times n, \ldots, E_k = p_k \times n.$$

The χ^2 statistic is

$$\chi^2 = \sum \frac{(O-E)^2}{E}$$

(2)

$$= \frac{(O_1-E_1)^2}{E_1} + \frac{(O_2-E_2)^2}{E_2} + \cdots + \frac{(O_k-E_k)^2}{E_k}.$$

The χ^2 statistic is to be referred to percentiles of the chi-squared distribution with $k-1$ degrees of freedom. That is, in this testing problem where the null hypothesis specifies the probabilities of k categories, the degrees of freedom parameter is $k-1$, one less than the number of categories.

The χ^2 procedure at the (approximate) α level is

(3)

$$\text{reject } H_0 \quad \text{if } \chi^2 \ge \chi^2_{\alpha,k-1},$$

$$\text{accept } H_0 \quad \text{if } \chi^2 < \chi^2_{\alpha,k-1},$$

where the symbol $\chi^2_{\alpha,k-1}$ denotes the upper α percentile point of the chi-squared distribution with $k-1$ degrees of freedom. Values of these percentiles are given in Table C8.

For our genetics example, with $k=4$, we found [see equation (1)] $\chi^2 = 121$. From Table C8 we find $\chi^2_{.001,3} = 16.27$. Thus the $\alpha = .001$ test defined by (3) is

$$\text{reject } H_0 \quad \text{if } \chi^2 \ge 16.27,$$

$$\text{accept } H_0 \quad \text{if } \chi^2 < 16.27.$$

Since $\chi^2 = 121$ exceeds 16.27, we reject H_0 at the $\alpha = .001$ level. Thus the data indicate that the Mendelian theory does not apply here. [The actual P value is much less than .001, but can't be obtained from our (limited) Table C8.]

Random Numbers

Can the 100 digits in Table 1 of Chapter 2 be viewed as a random sample of size 100 from a large table of random digits (such as the Rand table cited in Comment 4 of Chapter 3)? Such a random sample would tend to have the observed frequencies of the digits "close to" the expected frequencies. Let p_1 denote the probability of selecting a 0 from a large table of random numbers, p_2 the probability of selecting a 1, and so forth so that p_{10} denotes the probability of selecting a 9. We now check if the "data" of Table 1 of Chapter 2 are consistent with the null hypothesis H_0:

$$H_0: p_1 = 1/10, p_2 = 1/10, \cdots, p_{10} = 1/10.$$

We have as the observed frequencies

$$O_1 = 10, O_2 = 13, O_3 = 8, O_4 = 9, O_5 = 8, O_6 = 12, O_7 = 9, O_8 = 11,$$
$$O_9 = 6, O_{10} = 14.$$

The corresponding expected frequencies are

$$E_1 = \frac{1}{10} \times 100 = 10, E_2 = \frac{1}{10} \times 100 = 10, \cdots, E_{10} = 10.$$

From (2) with $k = 10$ we find

$$\chi^2 = \frac{(10-10)^2}{10} + \frac{(13-10)^2}{10} + \frac{(8-10)^2}{10} + \frac{(9-10)^2}{10} + \frac{(8-10)^2}{10}$$
$$+ \frac{(12-10)^2}{10} + \frac{(9-10)^2}{10} + \frac{(11-10)^2}{10} + \frac{(6-10)^2}{10} + \frac{(14-10)^2}{10}$$

$$= 5.6.$$

The $\alpha = .10$ test defined by (3) is found, from Table C8 entered at nine degrees of freedom, to be

$$\text{reject } H_0 \quad \text{if } \chi^2 \geq 14.68,$$
$$\text{accept } H_0 \quad \text{if } \chi^2 < 14.68.$$

Since the observed value of χ^2 is $\chi^2 = 5.6$, which is less than the critical value 14.68, we accept H_0 at $\alpha = .10$. (From Table C8 we note that the P value is between .70 and .80).

Comments

1. Recall the dichotomous data problem of Chapter 7 and, in particular, the test of the null hypothesis $H_0: p = p_0$. H_0 asserts that the true probability p of success is equal to a specified value p_0. The test of H_0 based on the standardized statistic B^* [see equation (5′) of Chapter 7] is equivalent to a χ^2 test. To see this we make the following identifications. Let $k = 2$ so that there are only two categories. Call category 1 the success category and category 2 the failure category. Let O_1 be the number of observations in category 1 and O_2 the number of observations in category 2. Then

$$O_1 = B, O_2 = n - B,$$

where B is the number of successes in n trials. Similarly, the expected frequencies under H_0 are

(4) $$E_1 = p_0 \times n, \quad E_2 = (1-p_0) \times n.$$

Then, from (2),

(5) $$\chi^2 = \frac{(B-np_0)^2}{np_0} + \frac{[(n-B)-n(1-p_0)]^2}{n(1-p_0)}.$$

Direct algebra shows that the right-hand side of equation (5) is, in fact, $(B^*)^2$. That is, the χ^2 test, in this $k=2$ situation, is simply the two-sided binomial test of Chapter 7.

2. An alternative expression for χ^2, which is usually more convenient than expression (2) for computational purposes, is

(6) $$\chi^2 = \left(\Sigma \frac{O^2}{E} \right) - n$$

$$= \frac{O_1^2}{E_1} + \frac{O_2^2}{E_2} + \cdots + \frac{O_k^2}{E_k} - n.$$

3. The justification for referring values of χ^2 to Table C8 is that as n gets large, the exact null distribution of the statistic χ^2 tends to the chi-squared distribution with $k-1$ degrees of freedom. A rule of thumb for this "large sample" approximation to be satisfactory is that each expected frequency under H_0 (that is, each E) be at least 5.

4. Sometimes the observed frequencies can be suspiciously close to the expected frequencies. Suppose the outcomes for the genetics experiment of Table 1 had been:

Types	Observed frequencies
Purple and long	143
Purple and round	48
Red and long	47
Red and round	18

For these "data" the value of χ^2 is

$$\chi^2 = \frac{(143-144)^2}{144} + \frac{(48-48)^2}{48} + \frac{(47-48)^2}{48} + \frac{(18-16)^2}{16}$$

$$= .28.$$

From Table C8 we find $\chi^2_{.95,3} = .35$. Thus the chance of getting a chi-squared value as small or smaller than the "observed" value of .28, when H_0 is true, is less than

.05, or 5%. Such a small value of χ^2 indicates that the data fit the model too well. Although such close fits can happen, they should alert the analyst to the possibility of an accidental error, or even the possibility of data tampering.

PROBLEMS

1. For a genetics experiment in the crossbreeding of peas, Mendel obtained the following results in a sample from the second generation of seeds resulting from crossing yellow round peas and green wrinkled peas:

Yellow and round	Yellow and wrinkled	Green and round	Green and wrinkled
315	101	108	32

Do these data support the theory that asserts that these four types should occur with theoretical probabilities 9/16, 3/16, 3/16, and 1/16, respectively?

2. Verify directly, or illustrate with a set of data, that χ^2 as given by equation (5) is equal to $(B^*)^2$, where B^* is given by equation (5') of Chapter 7. (See Comment 1.)

3. M. Hollander tossed a die 120 times with the following results:

Number of dots on face that turned up:	1	2	3	4	5	6
Observed frequency:	19	21	26	18	23	13

Let p_1 denote the probability that the die will show a 1, p_2 the probability of a 2, ..., p_6 the probability of a 6. Test the hypothesis H_0 that the die is fair:

$$H_0: p_1 = 1/6, p_2 = 1/6, p_3 = 1/6, p_4 = 1/6, p_5 = 1/6, p_6 = 1/6.$$

4. Verify directly, or illustrate with a set of data, that formulas (2) and (6) for χ^2 are equivalent.

2. $r \times k$ TABLES: CHI-SQUARED TESTS OF HOMOGENEITY AND INDEPENDENCE

Cheliped Laterality in Blue Crabs

Hamilton, Nishimoto, and Halusky (1976) conducted a study in which blue crabs, *Callinectes sapidus* Rathbun, from five locations on the Gulf of Mexico and Atlantic Ocean coasts of Florida and juvenile crabs hand-reared in the laboratory, were examined for the incidence of cheliped laterality. This species possesses crusher and cutter chelipeds (claws), which can be distinguished by a large blunt tooth located in the crotch of the crusher. Williams (1974) stated that the crusher and cutter of *Callinectes* are "usually" on the right and left, respectively, a few crabs have two cutters, and almost none have two crushers. Hamilton, Nishimoto, and Halusky (1976) obtained laterality data to study Williams' claim and related questions. Table 2 contains a subset of their data.

TABLE 2. Cheliped Configurations of 1096 *Callinectes sapidus* from Five Florida Locations

	Sample Origin					
Configuration	St. Marks River	Goose Creek Bay	Dickson Bay	Trout River	Nassau Inlet	Totals
Crusher *R* (right) Cutter *L* (left)	229	157	246	84	139	855
Crusher *L* Cutter *R*	90	19	14	21	22	166
Cutter *L* and *R*	19	5	12	12	22	70
Crusher *L* and *R*	2	0	2	0	1	5
Totals:	340	181	274	117	184	1096

Source: P. V. Hamilton, R. T. Nishimoto, and J. G. Halusky (1976).

One question of interest is, are the proportions of the four types of cheliped configurations the same for all five locations or are there differences between locations? More formally, let

p_{11} = probability that a randomly selected crab from the St. Marks River will possess the cheliped configuration crusher *R*, cutter *L*

p_{21} = probability that a randomly selected crab from the St. Marks River will possess the cheliped configuration crusher *L*, cutter *R*

p_{31} = probability that a randomly selected crab from the St. Marks River will possess the cheliped configuration cutter *L* and *R*

p_{41} = probability that a randomly selected crab from the St. Marks River will possess the cheliped configuration crusher *L* and *R*.

Define $p_{12}, p_{22}, p_{32},$ and $p_{42},$ by replacing St. Marks River with Goose Creek Bay in the preceding definitions. That is, $p_{12}, p_{22}, p_{32},$ and p_{42} are the Goose Creek Bay probabilities for the four cheliped configurations. With similar definitions for the other three locations, the null hypothesis of homogeneity which asserts there is no difference between the five locations is

$$H_0: \begin{array}{l} p_{11} = p_{12} = p_{13} = p_{14} = p_{15}, \\[6pt] p_{21} = p_{22} = p_{23} = p_{24} = p_{25}, \\[6pt] p_{31} = p_{32} = p_{33} = p_{34} = p_{35}, \\[6pt] p_{41} = p_{42} = p_{43} = p_{44} = p_{45}. \end{array}$$

That is, all the probabilities in the same row are equal to each other.

Let us now recast Table 2, generalizing the notation of display (7) of Chapter 8.

	Sample 1	Sample 2	Sample 3	Sample 4	Sample 5	Totals
Category 1	O_{11}	O_{12}	O_{13}	O_{14}	O_{15}	$n_{1.}$
Category 2	O_{21}	O_{22}	O_{23}	O_{24}	O_{25}	$n_{2.}$
Category 3	O_{31}	O_{32}	O_{33}	O_{34}	O_{35}	$n_{3.}$
Category 4	O_{41}	O_{42}	O_{43}	O_{44}	O_{45}	$n_{4.}$
Totals:	$n_{.1}$	$n_{.2}$	$n_{.3}$	$n_{.4}$	$n_{.5}$	$n_{..}$

In this notation, $O_{11} = 229, O_{12} = 157, \ldots, n_{1.} = 855, n_{2.} = 166, \ldots, n_{..} = 1096.$

An argument similar to the one that led to the estimated E's given by equation (8) of Chapter 8 yields the (estimated) expected frequencies under H_0:

$$E_{11} = \frac{n_{1.} \times n_{.1}}{n_{..}},$$

$$E_{12} = \frac{n_{1.} \times n_{.2}}{n_{..}},$$

and so forth, $\ldots,$

$$E_{45} = \frac{n_{4.} \times n_{.5}}{n_{..}}.$$

More generally, each E is estimated by (abusing notation slightly)

(7)
$$E = \frac{(\text{row total}) \times (\text{column total})}{n_{..}}.$$

Thus the estimated expected frequencies under H_0 for the crab data are

$$E_{11} = \frac{855 \times 340}{1096} = 265.237,$$

$$E_{12} = \frac{855 \times 181}{1096} = 141.200,$$

.

.

.

$$E_{45} = \frac{5 \times 184}{1096} = .839.$$

We display the E's in Table 3.

We measure closeness of the observed frequencies (Table 2) to the estimated expected frequencies (Table 3) via the χ^2 statistic:

$$(8) \quad \chi^2 = \sum \frac{(O-E)^2}{E}$$

$$= \frac{(229-265.237)^2}{265.237} + \frac{(157-141.200)^2}{141.200} + \frac{(246-213.750)^2}{213.750} + \frac{(84-91.273)^2}{91.273}$$

$$+ \frac{(139-143.540)^2}{143.540} + \frac{(90-51.496)^2}{51.496} + \frac{(19-27.414)^2}{27.414} + \frac{(14-41.500)^2}{41.500}$$

$$+ \frac{(21-17.721)^2}{17.721} + \frac{(22-27.869)^2}{27.869} + \frac{(19-21.715)^2}{21.715} + \frac{(5-11.560)^2}{11.560}$$

$$+ \frac{(12-17.500)^2}{17.500} + \frac{(12-7.473)^2}{7.473} + \frac{(22-11.752)^2}{11.752} + \frac{(2-1.551)^2}{1.551}$$

$$+ \frac{(0-.826)^2}{.826} + \frac{(2-1.250)^2}{1.250} + \frac{(0-.534)^2}{.534} + \frac{(1-.839)^2}{.839}.$$

An equivalent formula for χ^2, which is more convenient for computational purposes, is

$$(9) \qquad \chi^2 = \left(\sum \frac{O^2}{E} \right) - n_{..}.$$

Note that the $\sum$ in (8) and (9) means that the summation is over all $r \times k$ cells. For the blue crab data, $r \times k = 4 \times 5 = 20$, and thus 20 terms are summed to yield the

TABLE 3.   Estimated Expected Frequencies for the Cheliped Configurations

	Sample Origin				
Configuration	St. Marks River	Goose Creek Bay	Dickson Bay	Trout River	Nassau Inlet
Crusher R Cutter L	265.237	141.200	213.750	91.273	143.540
Crusher L Cutter R	51.496	27.414	41.500	17.721	27.869
Cutter L and R	21.715	11.560	17.500	7.473	11.752
Crusher L and R	1.551	0.826	1.250	· 0.534	0.839

value of χ^2. Using (9) we find

$$\chi^2 = \frac{(229)^2}{265.237} + \frac{(157)^2}{141.200} + \frac{(246)^2}{213.750} + \frac{(84)^2}{91.273} + \frac{(139)^2}{143.540}$$

$$+ \frac{(90)^2}{51.496} + \frac{(19)^2}{27.414} + \frac{(14)^2}{41.500} + \frac{(21)^2}{17.721} + \frac{(22)^2}{27.869}$$

$$+ \frac{(19)^2}{21.715} + \frac{(5)^2}{11.560} + \frac{(12)^2}{17.500} + \frac{(12)^2}{7.473} + \frac{(22)^2}{11.752}$$

$$+ \frac{(2)^2}{1.551} + \frac{(0)^2}{.826} + \frac{(2)^2}{1.250} + \frac{(0)^2}{.534} + \frac{(1)^2}{.839} - 1096$$

$$= 83.2.$$

For the case of an $r \times k$ table, the χ^2 statistic is to be referred to percentiles of the chi-squared distribution with $(r-1) \times (k-1)$ degrees of freedom. The hypothesis of homogeneity is to be rejected for "large" values of χ^2. Specifically, the χ^2 procedure, at the (approximate) α level, is

(10)

$$\text{reject } H_0 \quad \text{if } \chi^2 \geq \chi^2_{\alpha,(r-1)(k-1)},$$

$$\text{accept } H_0 \quad \text{if } \chi^2 < \chi^2_{\alpha,(r-1)(k-1)},$$

where the symbol $\chi^2_{\alpha,(r-1)(k-1)}$ denotes the upper α percentile point of the chi-squared distribution with $(r-1) \times (k-1)$ degrees of freedom. Values of these percentiles are given in Table C8.

For our blue crab example, with $r=4$ and $k=5$, we found $\chi^2 = 83.2$. From Table C8 we find $\chi^2_{.001,12} = 32.91$. Thus the $\alpha = .001$ test defined by (10) is

$$\text{reject } H_0 \quad \text{if } \chi^2 \geq 32.91,$$

$$\text{accept } H_0 \quad \text{if } \chi^2 < 32.91.$$

Since $\chi^2 = 83.2$ exceeds 32.91, we reject H_0 at the $\alpha = .001$ level and conclude that there are differences in cheliped laterality among the five locations. Note that the P value is much less than .001, but cannot be obtained from Table C8, which only gives selected percentiles.

Relationship Between Premarital Sexual Values and Region of the Country Where Reared

The following problem is from Zahn and van Belle (1970) and is based on a study by Greene, Curtis, Hopkin, and Strong (1970). A group of 1537 southern college students enrolled in introductory courses were surveyed by questionnaire in a study of premarital sexual values. Of interest was how these values related to those values they projected to their parents, and to several background variables measuring such things as sex, race, age, socioeconomic status, region where reared, and church attendance. The students were asked to rate as to acceptability

TABLE 4. 1407 Students Classified by PSI Rating and Region Where Reared

Region Where Reared	Very Acceptable	Somewhat Acceptable	Somewhat Unacceptable	Very Unacceptable	Totals
New England and the East	82	72	49	36	239
South	201	251	122	149	723
Midwest	30	20	26	13	89
West	169	96	36	28	329
Outside U.S.	13	7	4	3	27
Totals:	495	446	237	229	1407

Source: D. A. Zahn and G. van Belle (1970).

each of 12 sexual acts. The data in Table 4 relate to the question: does there exist an association between the variables "region where reared" and "self-rating of premarital sexual intercourse (*PSI*)"? Because of incomplete questionnaires, the total in the table is only 1407.

The null hypothesis H_0 asserts that the two variables "region where reared" and "rating of PSI" are independent. We now test this hypothesis using the χ^2 statistic defined by (8) [or by the equivalent formula (9)]. Recall that in Chapter 8 we found that for the 2×2 table, the expected frequencies (the E's) under the hypothesis of independence are estimated by the same values we use to estimate the expected frequencies under the hypothesis of homogeneity. It can be shown that this is also true of the $r \times k$ table (see Problem 9). Thus the expected frequency E for a given cell in the contingency table is again estimated by equation (7), and thus formulas (8) and (9), for the χ^2 test of homogeneity, are also valid for the test of the independence hypothesis. Computing χ^2, via (9), for the Table 4 data, yields

$$\chi^2 = \frac{(82)^2}{84.083} + \frac{(72)^2}{75.760} + \frac{(49)^2}{40.258} + \frac{(36)^2}{38.899}$$

$$+ \frac{(201)^2}{254.360} + \frac{(251)^2}{229.181} + \frac{(122)^2}{121.785} + \frac{(149)^2}{117.674}$$

$$+ \frac{(30)^2}{31.311} + \frac{(20)^2}{28.212} + \frac{(26)^2}{14.991} + \frac{(13)^2}{14.485}$$

$$+ \frac{(169)^2}{115.746} + \frac{(96)^2}{104.289} + \frac{(36)^2}{55.418} + \frac{(28)^2}{53.547}$$

$$+ \frac{(13)^2}{9.499} + \frac{(7)^2}{8.559} + \frac{(4)^2}{4.548} + \frac{(3)^2}{4.394} - 1407$$

$$= 80.9.$$

For the case of an $r \times k$ table, the χ^2 statistic is referred to percentiles of the chi-squared distribution with $(r-1) \times (k-1)$ degrees of freedom. The hypothesis H_0 of independence is rejected for large values of χ^2. Specifically, the χ^2 procedure at the (approximate) α level is

(11)
$$\text{reject } H_0 \quad \text{if } \chi^2 \geq \chi^2_{\alpha,(r-1)(k-1)},$$
$$\text{accept } H_0 \quad \text{if } \chi^2 < \chi^2_{\alpha,(r-1)(k-1)},$$

where $\chi^2_{\alpha,(r-1)(k-1)}$ is the upper α percentile point of the chi-squared distribution with $(r-1) \times (k-1)$ degrees of freedom. Values of these percentiles are given in Table C8.

For the questionnaire data, with $r = 5$ and $k = 4$, we found $\chi^2 = 80.9$. From Table C8 we find $\chi^2_{.001,12} = 32.91$. Thus the $\alpha = .001$ test defined by (11) is

$$\text{reject } H_0 \quad \text{if } \chi^2 \geq 32.91,$$

$$\text{accept } H_0 \quad \text{if } \chi^2 < 32.91.$$

Since $\chi^2 = 80.9$ exceeds 32.91, we reject H_0 at the $\alpha = .001$ level and conclude that the variables "region where reared" and "PSI rating" are not independent. Note that the P value is much less than .001, but the exact P value cannot be obtained from Table C8, which only provides selected percentiles.

Comments

5.   The $r \times k$ chi-squared tests of homogeneity and independence discussed in this section reduce, when $r = 2$, $k = 2$, respectively, to the 2×2 chi-squared tests of homogeneity and independence introduced in Section 2 of Chapter 8.

6.   The justification for referring values of χ^2 [see equation (8) or (9)] to Table C8 is based on a large sample approximation. The usual rule of thumb for this large sample approximation to be satisfactory is that each expected frequency under H_0 be at least 5. (Note that this requirement is violated with the blue crab data of Table 2. See the estimated expected frequencies given in Table 3. Also see Problem 6.)

PROBLEMS

5.   Farkas and Shorey (1972) have investigated the mechanism by which a small moth species, the pink bollworm *Pectinophora gossypiella* steer toward an odor source. Farkas and Shorey used a Plexiglas flight tunnel (see their paper for details) to investigate if males require a sense of wind direction when steering toward the odor source. The odor source was sex pheromone, prepared by extracting female abdomen tips in ether. Three experimental conditions were considered: condition 1, a pheromone (odor) plume in moving air; condition 2, a pheromone plume in still air; and condition 3, no pheromone plume in still air. The males were considered to be orienting within the pheromone plume if they flew in an upward direction through a square wire hoop (30 cm²) suspended in the center of the flight tunnel 30 cm downward from the pheromone source. The data are given in Table 5.
   Does the chance of flying through the hoop differ significantly in the three conditions?

TABLE 5. **Number of Male Moths Flying Through the Hoop and Outside of the Hoop for the Three Test Conditions**

	Conditions		
	Pheromone Plume in Moving Air	Pheromone Plume in Still Air	No Pheromone Plume in Still Air
Number flying through hoop	17	16	3
Number flying outside hoop	3	4	17

Source: S. R. Farkas and H. H. Shorey (1972).

6. Note that the observed frequencies (the O's) for row four of Table 2 are very small and the corresponding E's (see Table 3) are less than 5. Since the two configurations "cutter L and R" and "crusher L and R" are rare, it is reasonable to combine the last two rows of Table 2 to obtain Table 6.

TABLE 6. **Cheliped Configurations**

	Sample Origin					
Configuration	St. Marks River	Goose Creek Bay	Dickson Bay	Trout River	Nassau Inlet	Totals
Crusher R Cutter L	229	157	246	84	139	855
Crusher L Cutter R	90	19	14	21	22	166
Other	21	5	14	12	23	75
Totals:	340	181	274	117	184	1096

Source: P. V. Hamilton, R. T. Nishimoto, and J. G. Halusky (1976).

Compute χ^2 for this modified table and compare the result with the analysis in the text.

7. Mann (1970), in a study of the social psychology of waiting lines, interviewed 66 small boys in a line for free Batman shirts outside the Grosvenor Theater in downtown Melbourne. Mann states: "To attract a large juvenile audience the managment had announced that Batman T-shirts would be given away to the first 25 arrivals at the daily

10:00 A.M. show. At 7:00 o'clock each morning a queue consisting almost entirely of twelve-year old boys formed outside the theater. Because the queue was short and the critical point was near the front, this was a fine opportunity to determine whether systematic distortions would occur even when it was possible for judgement to be fairly accurate." The data are given in Table 7.

TABLE 7. Estimates of Position in Line as a Function of Actual Position in Line

Estimates of Number Ahead in Line	Actual Position	
	(earlycomers) 1–25	(latecomers) 26–50
Over-estimate	14	6
Correct	17	0
Under-estimate	9	20

Source: L. Mann (1970).

Do the latecomers and earlycomers tend to estimate their position with the same degree of optimism? Perform an appropriate hypothesis test.

8. Return to Table 5 of Chapter 2 and use the chi-squared statistic to test whether there exists a relationship or association between month of birth and month of death.

•9. Consider the questionnaire data of Table 4. Let A_1 denote the event "student's PSI rating is very acceptable," let A_2 denote the event "student's PSI rating is somewhat acceptable", and so forth so that A_4 denotes the event "student's PSI rating is very unacceptable." Similarly, let C_1 denote the event "student was reared in the region New England and the East," let C_2 denote the event "student was reared in the South," and so forth so that C_5 denotes the event "student was reared outside the U.S." The precise meaning of the assertion that the variables "region where reared" and "PSI rating" are independent is that:

events A_1 and C_1 are independent,

events A_1 and C_2 are independent,

$$\cdot \qquad \cdot$$
$$\cdot \qquad \cdot$$
$$\cdot \qquad \cdot$$

events A_1 and C_5 are independent,

events A_2 and C_1 are independent,

$$\begin{matrix} \cdot & & \cdot \\ \cdot & & \cdot \\ \cdot & & \cdot \end{matrix}$$

events A_4 and C_5 are independent.

Thus for the variables "rating of PSI" and "region where reared" to be independent, we must have:

(12)
$$\begin{aligned} Pr(A_1 \text{ and } C_1) &= Pr(A_1) \cdot Pr(C_1), \\ Pr(A_1 \text{ and } C_2) &= Pr(A_1) \cdot Pr(C_2), \\ \cdot \quad & \quad \cdot \\ \cdot \quad & \quad \cdot \\ \cdot \quad & \quad \cdot \\ Pr(A_4 \text{ and } C_5) &= Pr(A_4) \cdot Pr(C_5). \end{aligned}$$

That is, the $rk = (5)(4) = 20$ equations given by (12) must be true. Use this to show that, under the hypothesis of independence, E_{11} can be estimated by

$$E_{11} = \frac{n_{1.} \times n_{.1}}{n_{..}},$$

and, more generally, each E can be estimated by

$$E = \frac{(\text{row total}) \times (\text{column total})}{n_{..}}.$$

---

3.   A TEST FOR TREND IN PROPORTIONS

One-Year Survival Rates of Heart Transplant Patients—Is There an Increasing Trend?

Table 8, obtainable from the data of Table 4 of Chapter 1, contains, for each of the years 1968, 1969, 1970, and 1971, the number of patients who were accepted as candidates in the Stanford heart transplant program in that year and went on to receive a transplant. It also provides, for each year, the number of those who survived at least 1 year after their transplant.

The 1-year survival rates for the years 1968, 1969, 1970, and 1971 are, respectively,

$$\hat{p}_1 = 2/9 = .22, \quad \hat{p}_2 = 6/11 = .55, \quad \hat{p}_3 = 2/6 = .33, \quad \hat{p}_4 = 6/13 = .46.$$

TABLE 8.   Relationship Between Year of Acceptance and 1-Year Posttransplant Survival Rate

	Year				
	1968	1969	1970	1971	Totals
Number who survived at least 1 year after transplant	2	6	2	6	16
Number who did not survive at least 1 year after transplant	7	5	4	7	23
Totals:	9	11	6	13	39

Source: E. Dong (1973).

An increasing trend in the $\hat{p}$'s could possibly reflect improvement in surgical techniques, in the methods of selecting transplant patients, or in monitoring and caring for the posttransplant patients. Do these data indicate a significant increasing trend?

Testing for a Trend

The null hypothesis is that of no trend. Call this hypothesis H_0. Formally, we can write this null hypothesis as

$$H_0: p_1 = p_2 = \cdots = p_k,$$

where the p's are the population proportions. Employing a procedure discussed by Armitage (1955), we use the statistic S defined by equation (14) to test H_0 against the alternative

$$H_I: p_1 \leq p_2 \leq \cdots \leq p_k \quad \text{(where at least one of the inequalities is strict).}$$

The alternative hypothesis H_I corresponds to an increasing trend in the p's.

   In the Table 8 data there are four rates. To treat the general case of k rates we use display (13).

(13)

		Columns			
		1	2	... k	Totals
Rows	1	O_{11}	O_{12}	... O_{1k}	$n_{1.}$
	2	O_{21}	O_{22}	... O_{2k}	$n_{2.}$
Totals:		$n_{.1}$	$n_{.2}$	... $n_{.k}$	$n_{..}$

Set

(14)
$$S = C - D,$$

where

$$C = O_{21}(O_{12} + O_{13} + \cdots + O_{1k}) + O_{22}(O_{13} + \cdots + O_{1k})$$
$$+ \cdots + O_{2,k-1}(O_{1k}),$$

and

$$D = O_{11}(O_{22} + O_{23} + \cdots + O_{2k}) + O_{12}(O_{23} + \cdots + O_{2k})$$
$$+ \cdots + O_{1,k-1}(O_{2k}).$$

Note that C is simply the sum of products of each observed frequency in the second row with the observed frequencies above and to the right of it. Similarly, D is the sum of products of each observed frequency in the first row with these below and to the right of it. If the alternative hypothesis H_I corresponding to an increasing trend is true, then C would tend to be larger than D, and S would tend to be larger than 0. Analogously, we can use S to test H_0 against a decreasing trend alternative

$$H_{I'}: p_1 \geq p_2 \geq \cdots \geq p_k \quad \text{(where at least one of the inequalities is strict).}$$

In the presence of a decreasing trend, C would tend to be smaller than D, and S would tend to be smaller than 0. Under the hypothesis H_0 of no trend, the expected value of S is zero and the variance of S is

(15)
$$V = \frac{n_{1.}n_{2.}(n_{..}^3 - \sum n_{.i}^3)}{3n_{..}(n_{..} - 1)},$$

where

$$\sum n_{.i}^3 = n_{.1}^3 + n_{.2}^3 + \cdots + n_{.k}^3.$$

A procedure, with an approximate Type-I error probability of α, is

(16)
reject H_0, in favor of an increasing trend, if $S^* \geq z_\alpha$,

accept H_0 if $S^* < z_\alpha$,

where S^*, the properly standardized version of S, is

(17)
$$S^* = \frac{S}{\sqrt{V}}.$$

Similarly, to test H_0 versus a decreasing trend at the approximate α level,

(18)
$$\text{reject } H_0, \text{ in favor of a decreasing trend, if } S^* \leq -z_\alpha,$$
$$\text{accept } H_0 \quad \text{if } S^* > -z_\alpha.$$

A two-sided test of H_0 versus H_I and $H_{I'}$, at the approximate α level, is

(19)
$$\text{reject } H_0 \quad \text{if } S^* \geq z_{\alpha/2} \text{ or } \quad \text{if } S^* \leq -z_{\alpha/2},$$
$$\text{accept } H_0 \quad \text{if } -z_{\alpha/2} < S^* < z_{\alpha/2}.$$

For the data of Table 8, procedure (16) is appropriate. With $k = 4$ we have

$$C = 7(6+2+6) + 5(2+6) + 4(6) = 162,$$
$$D = 2(5+4+7) + 6(4+7) + 2(7) = 112.$$

Thus

$$S = C - D = 162 - 112 = 50.$$

Also, $n_{1.} = 16$, $n_{..} = 39$, and

$$\sum n_{.i}^3 = (9)^3 + (11)^3 + (6)^3 + (13)^3 = 4473.$$

From (15) we then find

$$V = \frac{16(23)\{(39)^3 - 4473\}}{3(39)(38)} = 4539.7.$$

Thus

$$S^* = \frac{50}{\sqrt{4539.7}} = .74.$$

For example take $\alpha = .05$. Then $z_\alpha = 1.65$ and, since $S^* = .74 < 1.65$, procedure (16) leads to acceptance of H_0. A one-sided P value of .23 is obtained by referring $S^* = .74$ to Table C2.

We conclude that although the data suggest an increasing (since S is positive) trend in 1-year survival rates for heart transplant patients, there is not sufficient evidence to reject the hypothesis of no trend.

Discussion Question 1.   Can you criticize the choice of the 1-year posttransplant survival rates as a measure of the success of the transplant program?

Discussion Question 2.   Let $\hat{p}_1 = O_{11}/n_{.1}$, $\hat{p}_2 = O_{12}/n_{.2}, \ldots, \hat{p}_k = O_{1k}/n_{.k}$. Do you think that the p's must be strictly increasing for one to be able to reject H_0 with procedure (16)?

Comments

7.   The χ^2 test of homogeneity, considered in Section 2 for the $r \times k$ table, can be applied to the $2 \times k$ contingency table of display (13). Using equation (9) we find for the data of Table 8

$$\chi^2 = \frac{(2)^2}{3.692} + \frac{(6)^2}{4.513} + \frac{(2)^2}{2.462} + \frac{(6)^2}{5.333}$$

$$+ \frac{(7)^2}{5.308} + \frac{(5)^2}{6.487} + \frac{(4)^2}{3.538} + \frac{(7)^2}{7.667}$$

$$- 39$$

$$= 2.43.$$

Since $r = 2$, $k = 4$, we refer χ^2 to the chi-squared distribution with $(2-1) \times (4-1) = 3$ degrees of freedom. From Table C8 we note that our observed value of $\chi^2 = 2.43$ yields a P value that is between .40 and .50. This is in agreement with the test based on S^*, which yielded a one-sided P value of .23. The chi-squared test is two-sided. The two-sided P value of the S^* test is .46. It is clear that both tests show there is insufficient evidence to reject H_0.

It is important to note here that although the statistics S^* and χ^2 are used to test the same hypothesis—the hypothesis of equal population proportions—they are designed to detect different types of alternatives. The test that rejects for large values of S^* is a one-sided test of H_0 versus H_I. The test that rejects for small values of S^* is a one-sided test of H_0 versus $H_{I'}$. The two-sided test that rejects for both large and small values of S^* is designed to detect both increasing and decreasing trends. However, the χ^2 test is designed to detect any deviation from H_0, including alternatives such as $p_1 = p_3 = p_4 = p$ (say), and $p_2 \neq p$, where there is no "trend." (In Problem 14, the reader is asked to produce contingency tables in which the two-sided S^* test and the χ^2 test lead to different conclusions.)

8.   The test procedure based on S^* introduced in this chapter is discussed by Armitage (1955). For theoretical details, see Armitage's paper, in which other tests for linear trends are also considered. Chapter 9 of Fleiss (1973) contains a good exposition of techniques for comparing proportions. Killion and Zahn (1976) present a bibliography of work published on contingency tables from 1900 to 1974.

Summary of Statistical Concepts

In this chapter the reader should have learned how to:
(i)   Test a hypothesis about k probabilities associated with k categories. The test is based on a χ^2 statistic with $k-1$ degrees of freedom.
(ii)   Use an appropriate χ^2 statistic for tests of homogeneity and independence in $r \times k$ tables. The statistic has $(r-1) \times (k-1)$ degrees of freedom.
(iii)   Test for a trend in proportions in a $2 \times k$ table.

PROBLEMS

10.   Table 9 is based on data of Dr. M. Menzel and can be found in van Belle (1971).

TABLE 9.   Number of Chromosomes of Type-I in *Hibiscus surattensis-sudanensis-breckeridgea*

Cell	Chromosomes of Type-1	Chromosomes Not of Type-1	Total
1	29	83	112
2	37	86	123
3	34	91	125
4	27	102	129
5	36	93	129
6	42	88	130
7	41	89	130
8	28	103	131
9	42	90	132
10	30	102	132
11	32	100	132
12	35	98	133
13	38	95	133
14	28	105	133
15	41	92	133
16	47	86	133
17	46	87	133
18	46	88	134
19	38	96	134

TABLE 9—continued

Cell	Chromosomes of Type-1	Chromosomes Not of Type-1	Total
20	27	107	134
21	39	95	134
22	45	89	134
23	33	102	135
24	41	95	136
25	38	98	136
26	28	108	136
27	41	95	136
28	36	100	136
29	32	104	136
30	36	100	136
31	37	100	137
32	42	95	137
33	35	102	137
34	29	108	137
35	48	90	138
36	31	108	139
37	42	99	141
Totals:	1357	3569	4926

Source: M. Menzel (1971).

Do these data provide evidence of a trend in the proportion of Type-I chromosomes as the total number of chromosomes per cell increases? Consider the data as a 2×15 table in which all cells of equal numbers of chromosomes are grouped together. For example, cells 12–17 all have 133 chromosomes, thus they are grouped into one cell with 235 chromosomes of Type-I and 563 chromosomes not of Type-I.

11. Table 10 can be found in Armitage (1955). It is based on data summarized in Holmes and Williams (1954).

Note that the carrier rates, for the categories $+$, $++$, and $+++$ are, respectively,

$$\hat{p}_1 = 19/516 = .0368, \hat{p}_2 = 29/589 = .0492, \hat{p}_3 = 24/293 = .0819.$$

Use the test based on S^* to see if you can conclude there is a significant trend in the carrier rates as the tonsil sizes increase. Apply the χ^2 statistic (see Comment 7) to these data and interpret the result.

TABLE 10.   Relationship Between Nasal Carrier Rate for *Streptococcus pyogenes* and Size of Tonsils Among 1398 Children Aged 0–15 Years

	Present, but Not Enlarged +	Enlarged Tonsils ++	+++	Totals
Carriers	19	29	24	72
Noncarriers	497	560	269	1326
Totals:	516	589	293	1398

Source: M. C. Holmes and R. E. O. Williams (1954).

12.   The data in Table 11 are from the National Center for Health Statistics (1970, Tables 1 and D). For each of seven age intervals, the table gives the number of men sampled in that interval and the number reporting perspiring hands.

TABLE 11.   Prevalence of Reported Perspiring Hands Among Adult Men by Age: United States, 1960–1962

	Ages 18–24	25–34	35–44	45–54	55–64	65–74	75–79	Totals
Number reporting perspiring hands	95	169	124	80	46	21	2	537
Number not reporting perspiring hands	316	506	579	467	372	244	70	2554
Totals:	411	675	703	547	418	265	72	3091

Source: National Center for Health Statistics (1970).

Do these data indicate a significant decreasing trend in the prevalence of perspiring hands as age increases?

13.   The data in Table 12, considered in Fleiss (1973), are from the National Center for Health Statistics (1970, Tables 1 and 6). For each of seven age intervals, the table gives the number of women sampled in that interval and the number reporting insomnia.

Do these data indicate a significant increasing trend in the prevalence of insomnia as age increases?

•14.   Construct a 2×4 contingency table in which a two-sided test based on S^* leads to rejection of H_0 at $\alpha = .05$ but where a two-sided test based on χ^2 (see Comment 7) leads to acceptance of H_0 at $\alpha = .05$. Now construct a different 2×4 table in which at $\alpha = .05$ the

TABLE 12. Prevalence of Reported Insomnia Among Adult Women by Age: United States, 1960–1962

| | Ages | | | | | | | |
	18–24	25–34	35–44	45–54	55–64	65–74	75–79	Totals
Number reporting insomnia	150	250	264	302	238	176	36	1416
Number not reporting insomnia	384	496	520	403	205	123	34	2165
Totals:	534	746	784	705	443	299	70	3581

Source: National Center for Health Statistics (1970).

two-sided test based on S^* leads to acceptance but the χ^2 test leads to rejection. Interpret your results.

15. Return to the Table 10 data of Problem 11, but collapse the 2×3 table into a 2×2 table by combining the ++ and +++ categories into a single "enlarged" category. Apply the χ^2 test for 2×2 tables (Chapter 8) and compare your results with those of Problem 11.

16. Apply the χ^2 test for $2 \times k$ tables (see Comment 7) to the data of Table 11. Compare your results with those of Problem 12.

17. Apply the χ^2 test for $2 \times k$ tables (see Comment 7) to the data of Table 12. Compare your results with those of Problem 13.

Supplementary Problems

1. Table 13, compiled from Table 5 of Chapter 2, classifies 348 notable Americans by birth month.

TABLE 13. Birth Months of 348 Notable Americans

Jan.	Feb.	Mar.	Apr.	May	June	July	Aug.	Sept.	Oct.	Nov.	Dec.
38	32	29	30	19	17	34	24	31	26	36	32

Source: D. P. Phillips (1972).

If all months are equally likely, we would expect about $\frac{1}{12}$ of all births to be January, $\frac{1}{12}$ in February, and so forth. Test the equally likely hypothesis.

2. Consider the data of Table 5, Chapter 8. Test the hypothesis that smoking history and myocardial infarction are independent, using the χ^2 statistic for a 6×2 table. Compare your results with those of Supplementary Problem 3, Chapter 8, where the analysis was to be based on a χ^2 statistic for a 2×2 table.

3. Table 14 is based on data from the study by Greene, Curtis, Hopkin, and Strong (1970) alluded to in Section 2. These data are cited by Zahn and van Belle (1970). The subjects are classified by sex and family income. Test for independence of sex and family income.

TABLE 14.   1258 Subjects Classified According to Sex and Family Income

Sex	Family Income (in thousands)							Totals
	<4	4–6	6–8	8–10	10–12	12–14	>14	
Male	30	74	92	107	86	77	177	643
Female	29	53	60	83	112	96	182	615
Totals:	59	127	152	190	198	173	359	1258

Source: D. A. Zahn and G. van Belle (1970).

4. Table 15 is based on the study by Hamilton, Nishimoto, and Halusky (1976) described in Section 2. The 340 blue crabs from the St. Marks River are classified by sex and cheliped configuration. Test for independence of sex and configuration.

TABLE 15.   340 Blue Crabs Classified According to Sex and Cheliped Configuration

Sex	Configuration				Totals
	Crusher R Cutter L	Crusher L Cutter R	Cutter L and R	Crusher L and R	
Males	148	55	12	0	215
Females	81	35	7	2	125
Totals:	229	90	19	2	340

Source: P. V. Hamilton, R. T. Nishimoto, and J. G. Halusky (1976).

•5. Consider the $2 \times k$ table of display (13). Let $\hat{p}_1 = O_{11}/n_{.1}$, $\hat{p}_2 = O_{12}/n_{.2}, \ldots, \hat{p}_k = O_{1k}/n_{.k}$ and let $\hat{p} = n_{1.}/n_{..}$. Show that the χ^2 statistic can be written as

$$\chi^2 = \frac{n_{.1}(\hat{p}_1 - \hat{p})^2 + n_{.2}(\hat{p}_2 - \hat{p})^2 + \cdots + n_{.k}(\hat{p}_k - \hat{p})^2}{\hat{p}(1 - \hat{p})}.$$

6. In the work of Goode and Coursey (1976) on the relationship between mononucleosis and tonsillectomy, described in Supplementary Problem 7 of Chapter 8, the investigators thought the proportion of Stanford students who had had a tonsillectomy might vary with age, both because the longer one lives the greater the chance of having any operation and because the medical opinion on the need for tonsillectomy has changed over the years. For the control group used by Goode and Coursey, the numbers of students with and without tonsillectomy can be obtained from Table 16. Is there any evidence in their data for a changing rate over the 7-year span studied? Apply the S^* test for trend. Also apply the 2×7 chi-squared test, with six degrees of freedom, designed to detect any deviations from the null hypothesis. Compare the results.

TABLE 16. Tonsillectomy Rates for Various Ages

	Age in Years						
	18	19	20	21	22	23	24
Tonsillectomy	17	26	34	48	45	29	36
Number of students	49	96	112	139	118	66	75

Source: R. L. Goode and D. L. Coursey (1976).

7. Jacobs (1975), in the work cited in Supplementary Problem 1 of Chapter 8, also cited a third group of 1500 males whom she and her colleagues had studied, in which they had found only one XYY abnormality. Use a chi-squared statistic to test the hypothesis that the rate of XYY abnormality is the same in all three populations sampled. Here a test for trend in proportions is not appropriate. (Explain.)

8. Zapp (1975) reported the results of a long-term animal experiment aimed at evaluating the results of a new and powerful solvent called HMPA (hexamethylphosphoramide) which is finding increasing industrial use. Zapp tested four groups of rats by exposing them to different levels of atmospheric HMPA. His results, by the eighth month, are given in Table 17. Test the null hypothesis that the cancer rate is unrelated to the

TABLE 17. Cancer Rates for Various HMPA Concentrations

Concentration of HMPA (ppb)	Group Size	Cases of Squamous Cell Cancer of Nasal Cavity
0	240	0
50	240	0
400	240	7
4000	240	12

Source: J. A. Zapp, Jr. (1975).

concentration of HMPA. The strength of the indictment against HMPA is enhanced by the fact that the particular cancer observed is relatively rare in these animals, and if the occurrences are not due to HMPA, another explanation is needed.

9. Becker, Kaback, Rosenstock, and Ruth (1975) studied samples of 500 participants and 500 nonparticipants in a screening program for Tay–Sachs disease in the Baltimore–Washington area. This disease is genetic in origin and particularly prevalent among Jews of Ashkenazi descent. The purpose was to determine the factors important in predicting who will participate in public health screening programs. One common and appealing theory is that those people who feel most susceptible will tend to avail themselves of the program. Becker *et al.* reported the Table 18 results among those subjects desiring additional children.

TABLE 18.   Levels of Perceived Susceptibility Versus Participation

	Level of Perceived Susceptibility			
Participation	Low	Moderate	High	Totals
Nonparticipants	53	18	13	84
Participants	202	113	64	379

Source: M. H. Becker, M. M. Kaback, I. M. Rosenstock, and M. V. Ruth (1975).

Is there significant evidence here of a trend toward greater participation among those who feel more susceptible?

10. At Stanford University Hospital, all prescriptions for patients are entered in a computer, where the prescription is compared with previous drug prescriptions for the

TABLE 19.   Reaction of Physicians to Drug Interaction Reporting System—Did the Reports Ever Cause You to Modify a Prescription?

Physician Category	Number Who Received Reports	Percent Responding Positively
Interns	25	36.0
Residents	42	19.0
Full-time faculty	16	6.3
Faculty/community physicians	72	36.1
Non-faculty/community physicians	15	13.3

Source: J. Morrell, M. Podlone, and S. N. Cohen (1977).

patient. If a comparison with a computer-stored list of drug interactions reveals that the patient has been prescribed a combination of drugs of potential harm to the patient, the physician in charge is notified. Morrell, Podlone and Cohen (1977) performed a survey of physicians to evaluate reactions to the drug interaction reporting system. The data in Table 19 are in answer to one of the many questions asked by Morrell *et al.* How strong is the evidence that these several types of physicians were influenced in varying degrees by the drug interaction reporting system?

REFERENCES

*Armitage, P. (1955). Tests for linear trends in proportions and frequencies. *Biometrics* **11,** 375–386.

Beadle, G. W. and Beadle, M. (1966). *The Language of Life.* Doubleday, Garden City, New York.

Becker, M. H., Kaback, M. M., Rosenstock, I. M., and Ruth, M. V. (1975). Some influences in public participation in a genetic screening program. *J. Community Health* **1,** 3–14.

Dong, E. (1973). Personal communication.

Farkas, S. R. and Shorey, H. H. (1972). Chemical trail-following by flying insects: A mechanism for orientation to a distant odor source. *Science* **178,** 67–68.

Fleiss, J. L. (1973). *Statistical Methods for Rates and Proportions.* Wiley, New York.

Goode, R. L. and Coursey, D. L. (1976). Data submitted to R. G. Miller, Jr. for statistical evaluation. Data can be found in the article "Combining 2×2 contingency tables" by R. G. Miller, Jr., appearing in *Biostatistics Case Book, Volume One,* by Brown, B. Wm. Jr., Efron, B., Hyde, J., Leurgans, S., Miller, R. G., Jr., Moses, L. E., Wong, S., and Wolfe, R. Stanford University Technical Report No. 21, 1976.

Greene, J. T., Curtis, J. H., Hopkin, D. T., and Strong, L. D. (1970). An investigation of premarital sexual values and their implications for family life education: A preliminary report. Abstract available in *The Proceedings of the National Council on Family Relations,* 1970 (Chicago).

Hamilton, P. V., Nishimoto, R. T., and Halusky, J. G. (1976). Cheliped laterality in *Callinectes sapidus* (Crustacea: Portunidae). *Biol. Bull.* **150,** 393–401.

Holmes, M. C. and Williams, R. E. O. (1954). The distribution of carriers of *Streptococcus pyogenes* among 2,413 healthy children. *J. Hyg., Camb.* **52,** 165–179.

Jacobs, P. A. (1975). XYY genotype. (Letter to Science.) *Science* **189,** 1044.

Killion, R. A. and Zahn, D. A. (1976). A bibliography of contingency table literature: 1900 to 1974. *International Statistical Review* **44,** 71–112.

Mann, L. (1970). The social psychology of waiting lines. *Amer. Sci.* **58,** 390–398.

Menzel, M. (1971). Personal communication.

Morrell, J., Podlone, M., and Cohen, S. N. (1977). Receptivity of physicians in a teaching hospital to a computerized drug interaction monitoring and reporting system. *Medical Care* **15,** 68–78.

National Center for Health Statistics (1970). Selected symptoms of psychological distress in the United States. *Data from National Health Survey,* Series 11, No. 37.

Phillips, D. P. (1972). Deathday and birthday: An unexpected connection. In: *Statistics: A Guide to the Unknown,* Tanur, J. M. *et al.* (Eds.), pp. 52–65. Holden-Day, San Francisco.

van Belle, G. (1971). Design and analysis of experiments in biology: Notes for a short course. Statistical Consulting Center Report. Florida State University.

Williams, A. B. (1974). The swimming crabs of the Genus *Callinectes* (Decapoda: Portunidae). *Fish. Bull.* **72,** 685–798.

Zahn, D. A. and van Belle, G. (1970). Tests and measures of association in the social sciences. Notes for a short course. Statistical Consulting Center Report. Florida State University.

Zapp, J. A., Jr. (1975). HMPA: A possible carcinogen. (Letter to *Science*.) *Science* **190**, 422.

CHAPTER 10

Analysis of k-Sample Problems

In this chapter we discuss two study designs in which two or more treatments can be compared simultaneously—the **completely randomized experiment** and the **randomized blocks experiment**. The completely randomized experiment is the generalization of the two-sample problem of Chapter 5, to $k(\geq 2)$ samples. The randomized blocks experiment generalizes the paired replicates of Chapter 4 to the case in which on each unit (subject, day, plot of land, etc.) we obtain $k(\geq 2)$ observations. (The units are called blocks, and we let n denote the number of blocks.)

Both the completely randomized experiment and the randomized blocks experiment offer the opportunities to make multiple inferences from a single set of data, and suitable procedures for controlling the added risk of errors in inference are presented. The display of the analysis of variance calculations for each type of experiment is also discussed.

1.	THE COMPLETELY RANDOMIZED EXPERIMENT

An Experiment with Four Treatments

In Chapter 1 an experiment by Zelazo and his colleagues (1972) was described in which infants were randomized to one of four treatments in an effort to determine if certain exercises would accelerate the time at which infants learn to walk by themselves. The nature of each treatment is described in Chapter 1, and the data obtained in the experiment are reproduced here in Table 1. Note that the treatment group of primary interest, namely, the active-exercise group, did indeed seem to walk earlier than any of the three control groups, but the overlap is considerable, and there is certainly some question as to whether the differences might not have come about simply through the imbalances inherent in random assignments of the infants to the treatment regimens.

TABLE 1.   Ages of Infants for Walking Alone (months)

1 Active-exercise Group	2 Passive-exercise Group	3 No-exercise Group	4 8-week Control Group
9.00	11.00	11.50	13.25
9.50	10.00	12.00	11.50
9.75	10.00	9.00	12.00
10.00	11.75	11.50	13.50
13.00	10.50	13.25	11.50
9.50	15.00	13.00	

Source: P. R. Zelazo, N. A. Zelazo, and S. Kolb (1972).

A preliminary step in the statistical evaluation of the data is calculation of the sample means and sample standard deviations for the several groups. For group 1, we find

$$\bar{X}_1 = \frac{9.00+9.50+9.75+10.00+13.00+9.50}{6} = \frac{60.75}{6} = 10.125,$$

and [see equation (3) of Chapter 5]

$$S_1 = \sqrt{\frac{(9)^2 + (9.5)^2 + (9.75)^2 + (10)^2 + (13)^2 + (9.5)^2 - \dfrac{(60.75)^2}{6}}{5}}$$

$$= 1.4470.$$

The reader should perform the corresponding calculations for the remaining three groups. The results are summarized as follows:

i:	1	2	3	4
$\bar{X}_i$:	10.1250	11.3750	11.7083	12.3500
s_i:	1.4470	1.8957	1.5200	.9618

Pooling the Sample Variances

Suppose we have k independent random samples. The parent populations are assumed to be normal populations, with "treatment" means $\mu_1, \mu_2, \ldots, \mu_k$ and common variance σ^2. One approach to comparing the means of the several treatments would be to calculate the confidence interval or t test for each pairwise comparison that interests us. In particular, in the $k = 4$ "walking" example, we would like to compare each of the three control groups with the mean for the active-exercise group. However, you will remember from Chapter 5 that this would mean pooling the sample variances of the two samples being compared to obtain an estimate of the assumed common population variance, and this pooling would then be repeated for each pair of samples to be compared. In fact, there is a substantial gain in convenience and in degrees of freedom for estimating the population variance if the sample variances are all pooled together to obtain one estimate of the population variance. The resultant pooled standard deviation is denoted by s_p, and calculated, in the case of k samples, as

$$s_p = \sqrt{\frac{(n_1 - 1)s_1^2 + (n_2 - 1)s_2^2 + \cdots + (n_k - 1)s_k^2}{(n_1 - 1) + (n_2 - 1) + \cdots + (n_k - 1)}}$$

$$= \sqrt{\frac{(n_1 - 1)s_1^2 + (n_2 - 1)s_2^2 + \cdots + (n_k - 1)s_k^2}{N - k}},$$

where n_1 is the size of the group 1 sample, n_2 is the size of the group 2 sample, $\ldots$, n_k is the size of the group k sample, and

$$N = n_1 + n_2 + \cdots + n_k.$$

For the data in Table 1 we have $k = 4$, $n_1 = n_2 = n_3 = 6$, $n_4 = 5$, $N = 23$, and

$$s_p = \sqrt{\frac{5(1.4470)^2 + 5(1.8957)^2 + 5(1.5200)^2 + 4(.9618)^2}{19}}$$

$$= 1.5164.$$

Thus s_p is obtained by calculating a weighted average of the sample variances, weighting by the degrees of freedom per sample, and then taking the square root. The degrees of freedom associated with s_p is the sum of the degrees of freedom for all samples. For k samples, with sample sizes $n_1, n_2, \ldots, n_k$, the degrees of freedom associated with s_p is $n_1 + n_2 + \cdots + n_k - k$.

The estimated standard deviation for each mean is then calculated by using the pooled standard deviation and dividing it by the square root of the sample size. Thus the standard deviation for $\bar{X}_4$, the sample mean of the 8-week control group, would be estimated as:

$$\widehat{SD}(\bar{X}_4) = \frac{s_p}{\sqrt{5}} = \frac{1.5164}{\sqrt{5}} = .6782.$$

Thus the estimated standard deviations of $\bar{X}_1$, $\bar{X}_2$, $\bar{X}_3$, $\bar{X}_4$ are

$$\widehat{SD}(\bar{X}_1) = .6191, \widehat{SD}(\bar{X}_2) = .6191, \widehat{SD}(\bar{X}_3) = .6191, \widehat{SD}(\bar{X}_4) = .6782.$$

The standard deviation of a difference between any two sample means can be estimated by squaring, adding, and taking the square root, as illustrated here for the case of $\bar{X}_4 - \bar{X}_1$:

$$\widehat{SD}(\bar{X}_4 - \bar{X}_1) = \sqrt{[\widehat{SD}(\bar{X}_4)]^2 + [\widehat{SD}(\bar{X}_1)]^2}$$

$$= \sqrt{\frac{s_p^2}{n_4} + \frac{s_p^2}{n_1}} = s_p\sqrt{\frac{1}{n_4} + \frac{1}{n_1}}.$$

For the Table 1 data we find

$$\widehat{SD}(\bar{X}_4 - \bar{X}_1) = 1.5164\sqrt{\frac{1}{5} + \frac{1}{6}} = .9182.$$

Using these estimates for the standard deviation of differences among the sample means, confidence intervals or t tests can be calculated. For the three differences of principal interest in the "walking" study, we have the following 95% confidence intervals for differences between population means:

$$\text{for } \mu_2 - \mu_1: \quad (\bar{X}_2 - \bar{X}_1) \pm t_{.025,19} s_p \sqrt{\frac{1}{6} + \frac{1}{6}}$$

$$= 1.2500 \pm (2.093)(1.5164)\sqrt{\frac{1}{6} + \frac{1}{6}}$$

$$= 1.2500 \pm 1.8324 = -.58 \text{ and } 3.08$$

$$\text{for } \mu_3 - \mu_1: \quad 1.5833 \pm 1.8324 = -.25 \text{ and } 3.42$$

$$\text{for } \mu_4 - \mu_1: \quad 2.2250 \pm 1.9218 = .30 \text{ and } 4.15.$$

Note that the degrees of freedom for the t distribution here is 19 for all intervals, since the pooled estimate, s_p, used for all intervals is the same and is based on all the sample data. This pooling of all the sample data to increase the degrees of freedom is a real advantage when the sample sizes for the individual samples are small. For the last interval above, the degrees of freedom would have been 9 instead of 19 if the other two samples had not been used. As in Chapter 5, *the pooling assumes equal variation in all treatments*, but this assumption is often (approximately) satisfied in practice.

The confidence interval for $\mu_4 - \mu_1$ excludes zero, suggesting that the evidence does justify the conclusion that the treatment does help the infants learn to walk.

Test statistics and P values can be obtained that parallel the confidence interval procedures. For the null hypothesis that $\mu_4 = \mu_1$, the appropriate test statistic is

$$\frac{\bar{X}_4 - \bar{X}_1}{s_p \sqrt{\frac{1}{n_4} + \frac{1}{n_1}}};$$

significantly large values indicate $\mu_4 > \mu_1$, significantly small values indicate $\mu_4 < \mu_1$.

For the Table 1 data we find

$$\frac{\bar{X}_4 - \bar{X}_1}{s_p \sqrt{\frac{1}{6} + \frac{1}{5}}} = \frac{2.2250}{.9182} = 2.42.$$

From Table C5 entered at 19 degrees of freedom, we find that the one-sided P value is between .01 and .025.

Contrasts Among Several Samples

The completely randomized experiment involving more than two treatments allows the comparison of combinations of means. For example, suppose we want to compare the active-exercise group in Table 1 with the average effect in the other three groups. We could calculate:

$$\frac{\bar{X}_2+\bar{X}_3+\bar{X}_4}{3}-\bar{X}_1=\frac{11.3750+11.7083+12.3500}{3}-10.1250=1.6861.$$

This combination of the sample means is called a contrast. A **contrast in the sample means** is any combination of the sample means that has the property that the coefficients (which multiply the sample means) sum to zero. Note that in the preceding case, the coefficients of $\bar{X}_1$, $\bar{X}_2$, $\bar{X}_3$, $\bar{X}_4$ are, respectively, -1, $+1/3$, $+1/3$, and $+1/3$. A difference between two sample means, for example, $\bar{X}_4-\bar{X}_1$, is a contrast, since the coefficients are $+1, -1$, with zeros for the remaining sample means.

More generally, if we express a contrast as

$$\sum a\bar{X} = a_1\bar{X}_1 + a_2\bar{X}_2 + \cdots + a_k\bar{X}_k,$$

where k is the number of treatments in the experiment and $a_1, a_2, \ldots, a_k$ are the coefficients (remember that some a's may be zero, but, in any case, the *sum* of the a's must be zero), then the standard deviation of the contrast is estimated by:

(1)
$$\widehat{SD}(\sum a\bar{X}) = s_p\sqrt{\frac{a_1^2}{n_1}+\frac{a_2^2}{n_2}+\cdots+\frac{a_k^2}{n_k}},$$

where $n_1, n_2, \ldots, n_k$ are the respective sample sizes, and s_p is the pooled standard deviation.

Note that the statistic $\sum a\bar{X}$ should be viewed as an estimator of the corresponding **contrast in the population means**

$$\sum a\mu = a_1\mu_1 + a_2\mu_2 + \cdots + a_k\mu_k.$$

For the contrast comparing $\bar{X}_1$ with the mean of the three control groups we have, from (1),

$$\widehat{SD}\left(\frac{\bar{X}_2+\bar{X}_3+\bar{X}_4}{3}-\bar{X}_1\right) = 1.5164\sqrt{\frac{(-1)^2}{6}+\frac{(1/3)^2}{6}+\frac{(1/3)^2}{6}+\frac{(1/3)^2}{5}}$$

$$= .7208.$$

If we wanted to test the null hypothesis that $\mu_1 = (\mu_2 + \mu_3 + \mu_4)/3$, the appropriate test statistic would be

$$\frac{\dfrac{\bar{X}_2 + \bar{X}_3 + \bar{X}_4}{3} - \bar{X}_1}{\widehat{SD}\left(\dfrac{\bar{X}_2 + \bar{X}_3 + \bar{X}_4}{3} - \bar{X}_1\right)} = \frac{1.6861}{.7208} = 2.34.$$

Referring the value 2.34 to the t table (Table C5), entered at 19 degrees of freedom, we find that the one-sided P value is between .01 and .025. Thus even when the control groups that reflect the effects of handling and testing are pooled with the pure control group, the effect of active exercise is statistically significant.

The confidence interval for $[(\mu_2 + \mu_3 + \mu_4)/3] - \mu_1$, at a confidence coefficient of .95, is

$$\left(\frac{\bar{X}_2 + \bar{X}_3 + \bar{X}_4}{3} - \bar{X}_1\right) \pm (t_{.025,19})\widehat{SD}\left(\frac{\bar{X}_2 + \bar{X}_3 + \bar{X}_4}{3} - \bar{X}_1\right)$$

$$= 1.6861 \pm (2.093)(.7208)$$

$$= 1.6861 \pm 1.5086 = .18 \text{ and } 3.19.$$

More generally, to test the hypothesis that the contrast $\sum a\mu$ is 0, refer the statistic

$$\frac{\sum a\bar{X}}{\widehat{SD}(\sum a\bar{X})}$$

to the t distribution with $N - k$ degrees of freedom. Significantly large values indicate $\sum a\mu > 0$; significantly small values indicate $\sum a\mu < 0$.

A $1 - \alpha$ confidence interval for $\sum a\mu$ is

(2) $$\sum a\bar{X} \pm (t_{\alpha/2,N-k})\widehat{SD}(\sum a\bar{X}),$$

where $\widehat{SD}(\sum a\bar{X})$ is given by (1), and $t_{\alpha/2,N-k}$ is the upper $\alpha/2$ percentile point of the t distribution with $N - k$ degrees of freedom.

Multiple Inference: Dangers and Safeguards

When a number of treatment groups are included in the same experiment, there is an added danger of error in inference that is sometimes not appreciated. In the experiment under consideration here, the four samples offer the opportunity to make many comparisons among the sample means. We have focused on three differences among the four means, namely $\bar{X}_4 - \bar{X}_1$, $\bar{X}_3 - \bar{X}_1$ and $\bar{X}_2 - \bar{X}_1$, but there are three others we could have checked: $\bar{X}_4 - \bar{X}_3$, $\bar{X}_4 - \bar{X}_2$, and $\bar{X}_3 - \bar{X}_2$. In addition, we examined one more complicated contrast, but this is only one among an infinity of contrasts we could have chosen to analyze.

When a set of data presents such a wealth of opportunities for inference, the possibilities of errors in inference increase with the number of inferences made. If there is a 5% chance of each confidence interval being wrong, it is clear that the confidence to be placed in all inferences being correct may be considerably less than 95%. Similarly, if there is a 5% probability of a Type-I error in testing one null hypothesis, there may be considerably more risk in testing a number of null hypotheses, each at the 5% level.

It can be shown that one way to guard against the risk of multiple inference is to perform each test at a more stringent level. If five hypotheses are to be tested, each should be tested at the 1% level, and this will assure that the probability of making at least one Type-I error is held to less than $5 \times 1\% = 5\%$, simultaneously over all five tests. Using analogous reasoning, if five confidence intervals are to be calculated from the data, and if an overall confidence of 95% is desired, each interval should be a 99% confidence interval. More generally, if an overall level of α is desired and m tests are to be made, one can perform each test at level $\alpha^* = \alpha/m$. If an overall confidence coefficient of $1 - \alpha$ is desired and m intervals are to be computed, one can set the individual confidence coefficients to be $1 - (\alpha/m)$. (See Comment 2.)

Taking the risk of simultaneous inference into account in evaluating the Zelazo *et al.* study, we realize that the data may not be as persuasive as our analyses have suggested thus far.

Selection of differences, and especially interesting contrasts, to test *after* examination of the data presents the most acute risks of error in inference. Examining the data post hoc for interesting facets might be viewed as a sifting of a multitude, a very large number of possible contrasts to find those most significant. One might suppose that this post hoc examination of the data, though essential to the generation of theories for future test, does not admit of any statistical control of possible errors of inference. However, the following procedure of Scheffé (1959, p. 67) does control the overall error probability.

If a very large (not necessarily specified, even infinite) number of confidence intervals are to be generated, the simultaneous confidence coefficient for all the intervals is at least 95% if the intervals for the contrasts are calculated as follows:

For the contrast $\sum a\mu = a_1\mu_1 + a_2\mu_2 + \cdots + a_k\mu_k$, use the interval

$$\sum a\bar{X} \pm \sqrt{(k-1)F_{.05, k-1, N-k}} \cdot \widehat{SD}(\sum a\bar{X}),$$

where

$$\sum a\bar{X} = a_1\bar{X}_1 + a_2\bar{X}_2 + \cdots + a_k\bar{X}_k,$$

$\widehat{SD}(\sum a\bar{X})$ being given by (1), and where $F_{.05,k-1,N-k}$ is the upper .05 percentile of an F distribution with $k-1$ degrees of freedom in the numerator and $N-k$ degrees of freedom in the denominator.

Recall (see Section 4 of Chapter 5) the F distributions (see Table C9) are identified by two positive integers, rather than one as with the t distribution. The first integer, ν_1, is called the degrees of freedom of the numerator, and the second integer, ν_2, is called the degrees of freedom of the denominator.

To obtain an overall confidence coefficient of $1-\alpha$, Scheffé has proved the following theorem:

> *The probability is $1-\alpha$ that the values of all contrasts of the form $\sum a\mu$ simultaneously satisfy the inequalities*

(3)
$$\sum a\bar{X} - \sqrt{(k-1)F_{\alpha,k-1,N-k}} \cdot \widehat{SD}(\sum a\bar{X}) \leq \sum a\mu$$
$$\leq \sum a\bar{X} + \sqrt{(k-1)F_{\alpha,k-1,N-k}} \cdot \widehat{SD}(\sum a\bar{X})$$

> *where $F_{\alpha,k-1,N-k}$ is the upper α percentile point of an F distribution with $k-1$ degrees of freedom in the numerator and $N-k$ degrees of freedom in the denominator.*

Let us compare the confidence intervals given by (2) and (3). Both intervals are of the form

$$\sum a\bar{X} \pm c \cdot \widehat{SD}(\sum a\bar{X})$$

but for interval (2) the constant c is $t_{\alpha/2,N-k}$ and for interval (3) the constant c is $\sqrt{(k-1)F_{\alpha,k-1,N-k}}$. The reader can use Tables C5 and C9 to see that $\sqrt{(k-1)F_{\alpha,k-1,N-k}}$ is larger than $t_{\alpha/2,N-k}$. This is because the Scheffé intervals provide simultaneous $100(1-\alpha)\%$ protection for all intervals that can be calculated, rather than just for a single confidence interval.

If we regard our estimate of the contrast $[(\mu_2+\mu_3+\mu_4)/3]-\mu_1$ among the four means for the Zelazo *et al.* data as an idea stemming from post hoc examination of the data, then the multiplier for the standard deviation of the contrast should be $\sqrt{3F_{.05,3,19}} \approx \sqrt{3(3.14)} = 3.07$, rather than $t_{.025,19} = 2.093$, the multiplier appropriate if the single contrast had been of interest. (Note that Table C9 gives $F_{.05,3,15} = 3.29$ and $F_{.05,3,20} = 3.10$ but does not give a value for $F_{.05,3,19}$ since there are no entries for $\nu_2 = 19$ degrees of freedom. Interpolating on the denominator degrees of freedom ν_2 yields $F_{.05,3,19} \approx 3.14$.) Thus the interval estimate is

$$1.6861 \pm (3.07)(.7208)$$

$$= 1.6861 \pm 2.2129$$

$$= -.53 \text{ and } 3.90.$$

The Omnibus Hypothesis and the Analysis of Variance

In most applications involving more than two samples, the investigator has some specific contrasts in mind for estimation or test. Usually the contrasts are simple differences among various pairs of treatments, as in the study by Zelazo *et al.*, in which the comparison of the active-exercise group with each of the remaining three groups determined the three contrasts of principal interest.

In some studies, however, it is of interest to test the omnibus hypothesis that there are no differences at all among the treatment groups. For k groups this omnibus null hypothesis is

$$H_0: \mu_1 = \mu_2 = \mu_3 = \cdots = \mu_k.$$

Another equivalent way of stating this hypothesis is that all contrasts in the population means are zero. It is clear that if all population means are equal (to μ, say), then the common value μ can be factored out of any contrast $\sum a\mu$, leaving $\mu \sum a$, which equals zero since by definition of the contrast $\sum a = 0$.

There is a third way of stating the omnibus hypothesis, an approach that yields a test of the hypothesis. First, define a weighted mean $\bar{\mu}$ of the individual population means $\mu_1, \mu_2, \ldots, \mu_k$ as follows:

$$\bar{\mu} = \frac{n_1\mu_1 + n_2\mu_2 + \cdots + n_k\mu_k}{n_1 + n_2 + \cdots + n_k}.$$

Suppose we measure the variation among $\mu_1, \mu_2, \ldots, \mu_k$ by the quantity

$$(4) \quad A = \frac{\sum n_i(\mu_i - \bar{\mu})^2}{k-1} = \frac{n_1(\mu_1 - \bar{\mu})^2 + n_2(\mu_2 - \bar{\mu})^2 + \cdots + n_k(\mu_k - \bar{\mu})^2}{k-1}.$$

It is clear that if the omnibus hypothesis is true, $\mu_1, \mu_2, \ldots, \mu_k$ will all be equal and their common value will be $\bar{\mu}$; hence, if the omnibus hypothesis is true, A as defined by (4) will be zero. If $\mu_1, \mu_2, \ldots, \mu_k$ are not all equal, A will be greater than zero.

To test the omnibus hypothesis we calculate the sample value analogous to the parameter A. This quantity is called the mean square among sample means and denoted by MS_A:

(5) $\quad MS_A = \dfrac{\sum n_i(\bar{X}_i - \bar{\bar{X}})^2}{k-1} = \dfrac{n_1(\bar{X}_1 - \bar{\bar{X}})^2 + n_2(\bar{X}_2 - \bar{\bar{X}})^2 + \cdots + n_k(\bar{X}_k - \bar{\bar{X}})^2}{k-1}$

where

$$\bar{\bar{X}} = \frac{n_1\bar{X}_1 + n_2\bar{X}_2 + \cdots + n_k\bar{X}_k}{n_1 + n_2 + \cdots + n_k}.$$

For the data of Table 1 we have:

$$\bar{\bar{X}} = \frac{6(10.1250) + 6(11.3750) + 6(11.7083) + 5(12.3500)}{6+6+6+5}$$

$$= 11.3478,$$

and

$$MS_A = [6(10.1250 - 11.3478)^2 + 6(11.3750 - 11.3478)^2$$

$$+ 6(11.7083 - 11.3478)^2 + 5(12.3500 - 11.3478)^2]/3$$

$$= 4.9259.$$

It can be shown that if $A = 0$, MS_A and s_p^2 independently estimate σ^2, the common population variance for the k populations. Hence when H_0 is true, the test statistic,

(6) $$\mathscr{F} = \frac{MS_A}{s_p^2}$$

would be expected to have a value of approximately one. However, if A is not zero, MS_A tends to be larger than s_p^2.

For the data of Table 1, $s_p^2 = (1.5164)^2 = 2.2995$, and we have a test statistic value of

$$\mathscr{F} = \frac{4.9259}{2.2995} = 2.14.$$

Thus $\mathscr{F}$ is greater than unity, suggesting that MS_A is too large and that A may not be zero.

How large need $\mathscr{F}$ be before the omnibus hypothesis can be rejected? When the underlying k populations are normal with common variance and the omnibus hypothesis H_0 is true (so that the k population means are equal), then $\mathscr{F}$ will have an F distribution with $k-1$ degrees of freedom in the numerator and $n_1 + n_2 + \cdots + n_k - k$ degrees of freedom in the denominator. The α level test rejects H_0 if $\mathscr{F} \geq F_{\alpha, k-1, N-k}$.

For the data of Table 1, $k - 1 = 3$, $n_1 + n_2 + n_3 + n_4 - 4 = 19$, and from the F table (Table C9) we found (via interpolation) that $F_{.05,3,19} \approx 3.14$, and therefore, the P value, for our test statistic $\mathscr{F} = 2.14$, is considerably greater than .05. Thus although $\mathscr{F}$ is greater than unity, it is not large enough to be attributed to anything but random variation, and the hypothesis that $A = 0$ cannot be rejected.

There are more convenient computational formulas for arriving at the test statistic $\mathscr{F}$ for the omnibus hypothesis. If the column or sample totals in Table 1 are denoted by T_1, T_2, T_3, T_4, we have:

$$T_1 = 60.75$$
$$T_2 = 68.25$$
$$T_3 = 70.25$$
$$\frac{T_4 = 61.75}{T = 261.00}$$

Then,

$$MS_A = \left[\frac{T_1^2}{n_1} + \frac{T_2^2}{n_2} + \frac{T_3^2}{n_3} + \frac{T_4^2}{n_4} - \frac{T^2}{N} \right] \Big/ 3.$$

More generally, for the k-sample case,

(7)
$$MS_A = \left[\frac{T_1^2}{n_1} + \frac{T_2^2}{n_2} + \cdots + \frac{T_k^2}{n_k} - \frac{T^2}{N} \right] \Big/ (k - 1),$$

where

$$T = T_1 + T_2 + \cdots + T_k.$$

For the Table 1 data, we have

$$MS_A = \left[\frac{(60.75)^2}{6} + \frac{(68.25)^2}{6} + \frac{(70.25)^2}{6} + \frac{(61.75)^2}{5} - \frac{(261)^2}{23} \right] \Big/ 3$$

$$= 4.9259.$$

A formula for quick calculation of s_p^2 is

(8)
$$s_p^2 = \left[\sum X^2 - \left(\frac{T_1^2}{n_1} + \frac{T_2^2}{n_2} + \cdots + \frac{T_k^2}{n_k} \right) \right] \Big/ (N - k),$$

where $\sum X^2$ is the sum of the squares of all the observations.

For the Table 1 data, we find, using (8),

$$s_p^2 = \frac{3020.25 - 2976.56}{19} = 2.2995.$$

These calculations are often displayed in what is called an analysis of variance table, as shown in Table 2. For the data of Table 1, the analysis of variance table is given in Table 3. As a check on the sums of squares among and within samples, the total sum of squares can be calculated as follows:

$$\text{Total sum of squares} = \sum X^2 - \frac{T^2}{N}$$

$$= 3020.25 - \frac{(261)^2}{23} = 58.47.$$

TABLE 2.   Analysis of Variance of Completely Randomized Experiment

Source of Variation	Sum of Squares (SS)	Degrees of Freedom	Mean Squares (MS)	$\mathscr{F}$ statistic
Among samples	$(k-1) \cdot MS_A$	$k-1$	MS_A	$\mathscr{F} = \dfrac{MS_A}{s_p^2}$
Within samples	$(N-k)s_p^2$	$N-k$	s_p^2	
Total:	$\sum X^2 - \dfrac{T^2}{N}$	$N-1$		

TABLE 3.   Analysis of Variance Table for the Data of Table 1

Source of Variation	Sum of Squares (SS)	Degrees of Freedom	Mean Squares (MS)	$\mathscr{F}$ statistic
Among samples	14.78	3	4.9259	2.14
Within samples	43.69	19	2.2995	
Total:	58.47	22		

We note that the total sum of squares is simply the sum of the squared deviations of all observations around the grand mean, $\sum (X - \bar{X})^2$. Thus in this test of the omnibus hypothesis, the total sum of squares is analyzed or broken down into two parts, one measuring the variation within samples and the other the variation among samples. Hence we have the name *analysis of variance*, even though the analysis itself is directed at a hypothesis concerning the population means.

If the omnibus hypothesis cannot be rejected, this would seem to imply that no contrasts can be taken to be significantly different from zero and some investigators would be firm on this point. From this point of view Zelazo and his colleagues would not be justified in concluding that the active-exercise treatment was any better than the other three treatments under investigation, in terms of teaching infants to walk early. However, the test of the omnibus hypothesis is very conservative. It is equivalent to the Scheffé procedure, guarding against error for all possible contrasts. If a particular small set of contrasts is of very special interest the omnibus hypothesis should be disregarded, and tests of the specific contrasts should be done, adjusting the level of significance only to take into account the small number of inferences that are of interest.

Discussion Question 1. In an experiment in which subjects are randomly assigned to each of several treatments, why might the variation within the several treatment groups be expected to be about the same in the several groups if the omnibus hypothesis is true?

Comments

1.   Pooling the sample variances of all samples to obtain one estimate makes sense only if the population variances under the several treatments are equal. Of course, this cannot be known precisely, since only estimates of the population variances are available. In general, it is a good idea to pool unless the sample standard deviations differ markedly one from another, say, by factors of five and more. If such marked heterogeneity in variability does occur, it is sometimes possible to transform the measurements, for example, by taking logarithms of the individual data values, so as to achieve values on a new scale that are amenable to standard statistical techniques. Snedecor and Cochran (1967) discuss the question of transformations at an elementary level.

2.   The distinction between the error rate for a single inference and the rate at which one or more errors are made in drawing inferences from a given set of data is made by labeling the latter the **experimentwise error rate**. Suppose that m inferences are to be made from the same set of data, each with a probability α of error. [These may be m interval estimates, each at the $100(1-\alpha)\%$ confidence level, or m tests of hypothesis, each at the $100\alpha\%$ rejection level.] The experimentwise error rate is the probability that at least one (i.e., one or more) of the inferences is wrong. Let this probability be denoted by α^*. It can be shown that the probability $1-\alpha^*$ that all inferences are correct is at least as large as $1-m\alpha$.

For example, for two inferences, I_1 and I_2,

$$1 - \alpha^* = Pr(I_1 \text{ and } I_2 \text{ are correct})$$
$$= 1 - Pr(I_1 \text{ or } I_2 \text{ or both incorrect})$$
$$= 1 - [Pr(I_1 \text{ incorrect}) + Pr(I_2 \text{ incorrect})$$
$$- Pr(I_1 \text{ and } I_2 \text{ are incorrect})]$$
$$\geq 1 - [Pr(I_1 \text{ incorrect}) + Pr(I_2 \text{ incorrect})]$$
$$= 1 - (\alpha + \alpha) = 1 - 2\alpha.$$

Thus an experimentwise error rate of α^* for m inferences can be assured if each individual inference is made with error rate set at $\alpha = \alpha^*/m$.

3. Recall (see Comment 11 of Chapter 5) the F percentiles have the property that $F_{\alpha,\nu_1,\nu_2} = 1/F_{1-\alpha,\nu_2,\nu_1}$. For example, find in Table C9 that $F_{.975,4,6} = .109$ and $F_{.025,6,4} = 9.20$, and note that $.109 = 1/9.20$.

PROBLEMS

1. Dolkart, Halpern, and Perlman (1971) compared antibody responses in normal and alloxan diabetic mice. Their investigation was designed to study the circulating antibody response in alloxan diabetic, insulin-treated diabetic, and normal CF-1 mice injected with bovine serum albumin.

Only those animals treated with alloxan who had elevated serum glucose levels (250 mg/100 ml or higher) were included in the study, together with a group of normal animals. Animals were bled from the orbital sinus, and the serum analyzed for antigen binding capacity of BSA, glucose concentration, and serum proteins. BSA was iodinated with I-131, and the antigen-binding capacity of each serum sample was determined as micrograms of BSA nitrogen bound by 1 ml of undiluted serum. The results are given in Table 4.

Is there evidence of different nitrogen-binding capacities for the two diabetic groups relative to normal animals? Calculate three confidence intervals (for an overall confidence of 95%): one for each group separately compared to the normal group, and one for the contrast comparing the mean of the two diabetic groups with the normal group. Use t intervals and Scheffé intervals and compare the results. Which intervals are to be preferred in this problem? Explain.

2. The data of Table 5 are given by Dunnett (1964), who acknowledges Dr. G. Tonelli of Lederle Laboratories for providing the data. The experiment related to the effect of certain drugs on the fat content of the breast muscle in cockerels. Eighty cockerels were divided at random into four treatment groups. Group A birds were untreated controls, and groups B, C, and D received, respectively, stilbesterol and two levels of acetyl enheptin in their diets.

TABLE 4. Micrograms of BSA Nitrogen Bound per ml of Undiluted Mouse Serum on Day 39, Following Injection of 5 mg BSA Antigen into Each Animal on Day 0 and 28

Normal	Alloxan Diabetic	Alloxan Diabetic—Treated with Insulin
155.76	390.72	82.50
282.00	46.20	99.66
197.34	468.60	97.66
297.00	86.46	150.48
115.50	174.02	242.88
126.72	132.66	67.98
119.46	13.20	227.70
29.04	498.96	130.68
252.78	167.64	73.26
122.10	62.04	17.82
349.14	127.38	19.80
108.90	275.88	100.32
143.22	176.22	71.94
64.02	145.86	133.32
25.54	108.24	464.64
85.80	275.88	36.96
122.10	50.16	46.20
454.85	72.60	34.32
655.38		43.56
13.86		

Source: R. E. Dolkart, B. Halpern, and J. Perlman (1971).

Birds from each group were sacrificed at specified times for the purpose of making certain measurements. One variable of interest was the fat content of the breast muscle. Table 5 gives fat content in percentage fat of fresh tissue for the twenty cockerels sacrificed at 7 weeks.

Is there evidence of differences in fat content for the four groups? Calculate the analysis of variance table for the data, and test the omnibus hypothesis.

3. Suppose that in a pilot study comparing two drugs designed for systemic treatment of rheumatoid arthritis, 12 patients are randomly assigned to three groups: placebo, drug A, and drug B. After one week, the groups are compared on the basis of duration of morning stiffness. The results are given in Table 6.

Is there evidence of differences among the three treatments in terms of their effects on the duration of morning stiffness?

TABLE 5. Fat Content of Breast Muscle in Cockerels
Sacrificed at 7 Weeks
(percentage fat of fresh tissue)

A	B	C	D
2.41	2.48	2.96	2.15
2.46	1.46	2.05	2.63
3.17	2.96	1.60	2.38
2.87	2.73	1.47	2.93
2.86	2.84	2.23	2.80

Source: C. W. Dunnett (1964).

TABLE 6. Duration (hr) of Morning
Stiffness

Placebo	Drug A	Drug B
5.2	3.7	4.8
6.8	2.4	5.9
5.6	5.1	4.0
6.1	1.8	4.7

4. Verify that formulas (5) and (7) for MS_A are equivalent.

•5. Refer to Table 2 and verify directly that

sum of squares among samples + sum of squares within samples = total sum of squares.

2.	THE RANDOMIZED BLOCKS EXPERIMENT

A Study of Wheelchair Cushions

Sitting immobile for an extended period of time can lead to the development of pressure ulcers. This is a serious problem for patients confined to wheelchairs. It would appear that a seat cushion that diminishes pressure under the ischial tuberosities would be desirable.

Since elevated pressures between bony prominences can cause tissue ischemia, Souther, Carr, and Vistnes (1974) performed a study designed to quantify the ability of various wheelchair cushions to reduce pressure. Quantitative pressure measurements were made beneath the right ischial tuberosity with a surface pressure manometer developed by Henry, Goodman, and Hyman (1946). Ten normal adult volunteers were evaluated. Measurements were made using a standard wheelchair without a cushion (WC) and with 11 wheelchair cushions, Reston Floatation Pad (RF), Stryker Floatation Pad (SF), Spenco Skin Care Pad (SS), Bye-Bye Decubiti (BB), Comfort Cushion (CC), Jobst Hydro-Float Pad (JP), Dri Flote Wheelchair Pad (DP), Bio Flote Gel Pad (BF), Latex Foam

TABLE 7.   Wheelchair Cushions (1973 data)

Cushion	Manufacturer	Composition	Weight (lb)	Price ($)
Reston Floatation Pad	3M Corporation St. Paul, Minnesota	Liquid-filled microcell sponge	12	98
Stryker Floatation Pad	Stryker Corporation Kalamazoo, Michigan	Special gel	14	298
Spenco Skin Care Pad	Spenco Medical Corporation Waco, Texas	Silicone emulsion	17	50
Bye-Bye Decubiti	Ken McRight Supplies, Inc. Tulsa, Oklahoma	Air-filled rubber bladder	<5	14
Comfort Cushion	Lyn-Bar Enterprises Van Nuys, California	Water-filled Polyurethane foam	3	30
Jobst Hydro-Float Pad	Jobst Institute, Inc. Toledo, Ohio	Water-filled polyurethane foam	25	49
Dri Flote Wheelchair Pad	Bio Clinic Company North Hollywood, California	Foamed plastic	5	49
Bio Flote Gel Pad	Bio Clinic Company North Hollywood, California	Special gel	6	129
Latex Foam Rubber Pad	Everest & Jennings, Inc. Los Angeles, California	2-inch latex foam rubber	<5	7
Adaptaire Pad	Everest & Jennings, Inc. Los Angeles, California	Resin-filled polyurethane foam	5	65
Jobst Hydro-Float Wheelchair Cushion	Jobst Institute, Inc. Toledo, Ohio	Water-filled nylon bladder	60	89

Source: S. G. Souther, S. D. Carr, and L. M. Vistnes (1974).

Rubber Pad (*LF*), Adaptaire Pad (*AP*), and Jobst Hydro-Float Cushion (*JC*). Some characteristics of the various cushions are presented in Table 7. The results of the pressure measurements are to be found in Table 8.

TABLE 8. Wheelchair cushion pressures (mm Hg)

Subjects	Cushions											
	WC	*RF*	*SF*	*SS*	*BB*	*CC*	*JP*	*DP*	*BF*	*LF*	*AP*	*JC*
1	74	47	59	40	49	55	40	52	60	46	52	49
2	53	43	47	45	49	42	36	55	43	41	51	32
3	60	42	43	38	30	48	41	54	51	42	53	32
4	123	76	84	64	46	82	49	84	63	68	78	33
5	56	49	70	57	58	57	29	70	47	50	50	38
6	130	62	86	55	63	92	67	96	76	87	93	49
7	64	53	68	51	50	65	48	65	53	49	68	43
8	75	75	61	58	56	64	34	47	63	59	69	20
9	82	88	80	76	82	79	62	88	68	75	90	48
10	80	77	76	88	56	88	75	80	65	71	70	67

Source: S. G. Souther, S. D. Carr, and L. M. Vistnes (1974).

We assume that the order of presentation of the 12 treatments (11 cushions and the standard wheelchair without a cushion) to each subject is chosen at random in such a way that for any particular subject each of the 12! possible orders has equal probability of being chosen, and that the choice is independent of the choices of presentation orders for the other subjects. This is known as the randomized blocks design. For each subject it is also assumed that there is no carryover effect between treatments.

Analysis of Variance Table for the Randomized Blocks Design

The randomized blocks design is a very commonly used study design in biology and medicine. In the wheelchair study, this design offers the possibility of each subject serving as his own control in evaluating the 11 cushions. Obviously, the subjects differ in contour and weight and thus in the pressures exerted. The data in Table 8 clearly indicate that some patients show higher pressures than others across all types of cushions.

The randomized blocks experiment, in general, involves groups of experimental units or subjects, with the random assignment of treatments done separately within each group. For example, in an experiment involving rats as experimental subjects with four treatments, the rats may be divided into quartets homogeneous with regard to age and weight. Each quartet is a block and the treatments are randomly assigned to the members of each block. In the wheelchair cushion study the experimental units are the time periods for measurement of the pressure for each cushion, the cushions being randomly assigned to the times within each subject.

In evaluating the results of this study, first we can compute the mean pressure for each cushion.

Cushions (treatments)	Total	Mean
WC	797	79.7
RF	612	61.2
SF	674	67.4
SS	572	57.2
BB	539	53.9
CC	672	67.2
JP	481	48.1
DP	691	69.1
BF	589	58.9
LF	588	58.8
AP	674	67.4
JC	411	41.1
Overall	7300	60.83

Note that all 11 cushions achieved means that were less than that achieved without a cushion (WC). Cushions JC and JP showed the lowest mean pressures. To make inferences from these data, we need to estimate the reliability of these means, the reliability of differences among them, and the reliability of any other contrasts among the means that may be of interest.

It would be misleading to calculate the standard deviation of the 10 measurements for a given cushion as in the completely randomized experiment, because this would not take into account the blocking that was so carefully done in the study. Instead, the data can be analyzed by an analysis of variance technique that breaks down the total variance into a component due to treatments (cushions), a component due to blocks (subjects), and a remainder due to random variation within each block.

The sum of squares for treatments in a randomized blocks design with k treatments and n blocks [so that there are n measurements (one in each block) for each treatment] is

(9) $$SS(\text{treatments}) = \frac{\sum T_i^2}{n} - \frac{T^2}{nk} = \frac{(T_1^2 + T_2^2 + \cdots + T_k^2)}{n} - \frac{T^2}{nk}$$

where T_1 denotes the total (over n blocks) for treatment 1, T_2 the total for treatment 2, $\ldots$, T_k the total for treatment k, and

$$T = T_1 + T_2 + \cdots + T_k$$

is the grand total of all nk measurements.

For the cushion data of Table 8 we find, via (9), with $n = 10$, $k = 12$,

$$SS(\text{cushions}) = \frac{(T_1^2 + T_2^2 + \cdots + T_{12}^2)}{10} - \frac{T^2}{120}$$

$$= \frac{(797)^2 + (612)^2 + \cdots + (411)^2}{10} - \frac{(7300)^2}{120}$$

$$= \frac{4{,}558{,}022}{10} - \frac{53{,}290{,}000}{120}$$

$$= 11{,}718.87.$$

The sum of squares for subjects is calculated in parallel fashion, first obtaining the total (U, say) for each subject. Referring to Table 8, we obtain

Subjects (blocks)	Total
1	623
2	537
3	534
4	850
5	631
6	956
7	677
8	681
9	918
10	893
	7300

In a randomized blocks design with k treatments and n blocks, the sum of squares for blocks is given by

(10) $$SS(\text{blocks}) = \frac{\sum U_j^2}{k} - \frac{T^2}{nk} = \frac{(U_1^2 + U_2^2 + \cdots + U_n^2)}{k} - \frac{T^2}{nk},$$

where U_1 is the total for block 1, U_2 is the total for block 2, ..., U_n is the total for block n, and again

$$T = U_1 + U_2 + \cdots + U_n = T_1 + T_2 + \cdots + T_k$$

is the grand total of all nk measurements.

For the Table 8 data, $SS(\text{blocks})$ is found, via (10) with $n = 10$, $k = 12$, to be

$$SS(\text{subjects}) = \frac{(623)^2 + (537)^2 + \cdots + (893)^2}{12} - \frac{(7300)^2}{120}$$

$$= \frac{5,558,514}{12} - \frac{53,290,000}{120}$$

$$= 19,126.17.$$

The total sum of squares is given by

(11) $$SS(\text{total}) = \sum X^2 - \frac{T^2}{nk}$$

where $\sum X^2$ is obtained by squaring each measurement and summing the nk squared values.

For the Table 8 data,

$$SS(\text{total}) = (74)^2 + (47)^2 + \cdots + (67)^2 - \frac{(7300)^2}{120}$$

$$= 484,886 - \frac{53,290,000}{120}$$

$$= 40,802.67.$$

The sum of squares attributed to random variation within blocks, that is, not due to or explainable by differences among blocks or among treatments, can be obtained by subtraction. This sum of squares is called the sum of squares due to "error" or the sum of squares of the residuals:

(12) $$SS(\text{residuals}) = SS(\text{total}) - SS(\text{blocks}) - SS(\text{treatments}),$$

where $SS(\text{total})$, $SS(\text{blocks})$, and $SS(\text{treatments})$ are given, respectively, by formulas (11), (10), and (9).

For the Table 8 data we have

$$SS(\text{residuals}) = SS(\text{total}) - SS(\text{subjects}) - SS(\text{cushions})$$

$$= 40{,}802.67 - 19{,}126.17 - 11{,}718.87$$

$$= 9957.63.$$

As in the completely randomized experiment, the sum of squares calculations are often displayed in an analysis of variance table. The analysis of variance table for the randomized blocks experiment is given in Table 9.

TABLE 9. Analysis of Variance of Randomized Blocks Experiment

Source of Variation	Sum of Squares (SS)	Degrees of Freedom	Mean Squares (MS)	$\mathcal{F}$ statistics
Treatments	$\dfrac{\sum T_i^2}{n} - \dfrac{T^2}{nk}$	$k-1$	$\dfrac{SS(\text{treatments})}{k-1}$	$\mathcal{F}_T = \dfrac{MS(\text{treatments})}{MS(\text{residuals})}$
Blocks	$\dfrac{\sum U_j^2}{k} - \dfrac{T^2}{nk}$	$n-1$	$\dfrac{SS(\text{blocks})}{n-1}$	$\mathcal{F}_B = \dfrac{MS(\text{blocks})}{MS(\text{residuals})}$
Residuals	$SS(\text{total}) - SS(\text{blocks})$ $- SS(\text{treatments})$	$(k-1)(n-1)$	$\dfrac{SS(\text{residuals})}{(k-1)(n-1)}$	
Total:	$\sum X^2 - \dfrac{T^2}{nk}$	$nk-1$		

Note that in Table 9 the degrees of freedom for treatments is $k-1$ (one less than the number of treatments), the degrees of freedom for blocks is $n-1$ (one less than the number of blocks), and the degrees of freedom for the total is $nk-1$ (one less than the total number of observations). The degrees of freedom for residuals is $(nk-1)-(k-1)-(n-1)$ or equivalently, $(k-1)(n-1)$ (the product of the degrees of freedom for treatments and for blocks). The mean squares are simply the sums of squares divided by their respective degrees of freedom.

For the Table 8 cushion data the analysis of variance is given in Table 10.

One of the most useful entries in the analysis of variance table is the mean square for residuals. The square root of this mean square is an estimate of the standard deviation of the random variation in the individual observations within blocks (subtracting out variation due to treatments and to blocks). This estimate, denoted by s_p', is

(13) $$s_p' = \sqrt{MS(\text{residuals})},$$

TABLE 10.   Analysis of Variance for the Data of Table 8

Source of Variation	Sum of Squares (SS)	Degrees of Freedom	Mean Squares (MS)	$\mathscr{F}$ statistics
Cushions	11,718.9	11	1065.4	10.6
Subjects	19,126.2	9	2125.1	21.1
Residuals	9,957.6	99	100.6	
Total:	40,802.7	119		

and for the cushion data

$$s'_p = \sqrt{100.6} = 10.03.$$

This estimate is used in the next subsection to make comparisons among the treatments (cushions).

In addition, the analysis of variance table yields a test of the omnibus hypothesis

$$H_0: \text{The } k \text{ treatments are equivalent.}$$

The test statistic is the ratio of the mean square for treatments to the mean square for residuals:

(14)
$$\mathscr{F}_T = \frac{MS(\text{treatments})}{MS(\text{residuals})}.$$

The $\mathscr{F}_T$ ratio should be approximately unity if the null hypothesis is true. If H_0 is not true, MS(treatments) will tend to be larger than MS(residuals). Hence when $\mathscr{F}_T$ is significantly larger than 1, this suggests that H_0 is false.

How large need $\mathscr{F}_T$ be in order to reject H_0? If we assume the nk measurements have normal distributions with common variance σ^2, then, when H_0 is true, the statistic $\mathscr{F}_T$ has an F distribution with $k-1$ degrees of freedom in the numerator and $(k-1)(n-1)$ degrees of freedom in the denominator. The α level test rejects H_0 if $\mathscr{F}_T \geq F_{\alpha,k-1,(k-1)(n-1)}$.

For the Table 8 data the observed value $\mathscr{F}_T = 10.6$ is substantially larger than the value of 1 expected under the null hypothesis. Referring to Table C9, the table of F percentiles, we note the absence of entries corresponding to $v_1 = k - 1 = 11$, $v_2 = (k-1)(n-1) = 99$. But from Table C9 we find that $F_{.0005,11,60} = 3.69$, and $F_{.0005,11,120} = 3.34$. Thus $F_{.0005,11,99}$ is between 3.34 and 3.69, and since $\mathscr{F}_T = 10.6$ exceeds 3.69, H_0 can be rejected at α values (well) below .0005; that is, the P value is less than .0005. Thus there is strong evidence of differences among cushions.

The analog to the $\mathcal{F}_T$ test for treatment differences is the $\mathcal{F}_B$ test for block differences. For testing

$$H_0': \text{the } n \text{ blocks are equivalent,}$$

the test statistic is

$$(15) \qquad \mathcal{F}_B = \frac{MS(\text{blocks})}{MS(\text{residuals})}.$$

$\mathcal{F}_B$ is to be referred to the F distribution with $\nu_1 = n-1$ and $\nu_2 = (k-1)(n-1)$ degrees of freedom; significantly large values indicate H_0' is false.

The Souther *et al.* study was done to find out about possible differences between cushions. Subject differences are presumed, but, in any case, are of little substantive interest. Furthermore, we are assuming the cushions were randomized within subjects, so that although there is a firm basis in randomization for the $\mathcal{F}_T$ test for differences among treatments (cushions), a rationale for the analogous $\mathcal{F}_B$ test for differences among blocks (subjects) would be less directly founded on the actual experimental procedure used in the Souther *et al.* experiment.

Inferences About Contrasts

To estimate a contrast

$$\sum a\mu = a_1\mu_1 + a_2\mu_2 + \cdots + a_k\mu_k,$$

where $\mu_1, \mu_2, \ldots, \mu_k$ are the population means, the estimate is

$$\sum a\bar{X} = a_1\bar{X}_1 + a_2\bar{X}_2 + \cdots + a_k\bar{X}_k,$$

where

$$\bar{X}_1 = \frac{T_1}{n}, \bar{X}_2 = \frac{T_2}{n}, \ldots, \bar{X}_k = \frac{T_k}{n}$$

are the sample means.

The standard deviation of $\sum a\bar{X}$ is estimated by

$$(16) \qquad \widehat{SD}(\sum a\bar{X}) = s_p'\sqrt{\frac{a_1^2 + a_2^2 + \cdots + a_k^2}{n}},$$

where s_p' is given by (13).

To test the hypothesis that $\sum a\mu = 0$, refer the statistic

$$\frac{\sum a\bar{X}}{\widehat{SD}(\sum a\bar{X})}$$

to the t distribution with $(k-1)(n-1)$ degrees of freedom.

A $1-\alpha$ confidence interval for $\sum a\mu$ is given by

(17) $$\sum a\bar{X} \pm (t_{\alpha/2,(k-1)(n-1)})\widehat{SD}(\sum a\bar{X}),$$

where $t_{\alpha/2,(k-1)(n-1)}$ is the upper $\alpha/2$ percentile point of the t distribution with $(k-1)(n-1)$ degrees of freedom.

Consider, for simplicity, the contrast

$$\mu_{WC} - \mu_{RF}$$

for the wheelchair cushion problem. With $k = 12$, $n = 10$,

$$a_1 = 1, a_2 = -1, a_3 = a_4 = \cdots = a_{12} = 0,$$

we have

$$\sum a\bar{X} = 79.7 - 61.2 = 18.5,$$

and

$$\widehat{SD}(\sum a\bar{X}) = s'_p\sqrt{\frac{(1+1)}{10}} = (10.03)\sqrt{\frac{2}{10}} = 4.49.$$

The test statistic for the null hypothesis $\mu_{WC} = \mu_{RF}$ is

$$\frac{\sum a\bar{X}}{\widehat{SD}(\sum a\bar{X})} = \frac{18.5}{4.49} = 4.12.$$

This value of 4.12 should be referred to the t distribution with $(12-1)(10-1) = 99$ degrees of freedom. For so many degrees of freedom the normal distribution is adequate to approximate the t; that is, we can approximate $t_{\alpha,99}$ by z_α. Referring 4.12 to Table C2, we see the one-sided P value is less than .0002. Thus there is very strong evidence that $\mu_{WC} > \mu_{RF}$.

A 99% confidence interval for $\mu_{WC} - \mu_{RF}$ is (approximating $t_{.005,99}$ by $z_{.005} = 2.58$)

$$(79.7 - 61.2) \pm (2.58)(4.49) = 18.5 \pm 11.6.$$

Ninety-nine percent confidence intervals for the other 10 contrasts comparing a treatment cushion with the control are

$$\mu_{WC} - \mu_{SF} = 12.3 \pm 11.6$$

$$\mu_{WC} - \mu_{SS} = 22.5 \pm 11.6$$

$$\mu_{WC} - \mu_{BB} = 25.8 \pm 11.6$$

$$\mu_{WC} - \mu_{CC} = 12.5 \pm 11.6$$

$$\mu_{WC} - \mu_{JP} = 31.6 \pm 11.6$$

$$\mu_{WC} - \mu_{DP} = 10.6 \pm 11.6$$

$$\mu_{WC} - \mu_{BF} = 20.8 \pm 11.6$$

$$\mu_{WC} - \mu_{LF} = 20.9 \pm 11.6$$

$$\mu_{WC} - \mu_{AP} = 12.3 \pm 11.6$$

$$\mu_{WC} - \mu_{JC} = 38.6 \pm 11.6.$$

The eleven contrasts of the form

$$\mu_{WC} - \mu_{RF}, \mu_{WC} - \mu_{SF}, \ldots, \mu_{WC} - \mu_{JC}$$

were of interest from the beginning of the experiment. That is, the experiment was performed to see if the cushions reduced pressure, compared with no cushion at all (WC). Since we calculated 11 intervals, each with confidence coefficient .99, the overall confidence attached to the 11 intervals is approximately .89. [More generally, if we desire confidence intervals for m contrasts and seek an overall confidence of at least $1 - \alpha$, then the confidence coefficient for each individual interval should be $1 - (\alpha/m)$.] Note that a difference of zero can be ruled out at this 89% level of confidence for all but one cushion (DP), though three others (SF, CC, AP) almost enclose zero.

Scanning the data post hoc for interesting differences can be done in an analogous fashion to the Scheffé procedure (3) of Section 1. Scheffé's simultaneous confidence interval procedure for the randomized blocks experiment is based on the following theorem:

> The probability is $1 - \alpha$ that the values of all contrasts of the form $\sum a\mu$ simultaneously satisfy the inequalities

(18)
$$\sum a\bar{X} - \sqrt{(k-1)F_{\alpha, k-1, (k-1)(n-1)}} \cdot \widehat{SD}(\sum a\bar{X}) \leq \sum a\mu$$
$$\leq \sum a\bar{X} + \sqrt{(k-1)F_{\alpha, k-1, (k-1)(n-1)}} \cdot \widehat{SD}(\sum a\bar{X}),$$

where $F_{\alpha,k-1,(k-1)(n-1)}$ is the upper α percentile point of an F distribution with $k-1$ degrees of freedom in the numerator and $(k-1)(n-1)$ degrees of freedom in the denominator.

For example, suppose we note, on the basis of the data, that the pads made by the Jobst Institute (that is, *JP* and *JC*) seem superior to those made by the Bio Clinic Company (*DP* and *BF*). If we wish to estimate this contrast we should use the Scheffé procedure, since the contrast was suggested post hoc by examination of the data. The 95% confidence interval for the contrast

$$\left(\frac{\mu_{JP}+\mu_{JC}}{2}\right)-\left(\frac{\mu_{DP}+\mu_{BF}}{2}\right)$$

is, from (18),

$$\left(\frac{\bar{X}_{JP}+\bar{X}_{JC}}{2}\right)-\left(\frac{\bar{X}_{DP}+\bar{X}_{BF}}{2}\right)$$

$$\pm\sqrt{11(F_{.05,11,99})}\cdot s_p'\sqrt{\frac{(1/2)^2+(1/2)^2+(-1/2)^2+(-1/2)^2}{10}}.$$

Table C9 does not contain entries for v_2 (the denominator degrees of freedom) equal to 99, but from Table C9 we find $F_{.05,11,60}=1.95$ and $F_{.05,11,120}=1.87$, and linear interpolation on the denominator degrees of freedom gives $F_{.05,11,99}\approx1.90$. Thus the 95% confidence interval is

$$\left(\frac{48.1+41.1}{2}\right)-\left(\frac{69.1+58.9}{2}\right)\pm\sqrt{11(1.90)}\cdot(10.03)\cdot\sqrt{\frac{1}{10}}$$

$$=-19.4\pm14.5.$$

Note that this interval does not include the point 0. Thus even though the inference was suggested by the data, we can be highly confident that the two cushions made by the one manufacturer are better (in terms of pressure reduction) than those of the other, and did not simply fare better in the test due to random variation in the allocation of test times within test subjects.

Discussion Question 2.   The data seem to demonstrate conclusively that the pressures exerted by the 10 subjects in the experiment were eased by most of the cushions and that some cushions were more effective than others. But suppose that one of the manufacturers were to claim that the cushion tested under his name was not representative, that it must have been defective in some way. Would this nullify the conclusions?

Comments

4. If the variation is too large, so that fairly small treatment differences cannot be detected with a reasonable number of observations, it is proper to consider restricting the randomization to more homogeneous subcategories or blocks. The term block comes from agriculture, where these ideas were first used. For example, in comparing yields of four varieties of wheat it would be unwise to randomly allocate varieties to plots if a known fertility gradient exists in the field.

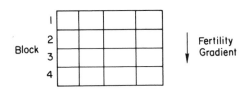

If varieties are randomly allocated to plots, the variation due to fertility is incorporated in the error. Thus it is better to randomly assign varieties within blocks. Then differences among varieties cannot be "explained" by the variation in fertility gradient. Hence the term "randomized blocks experiment." The object of the randomized blocks design is roughly to maximize the variation among blocks and minimize the variation within blocks.

5. The randomized blocks experiment is an example of the kind of design and statistical analysis than can be done to study a series of treatments while adjusting for or taking into account an important additional factor, namely, differences among blocks (subjects, days, plots of land, etc.). The area of statistics called design of experiments deals with these questions for a whole range of situations involving the possibilities of multiple treatment factors, multiple factors for adjustment or blocking, and various constraints on the study design. The elementary books by Ostle (1963) and Winer (1962) are particularly well-written expositions for the serious student who does not want a mathematical treatment of the subject.

In the randomized blocks design considered in this section, in each block there is one observation on each treatment. In some studies there may be enough units per block so that several units in each block can be assigned to each treatment. In other studies, there are fewer units per block than the number of treatments under study. Such a shortage requires care in balancing and a relatively complex analysis of the results.

Summary of Statistical Concepts

In this chapter the reader should have learned the following, for both the completely randomized experiment and the randomized blocks experiment:
(i)   To estimate any contrast among treatment means.
(ii)   To control the overall confidence level for all inferences when many confidence intervals are made, or when post hoc inferences are made.
(iii)   To calculate and display the analysis of variance table.
(iv)   To test the omnibus hypothesis of treatment equivalence using an appropriate $\mathscr{F}$ statistic.

PROBLEMS

6.   Recall the study on the transfer of information from the manual control system to the oculomotor control system described in Chapter 1. The basic data of Table 3, Chapter 1 are reproduced here as Table 11.
   The variables (Y's, say) in Table 11 are defined as

$$Y = \frac{\text{Saccades during test}}{\text{Saccades during playback}}.$$

TABLE 11.   Saccades During Test/Saccades During Playback

			Lags		
Subjects	1 (0 sec)	2 (0.18 sec)	3 (0.36 sec)	4 (0.72 sec)	5 (1.44 sec)
1	0.600	0.880	0.587	1.207	0.909
2	0.610	0.526	0.603	1.333	0.813
3	0.592	0.527	0.976	1.090	0.996
4	0.988	0.841	0.843	1.010	1.112
5	0.611	0.536	0.654	0.910	0.936
6	0.981	0.830	1.135	0.838	0.876
7	0.698	0.520	0.673	0.892	0.857
8	0.619	0.493	0.734	0.948	0.862
9	0.779	0.759	0.764	0.824	0.851
10	0.507	0.712	0.787	0.868	0.880

Source: R. W. Angel, M. Hollander, and M. Wesley (1973).

Is there evidence that the various lags affect the Y response? (This question can be answered via the $\mathscr{F}_T$ test. A slightly different approach, taking into account the natural ordering of the lags, was used in the paper by Angel *et al.*) At (roughly) what time lag do the Y values tend to be close to 1, indicating the hand movement information is no longer available to the brain?

7.   Graybiel *et al.* (1975) reported a series of experiments on the efficacy of various drugs in preventing motion sickness induced by a specially designed chair in a slow rotation room. The data of Table 12 are a subset of the data obtained in "Experiment 1" of Graybiel *et al.*

TABLE 12. Decrease in Motion Sickness Relative to Placebo

Subject	Scopolamine	Dimenhydrinate	Amphetamine
2	9.8	6.0	1.6
3	4.0	−3.0	3.0
4	1.6	0.4	−0.9
5	−1.0	−2.6	10.5
6	4.3	2.0	4.0
7	9.0	5.0	1.8
8	2.0	7.5	0.0
9	2.0	5.0	−1.0
10	1.0	1.2	1.0
11	0.0	−3.0	2.0

Source: A. Graybiel, C. D. Wood, J. Knepton, J. P. Hoche, and G. F. Perkins (1975).

Calculate the analysis of variance table for these data. Is there evidence of differences in efficacy among the drugs?

*8.   Using Scheffé's procedure, which provides simultaneous confidence intervals for all contrasts, makes it difficult for the data to indicate (as summarized by the Scheffé intervals) that population contrasts are not zero, when in fact they are not zero, and this difficulty becomes more severe as k increases. Explain. [Hint: Scheffé's procedure ensures that the probability is $1 - \alpha$ that *all* contrasts simultaneously satisfy inequality (18) (inequality (3) in the case of the completely randomized experiment).]

9.   Describe some situations in which the difficulty alluded to in Problem 8 would be an advantage to the public. (Hint: Consider a situation in which a manufacturer advertises on television that clinical trials have shown his product to be more effective than any of its competitors.)

10.   Define completely randomized experiment, randomized blocks experiment, contrast in sample means, contrast in population means, and experimentwise error rate.

Supplementary Problems

1. M. Crouch (1975), a medical student at Stanford, randomized subjects into three diet treatment groups, each treatment aimed at advising the subject on how to lower his somewhat elevated serum cholesterol level. After initial advice, one group received no follow-up information and help, one group received further contact by phone and mail, and a third group received more costly and time-consuming personal attention. Table 13 gives the changes in cholesterol after 1 year.

TABLE 13.   Changes (mgm%) in Cholesterol After 1 Year

Personal Follow-up	Mail/Phone Follow-up	No Follow-up
−25	−2	+10
−20	−4	+2
−27	−23	−19
−3	−41	−32
−12	−38	+26
−5	−28	+25
−17	−69	−1
−69	−6	−46
−22		+22
−21		+10

Source: M. Crouch (1975).

Calculate and display the analysis of variance table. Is the omnibus null hypothesis tenable? Calculate individual P values for each of the following three null hypotheses: Personal follow-up and no follow-up equally effective in lowering cholesterol; mail/phone follow-up and no follow-up equally effective; personal follow-up and mail/phone follow-up equally effective. Can any of these hypotheses be rejected at the 5% level, considering the necessary adjustment for making three inferences from the same set of data?

2. Crouch (1975) conducted the same kind of study as described in Supplementary Problem 1, but on a second set of subjects. The results were as follows:

Treatment i:	1 Personal Follow-up	2 Mail/Phone Follow-up	3 No Follow-up
n_i:	14	13	13
$\bar{X}_i$:	−17.00	−7.92	−10.62
s_i:	24.58	26.02	22.20

Are these results consistent with the results of Table 13? Explain. What are some possible explanations for the difference?

3. Graybiel *et al.* (1975), in the same study referred to in Problem 7 earlier in this chapter, reported additional results in comparing several antimotion sickness drugs. The Table 14 data are a subset of the data considered by Graybiel *et al.*

TABLE 14. Decrease in Motion Sickness Relative to Placebo

Subject	Scopolamine	Promethazine	Amphetamine	Ephedrine
12	4.0	1.0	0.0	1.0
13	11.5	4.0	0.0	−2.0
14	7.4	3.0	6.0	2.6
15	9.6	9.0	5.0	4.6
16	1.7	0.7	2.7	4.7
17	−1.0	2.0	2.0	3.0
18	3.2	4.0	1.4	2.5
19	0.0	2.0	0.0	0.0

Source: A. Graybiel, C. D. Wood, J. Knepton, J. P. Hoche, and G. F. Perkins (1975).

Calculate and display the analysis of variance table for this set of data. How strong is the evidence for differences in efficacy among these four drugs at the doses used? Calculate a Scheffé confidence interval for the contrast comparing scopolamine with the average effect of the other three.

4. Fouts (1973) described an experiment in which four chimpanzees (two male and two female) were each taught 10 signs of American Sign Language. The acquisition rates of the signs were measured by the number of minutes required in training to reach a criterion of five consecutive unprompted correct responses. The data are given in Table 15.

TABLE 15. Minutes to Criterion for Acquiring Signs of American Sign Language

Chimpanzee	Hat	Shoe	Fruit	Drink	More	Look	Key	Listen	String	Food
Bruno	178	60	177	36	225	345	40	2	287	14
Booee	78	14	80	15	10	115	10	12	129	80
Cindy	99	18	20	25	15	54	25	10	476	55
Thelma	297	20	195	18	24	420	40	15	372	190

(Columns grouped under heading: **Sign Taught:**)

Source: R. S. Fouts (1973).

Is there evidence of significant differences in the acquisition times for the various signs? Why might some signs be easier for a chimpanzee to acquire than other signs?

5. Thompson and Short (1969) have assessed mucociliary efficiency from the rate of removal of dust in normal subjects, subjects with obstructive airway disease, and subjects with asbestosis. Table 16 is based on a subset of the Thompson–Short data.

TABLE 16. Half-time of Mucociliary Clearance (hr)

	Subjects with	
Normal Subjects	Obstructive Airway Disease	Asbestosis
2.9	3.8	2.8
3.0	2.7	3.4
2.5	4.0	3.7
2.6	2.4	2.2
3.2		2.0

Source: M. L. Thomson and M. D. Short (1969).

Is there evidence of differences among the three groups?

6. Daly and Cooper (1967) considered the rate of stuttering adaptation under three conditions. Eighteen subjects (college-age stutterers) read each of three different passages five consecutive times. In one condition electroshock was administered during each moment of stuttering, and in another condition electroshock was administered immediately following each stuttered word. The remaining condition was a control with no electroshock administered. The percentage of stuttering behavior during each reading was recorded, and Table 17 presents for each subject a rate of adaptation score under each condition.

Is there evidence of differences between the control and treatments?

7. Armitage (1973; p. 222) gives the results of a randomized blocks experiment designed to compare the effects on the clotting time of plasma of four different methods of treating the plasma. There were eight subjects. Samples of plasma from the subjects were randomly assigned to the four treatments. The results are given in Table 18.

Test the hypothesis that the four treatments are equivalent with respect to their effects on the clotting times of plasma. Use Scheffé's procedure for the randomized blocks experiment to give a 95% confidence interval for $\mu_4 - \mu_1$.

8. If forced to pick one of the 11 cushions (see Tables 7 and 8), which would you select? Why? Can you make a case for two different choices?

TABLE 17. Adaptation Scores for College-age Stutterers

Subject	1 (No Shock)	2 (Shock Following)	3 (Shock During)
1	57	38	51
2	59	48	56
3	44	50	44
4	51	53	44
5	43	53	50
6	49	56	54
7	48	37	50
8	56	58	40
9	44	44	50
10	50	50	50
11	44	58	56
12	50	48	46
13	70	60	74
14	42	58	57
15	58	60	74
16	54	38	48
17	38	48	48
18	48	56	44

Source: D. A. Daly and E. B. Cooper (1967).

TABLE 18. Clotting Times (minutes) of Plasma

Subjects	Treatments			
	1	2	3	4
1	8.4	9.4	9.8	12.2
2	12.8	15.2	12.9	14.4
3	9.6	9.1	11.2	9.8
4	9.8	8.8	9.9	12.0
5	8.4	8.2	8.5	8.5
6	8.6	9.9	9.8	10.9
7	8.9	9.0	9.2	10.4
8	7.9	8.1	8.2	10.0

Source: P. Armitage (1973).

9. The paper by Souther *et al.* (1974) does not specify the manner in which the treatment orders were randomized for the subjects. The treatment orders should be selected in such a way that (i) for each subject all 12! possible orders (of presentation of the cushions) are equally likely to be picked, and (ii) the randomization for each subject is independent of the randomization for the other subjects. Explain how this could be done in practice.

10. The paper by Zelazo *et al.* (1972) does not indicate the manner in which the infants were assigned to the four groups, other than stating that six infants were assigned to each of the four groups. (Since there are only five 8-week control observations in Table 1, apparently, one infant in the 8-week control group was lost to dropout.) The infants should have been assigned at random so that each possible assignment of 6 infants to the active-exercise group, 6 infants to the passive-exercise group, 6 to the no-exercise group, and 6 to the 8-week control group was equally likely. Explain how this could be done in practice.

REFERENCES

Angel, R. W., Hollander, M., and Wesley, M. (1973). Hand-eye coordination: The role of "motor memory." *Perception and Psychophysics* **14**, 506–510.

Armitage, P. (1973). *Statistical Methods in Medical Research*. Wiley, New York.

Crouch, M. (1975). Personal communication.

Daly, D. A. and Cooper, E. B. (1967). Rate of stuttering adaptation under two electro-shock conditions. *Behav. Res. Ther.* **5**, 49–54.

Dolkart, R. E., Halpern, B., and Perlman, J. (1971). Comparison of antibody responses in normal and alloxan diabetic mice. *Diabetes* **20**, 162–167.

Dunnett, C. W. (1964). New tables for multiple comparisons with a control. *Biometrics* **20**, 482–491.

Fouts, R. S. (1973). Acquisition and testing of gestural signs in four young chimpanzees. *Science* **180**, 978–980.

Graybiel, A., Wood, C. D., Knepton, J., Hoche, J. P., and Perkins, G. F. (1975). Human assay of antimotion sickness drugs. *Aviation Space Environ. Med.* **46**, 1107–1118.

Henry, J. P., Goodman, J., and Hyman, C. (1946). Surface pressure measurements with the University of Southern California pressure suit. Report submitted to National Research Council, Committee on Aviation Medicine.

Ostle, B. (1963). *Statistics in Research*. Iowa State University Press, Ames, Iowa.

*Scheffé, H. (1959). *The Analysis of Variance*. Wiley, New York.

Snedecor, G. W. and Cochran, W. G. (1967). *Statistical Methods*. Iowa State University Press, Ames, Iowa.

Souther, S. G., Carr, S. D., and Vistnes, L. M. (1974). Wheelchair cushions to reduce pressure under bony prominences. *Arch. Phys. Med. Rehabil.* **55**, 460–464.

Thomson, M. L. and Short, M. D. (1969). Mucociliary function in health, chronic obstructive airway disease, and asbestosis. *J. Appl. Physiol.* **26**, 535–539.

Winer, B. J. (1962). *Statistical Principles in Experimental Design*. McGraw-Hill, New York.

Zelazo, P. R., Zelazo, N. A., and Kolb, S. (1972). "Walking" in the newborn. *Science* **176**, 314–315.

CHAPTER 11

Linear Regression and Correlation

In this chapter we consider the problem of relating one variable to another variable. This problem often arises in experimental work in which one wishes to study how changes in one variable affect another variable. The consideration of the distribution of one variable, when another is held fixed at each of several levels, constitutes a *regression* problem. When one studies the association between two measurements, neither one of which is held fixed by the experimenter, this constitutes a *correlation* problem. As one would expect, regression problems and methods are linked with correlation problems and methods, but there are important differences, both in technique and interpretation.

Section 1 considers regression problems for situations in which the relationship between the variables under study is linear—that is, the relationship can be expressed by a straight line. We show how to estimate the line by the method of *least squares*. Furthermore, we describe methods to obtain tests and confidence intervals for the intercept and slope of the line.

Section 2 introduces a popular measure of association called Pearson's product moment correlation coefficient (denoted by *r*), and the use and interpretation of *r* is discussed.

Section 3 contains the description of another popular measure of association between two variables—namely, Spearman's rank correlation coefficient (denoted by r_s). We discuss the use and interpretation of r_s, its relationship to Pearson's r, and certain advantages it enjoys over r. The introduction of ranks in Section 3 also provides a transition to the ranking methods covered in Chapters 12, 13, and 14.

1.	LINEAR REGRESSION

Measurement of Intracranial Pressure

The usual technique for measuring intracranial pressure on the brain requires access to the fluid itself through the base of the brain. Ream *et al.* (1975) have developed a safer method of indirect measurement through a hole in the skull cap. In a series of dog experiments Ream and his colleagues attempted to show that the new method is accurate (i.e., gives values close to the true value) and reliable (i.e., gives values that are nearly the same in repeated measurements). Table 1 shows the results obtained by Ream *et al.* for one dog. The pressure on the dog's brain

TABLE 1. Intracranial Pressures of a Dog, Measured by a Standard and an Experimental Technique (Measurements in mm Hg)

Standard, Direct Method (X)	Experimental, Indirect Method (Y)
9	6
12	10
28	27
72	67
30	25
38	35
76	75
26	27
52	53

Source: A. K. Ream, G. D. Silverberg, J. Rolfe, B. Wm. Brown, Jr., and J. Halpern (1975).

was varied experimentally in a randomized pattern, and after each change in pressure, measurements were taken by both the old and the new methods. Based on these data we would like to evaluate the accuracy and the reliability of the new technique, considering the measurements by the direct technique as our standard.

The Scatterplot

A basic aid for studying the relationship between two variables is a graph called the **scatterplot** (some statisticians prefer the term scatter diagram, which itself is often shortened to scattergram). In a scatterplot, each (X, Y) pair is plotted as a single point, with the X value on the horizontal axis and the Y value on the vertical axis. Figure 1 is a scatterplot for Ream's intracranial pressure data.

In Figure 1 we started the horizontal and vertical scales at zero because that was a convenient starting point for the intracranial pressure data. In general, the starting points can be different for the two axes, and should be chosen for convenience and clear representation of the particular set of data being plotted.

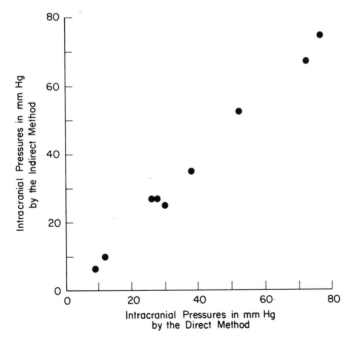

FIGURE 1. A scatterplot of the intracranial pressure data

From Figure 1 we see that although the indirect method is not in perfect agreement with the direct method, it does seem fairly accurate and reliable. We do need, however, methods for describing this agreement quantitatively and objectively. Such methods are described in the remaining subsections of Section 1.

Fitting a Straight Line to the Data

From Figure 1 it seems reasonable to describe the relationship between the indirect and direct measurements as approximating a straight line. In fact, we might assume that if we were able to take many more measurements on the dog at a variety of pressure levels, we would find that the average pressure obtained by the indirect method was a straight line or linear function of the direct pressure. In other words, we consider that the reason the points in the scatterplot do not lie on a perfectly straight line is that there is some variation due to animal and technical fluctuation, but these would average out to yield a straight line relationship in a very long series of measurements.

Let us denote the mean of the distribution of the Y variable for a given fixed value of X as $\mu_{Y \cdot X}$. In general, the curve generated by plotting $\mu_{Y \cdot X}$ against X is called the **regression curve of Y on X**. If we make the further assumption that this curve is really a straight line, then we can represent it by the equation

$$(1) \qquad \mu_{Y \cdot X} = A + BX.$$

When equation (1) holds we say there is a **linear regression of Y on X**. The values A and B are population parameters, which are to be estimated from the data.

Equation (1) specifies that the mean of the distribution of Y, for a fixed value of X, varies with X in a linear manner. However, we assume that the variance of the distribution of Y for a fixed value of X does not change when we change the fixed value X. Denoting this variance as $\sigma^2_{Y \cdot X}$, we assume that $\sigma^2_{Y \cdot X}$ is constant for all values of X. This is known as the assumption of **homogeneous variance**. Some writers call this the assumption of **homoscedasticity**.

We wish to estimate A and B, and hence estimate the straight line given by equation (1).

For the Ream *et al.* data, to estimate the underlying straight line from the nine pairs of measurements we have we could simply apply pencil and ruler to Figure 1, trying to draw in the best fitting straight line. However, this approach lacks the objectivity we desire in a statistical procedure. Furthermore, we would not have standard deviations for estimates of the intercept and slope read from such a line.

The standard approach to fitting a straight line to the data is to use the line that best fits the data by the criterion of **least squares**. For any line $Y = a + bX$, that is fitted to the data, the vertical distance of any data point (X_i, Y_i) from the line can be computed as $Y_i - (a + bX_i)$. A positive difference would mean that the point is above the line and a negative value that it is below. If these distances are squared and summed, the best fitting straight line is defined by that choice of a and b which yields the smallest sum of squares.

The least squares line for a particular set of data could be obtained by trial and error, selecting an intercept (a) and slope (b), computing the distance of each point from the trial line, obtaining the sum of squared distances, and repeating this for various choices of a and b until the least squares values are obtained, at least to a close approximation. Such a trial-and-error approach is not necessary, however, because it can be shown that the least squares estimates are given by the following expressions:

(2)
$$b = \frac{n \sum XY - (\sum X)(\sum Y)}{n \sum X^2 - (\sum X)^2},$$

$$a = \bar{Y} - b\bar{X},$$

where

$$\bar{X} = \frac{\sum X}{n}, \qquad \bar{Y} = \frac{\sum Y}{n},$$

and n denotes the number of (X, Y) pairs.

The calculation of the least squares estimates of the slope and intercept for the data of Table 1 are shown in Table 2. (Note that to calculate a and b we need only use columns 1–5 of Table 2.) The least squares line has slope $b = .9968$ and intercept $a = -1.88$. The least squares line is plotted with the data in Figure 2.

Estimation of the Standard Deviation about the Line

Recall our assumption that the standard deviation of the Y measurements at a fixed level of X is the same for various values of X. This standard deviation can be estimated by pooling the squared deviations of the several Y measurements from the estimated line. The estimate is

(3)
$$\hat{\sigma}_{Y \cdot X} = \sqrt{\frac{\sum [Y_i - (a + bX_i)]^2}{n - 2}}$$

$$= \sqrt{\frac{[Y_1 - (a + bX_1)]^2 + [Y_2 - (a + bX_2)]^2 + \cdots + [Y_n - (a + bX_n)]^2}{n - 2}}.$$

TABLE 2.   Calculations of a, b and $\hat{\sigma}_{Y \cdot X}$ for the Data of Table 1

X	Y	X^2	Y^2	XY	$a+bX$	$Y-(a+bX)$	$[Y-(a+bX)]^2$
9	6	81	36	54	7.09	−1.09	1.188
12	10	144	100	120	10.08	−0.08	0.006
28	27	784	729	756	26.03	0.97	0.941
72	67	5,184	4,489	4,824	69.89	−2.89	8.352
30	25	900	625	750	28.02	−3.02	9.120
38	35	1,444	1,225	1,330	36.00	−1.00	1.000
76	75	5,776	5,625	5,700	73.88	1.12	1.254
26	27	676	729	702	24.04	2.96	8.762
52	53	2,704	2,809	2,756	49.95	3.05	9.303
Σ 343	325	17,693	16,367	16,992		0.02†	39.93

$$b = [9(16,992)-(343)(325)]/[9(17,693)-(343)^2] = \frac{41,453}{41,588} = 0.9968$$

$$a = (325/9)-(0.9968)(343/9) = 36.11 - 37.99 = -1.88$$

$$\hat{\sigma}_{Y \cdot X} = \sqrt{\frac{39.93}{7}} = 2.39$$

†$\Sigma [Y_i - (a+bX_i)]$ will always equal 0; the 0.02 value in this table is due to round-off error.

Note that the sum of the n squared distances is divided by $n-2$. This is done for theoretical reasons. Roughly speaking, the loss of two degrees of freedom derives from the estimation of two parameters from the data in order to estimate the means of the Y distributions at the several levels of X. The estimator $\hat{\sigma}_{Y \cdot X}$ is often called the standard error of estimate. From equation (3) we see that $\hat{\sigma}_{Y \cdot X}$ is large when the vertical distances of the data points from the least squares line are large, and $\hat{\sigma}_{Y \cdot X}$ is small when those vertical distances are small.

The calculation of $\hat{\sigma}_{Y \cdot X}$ is shown in Table 2. The value of $\hat{\sigma}_{Y \cdot X}$ is 2.39. (Also see Comment 3.)

To this point we have not made an assumption about the form of the distribution of Y measurements at each X level. Suppose now that in addition to the assumptions of linearity of regression [equation (1)] and homogeneous variance, we also assume that at each X level the Y measurements have a normal distribution. Then, on the average, $100(1-\alpha)\%$ of the Y measurements will lie within $\pm(t_{\alpha/2, n-2})\hat{\sigma}_{Y \cdot X}$ of the regression line.

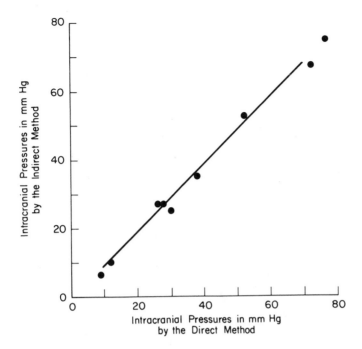

FIGURE 2. The least squares line for the intracranial pressure data.

Consider again the data of Table 1. Since $\hat{\sigma}_{Y \cdot X} = 2.39$, we can expect that 95% of the measurements of intracranial pressure by the indirect method will lie within $\pm(t_{.025,7})\hat{\sigma}_{Y \cdot X} = \pm(2.365)(2.39) = \pm 5.65$ mm Hg of the pressure as measured by the direct method. Hence the data for the dog suggest that the relationship between indirect and direct measurement methods is nearly the 45° line through the origin, that is, with intercept 0 and slope 1, with variation about the line of plus and minus 5.65 mm Hg or less.

Confidence Intervals for the Intercept *A* and the Slope *B*

In computing the estimates *a* and *b* of the intercept and slope, respectively, we did not need to make distributional assumptions about the *Y* variable at each fixed value of *X*. However, to obtain exact confidence intervals and tests for the population values *A* and *B* we now impose the additional assumption that for each *X* the distribution of *Y* is normal.

The linearity of regression assumption, the homogeneous variance assumption, and the normality assumption are portrayed in the conceptual regression model of Figure 3.

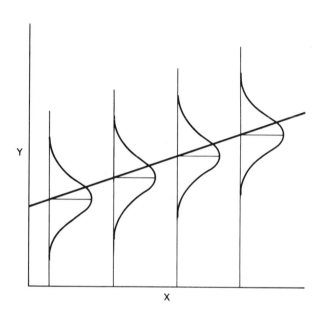

FIGURE 3. Conceptual regression model encompassing linearity of regression, homogeneity of variance, and normality of distributions.

Under the assumptions of linearity of regression, homogeneity of variances, and normality, confidence intervals for A and B are as follows. The confidence interval for A, with confidence coefficient $100(1-\alpha)\%$, is

$$(4) \qquad a \pm (t_{\alpha/2,n-2}) \cdot \widehat{SD}(a),$$

where $\widehat{SD}(a)$, the estimated standard deviation of a, is given by

$$(5) \qquad \widehat{SD}(a) = \hat{\sigma}_{Y \cdot X} \sqrt{\frac{1}{n} + \frac{n\bar{X}^2}{n \sum X^2 - (\sum X)^2}}.$$

The confidence interval for B, with confidence coefficient $100(1-\alpha)\%$, is

$$(6) \qquad b \pm (t_{\alpha/2,n-2}) \cdot \widehat{SD}(b),$$

where $\widehat{SD}(b)$, the estimated standard deviation of b, is given by

(7)
$$\widehat{SD}(b) = \frac{\hat{\sigma}_{Y\cdot X}}{\sqrt{\sum X^2 - \frac{(\sum X)^2}{n}}}.$$

In formulas (4) and (6), $t_{\alpha/2, n-2}$ is the upper $\alpha/2$ percentile point of student's t distribution with $n-2$ degrees of freedom. Selected t percentiles are given in Table C5.

Consider the intracranial pressure data and suppose our desired confidence is 99%. Then with $\alpha = .01$, $\alpha/2 = .005$, and $n-2 = 7$, we find from Table C5 that $t_{.005,7} = 3.499$. From (5) and Table 2 we obtain

$$\widehat{SD}(a) = (2.39)\sqrt{\frac{1}{9} + \frac{9(343/9)^2}{9(17,693) - (343)^2}}$$

$$= 1.56.$$

Then from (4), the 99% confidence interval for A is

$$-1.88 \pm (3.499) \cdot (1.56) = -1.88 \pm 5.46$$

$$= -7.34 \text{ and } 3.58.$$

A 99% confidence interval for B is computed in a similar manner using (6) and (7). For the intracranial pressure data we have, from (7) and Table 2,

$$\widehat{SD}(b) = \frac{(2.39)}{\sqrt{17,693 - \frac{(343)^2}{9}}} = .035.$$

Then from (6) the interval is found to be

$$.9968 \pm (3.499) \cdot (.035) = .9968 \pm .122$$

$$= .87 \text{ and } 1.12.$$

Hypothesis Tests for the Intercept *A* and the Slope *B*

We sometimes want to investigate whether the intercept A is equal to some hypothesized value A_0. We may also want to investigate whether the slope B is equal to some hypothesized value B_0. In practice, the most common hypothesis, in the case of straight-line data, is the hypothesis that B is zero, since this would mean that the line is perfectly horizontal and the mean of Y does not change with a change in the level of X (roughly speaking, Y does not depend on X).

For the Ream *et al.* data, the interesting question is whether the experimental indirect method "agrees" with the standard direct method on the average. Thus we wish to test whether the slope B is one and the intercept A is zero, so that the true line is simply $\mu_{Y \cdot X} = X$.

Under the assumptions of linearity of regression, homogeneity of variances, and normality, hypothesis tests for A and B are as follows.

At the α level, a two-sided test of the hypothesis $A = A_0$ against the alternative $A \neq A_0$, rejects whenever

$$(8) \qquad \left| \frac{a - A_0}{\widehat{SD}(a)} \right| > t_{\alpha/2, n-2}.$$

At the α level, a two-sided test of the hypothesis $B = B_0$ against the alternative $B \neq B_0$, rejects whenever

$$(9) \qquad \left| \frac{b - B_0}{\widehat{SD}(b)} \right| > t_{\alpha/2, n-2}.$$

The one-sided analogs of tests (8) and (9) are defined in Comment 6. For simultaneously testing $A = A_0$ and $B = B_0$, see Comment 7.

We illustrate the test procedure for the slope B on the intracranial pressure data. First consider a test of $B = 0$. Using the hypothesized value $B_0 = 0$ in (9) yields

$$\left| \frac{b - 0}{\widehat{SD}(b)} \right| = \left| \frac{.9968}{.035} \right| = 28.5.$$

Referring 28.5 to Table C5, entered at $n - 2 = 7$ degrees of freedom, shows that $P < .001$ (note that $t_{.0005,7} = 5.408$ and since 28.5 is greater than 5.408, we reject $B = 0$ at $\alpha = .001$). Thus there is little doubt that the new measurement technique is measuring something related to pressure as measured by the direct method.

As we indicated earlier, a more meaningful question for the Ream *et al.* data is whether or not $B = 1$. Testing the hypothesis that $B = 1$, via (9), we have

$$\left| \frac{b - 1}{\widehat{SD}(b)} \right| = \left| \frac{.9968 - 1}{.035} \right| = .091.$$

The two-sided $\alpha = .01$ test rejects if $|(b - 1) / \widehat{SD}(b)|$ exceeds $t_{.005,7} = 3.499$. Since $.091 < 3.499$, we accept the hypothesis $B = 1$. This could have been noted in the previous subsection, since there we saw that the 99% confidence interval for B, obtained via (6), contained the point 1.

To test $A = 0$, at $\alpha = .01$, we compute

$$\left|\frac{a-0}{\widehat{SD}(a)}\right| = \left|\frac{-1.88}{1.56}\right| = 1.21.$$

The two-sided $\alpha = .01$ test rejects the hypothesis $A = 0$ if $|(a - 0)/\widehat{SD}(a)|$ exceeds $t_{.005,7} = 3.499$. Since $1.21 < 3.499$, we accept the hypothesis $A = 0$. This also could have been noted in the previous subsection, since there we saw that the 99% confidence interval for A, obtained via (4), contained the point 0.

Based on the analysis of the Ream *et al.* data, we can conclude that the indirect method is in good agreement with direct measurements. From examination of the scatterplot, the relationship between the two methods seems quite linear, at least over the interval from $X = 10$ to $X = 70$. The scatter about the line, as measured by $\hat{\sigma}_{Y\cdot X}$, is fairly small. The estimated slope of .9968 indicates a one-to-one relationship in increments (slope) between the two methods. The intercept is close to zero, and this indicates that the indirect method neither overestimates nor underestimates the pressure by very much. Of course, all these conclusions must be tentative until they are confirmed on further animals. At the time of this writing Ream and his colleagues had completed a series of dog experiments and the method was consistent from dog to dog.

Confidence Band for the Regression line

In addition to estimating the parameters A and B, in the case of the Ream *et al.* data (and in many regression problems) there is also interest in estimation of the line itself, $A + BX$, at various values of X. Suppose we focus on $X = 30$ as a typical pressure level. Our estimate of the value of the line, that is, the mean of the distribution of Y measurements, at $X = 30$, denoted by $\mu_{Y\cdot 30}$, would be:

$$\hat{\mu}_{Y\cdot 30} = a + b \cdot 30 = -1.88 + (.9968)(30) = 28.02.$$

The standard deviation of $\hat{\mu}_{Y\cdot 30}$ is estimated as follows:

$$\widehat{SD}(\hat{\mu}_{Y\cdot 30}) = \hat{\sigma}_{Y\cdot X}\sqrt{\frac{1}{n} + \frac{n(30 - \bar{X})^2}{n\sum X^2 - (\sum X)^2}}.$$

For the Ream *et al.* data, we have:

$$\widehat{SD}(\hat{\mu}_{Y\cdot 30}) = 2.39\sqrt{\frac{1}{9} + \frac{9(30 - 38.11)^2}{9(17,693) - (343)^2}} = .85.$$

In general, a $100(1-\alpha)\%$ confidence interval for $\mu_{Y \cdot X_0}$, the ordinate of the regression line at a specified value $X = X_0$, is given by:

(10)
$$\hat{\mu}_{Y \cdot X_0} \pm (t_{\alpha/2, n-2}) \cdot \widehat{SD}(\hat{\mu}_{Y \cdot X_0})$$

$$= (a + bX_0) \pm (t_{\alpha/2, n-2})\hat{\sigma}_{Y \cdot X} \sqrt{\frac{1}{n} + \frac{n(X_0 - \bar{X})^2}{n \sum X^2 - (\sum X)^2}}.$$

For the Ream *et al.* data, a 95% confidence interval for $\mu_{Y \cdot 30}$ is, from (10),

$$28.02 \pm (t_{.025,7})(.85) = 28.02 \pm (2.365)(.85)$$

$$= 26.0 \text{ and } 30.0.$$

In practice, as in the intracranial pressure data of Ream *et al.*, one may not be satisfied with an interval estimate at a single level of X. In fact, one is tempted to calculate the interval given by expression (10) at many levels of X and then plot what would amount to a confidence band for the whole regression line over some wide interval of X levels. However, it is clear from our discussion in Chapter 10 (see, e.g., Comment 2 of Chapter 10) that if expression (10) is used to produce a 95% confidence interval for $\mu_{Y \cdot X}$ at $X = 30$, say, *and* a 95% confidence interval for $\mu_{Y \cdot X}$ at another level of X, say $X = 60$, then the joint confidence that *both* intervals contain the true line ordinates at these two levels of X is *less* than 95%. The problem of assuring satisfactory *joint* confidence would seem more acute, if not hopeless, if one desired a confidence interval for the line at many levels of X, perhaps even for a whole interval of X values. Clearly, the confidence interval at each X level would have to be wider, but how much wider? The following interval due to Scheffé, analogous to the Scheffé intervals of Chapter 10, solves the problem. Substitute for the t percentile $t_{\alpha/2, n-2}$ in (10), the multiplier $\sqrt{2F_{\alpha,2,n-2}}$ (where $F_{\alpha,2,n-2}$ is the upper α percentile point of an F distribution with two degrees of freedom in the numerator and $n-2$ degrees of freedom in the denominator) to obtain the interval:

(11)
$$(a + bX_0) \pm \sqrt{2F_{\alpha,2,n-2}} \cdot \hat{\sigma}_{Y \cdot X} \sqrt{\frac{1}{n} + \frac{n(X_0 - \bar{X})^2}{n \sum X^2 - (\sum X)^2}}.$$

One can calculate the interval given by (11) at any number of X_0 values, construct a confidence band for the regression line, and take the joint confidence that the true line is wholly within the band to be at least $100(1-\alpha)\%$.

For the data on intracranial pressure measurement, expression (11) can be used to construct a 95% confidence band for $\mu_{Y \cdot X}$ by calculating the limits successively at $X_0 = 0, 10, 20, 30, 40, 50, 60,$ and 70, plotting the results and then connecting the upper limits by a smooth curve and the lower limits by a smooth curve to form a confidence band for the line. For example, at $X_0 = 30$, we obtain from (11) with $\alpha = .05$ the interval

$$28.02 \pm \sqrt{2(4.74)} \cdot (.85) = 25.4 \text{ and } 30.6.$$

The value $F_{.05,2,7} = 4.74$ is obtained from Table C9.

The 95% confidence band for the regression line for the Ream *et al.* data is given in Figure 4. Note that the band gets wider as X moves away from $\bar{X}$. This is a consequence of the fact that the term $n(X_0 - \bar{X})^2$, appearing in the limits given by (11), assumes its minimum value of 0 when $X_0 = \bar{X}$ and increases as X_0 moves away from $\bar{X}$.

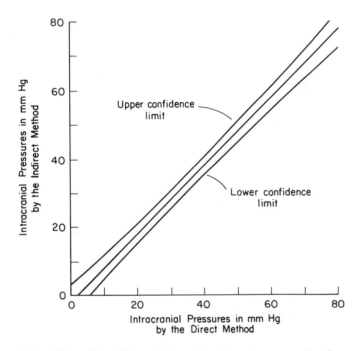

FIGURE 4. The 95% confidence band for the regression line for the intracranial pressure data.

It should be noted that expressions (10) and (11) give confidence intervals for the regression line at specified values of X, that is, for the *mean* of the Y values that might be generated at any specified X. Sometimes such an interval or band is mistakenly taken to be the interval for *individual* values of Y at an X, but the individual values vary about the mean with a standard deviation estimated by $\hat{\sigma}_{Y \cdot X}$. Thus individual values of Y can fall outside the interval estimate for the line itself.

If instead of a confidence interval for a mean value, one is more interested in obtaining a probability statement about a future observation, the following procedure can be used.

The probability is $1 - \alpha$ that a future observation Y_0 (say), corresponding to a specified value X_0, will lie in the interval

$$(a + bX_0) \pm (t_{\alpha/2, n-2}) \hat{\sigma}_{Y \cdot X} \sqrt{1 + \frac{1}{n} + \frac{n(X_0 - \bar{X})^2}{n \sum X^2 - (\sum X)^2}}.$$

Discussion Question 1. What are some possible objectives for fitting a least squares line to paired data?

Discussion Question 2. Can you describe an experiment where the hypothesis $B = -1$ arises in a natural way? Explain.

Comments

1. The calculations of Section 1 are slightly more complicated and more tedious than the techniques in most of the preceding chapters. They can be carried out with pencil, paper, and a simple desk or pocket calculator, as shown in Table 2, and the student should do the calculations in that manner in order to get a better feeling for the numbers, the relationships, and the meanings of the various statistics. In practice, one should certainly check to see if there are any calculators close by that do the calculations more automatically through internal programs or program cards. Further, every scientific computer center will have least squares programs, and the simple routines can be learned and used in less time than it would take to do the calculations (for a large set of data) by hand. It should be noted here that programming the calculations for a high-speed computer requires some professional judgment, and the handiest formulas for desk calculation sometimes lead to the worst rounding errors in high-speed electronic computers. Often special formulas are developed for the digital computer programs.

2. The calculational formulas [e.g., (2), (3), (5), and (7)] are all written as if the data were available as n pairs of X, Y values. Often, however, there are multiple values of Y at each level of X, and then care must be taken to include the multiple values of X in the calculations. For example, if two observations on Y have been

taken at $X = 30$, and one each at $X = 40$, 50, and 60, then the $\bar{X}$ in formula (2) is $\bar{X} = (30+30+40+50+60)/5 = 42$, not $\bar{X} = (30+40+50+60)/4 = 45$. Similarly, $\sum X^2$ includes the square of 30 twice and the squares of 40, 50, and 60 once.

3.   Formula (3), for $\hat{\sigma}_{Y \cdot X}$, is useful for pencil and paper calculations, and one of the better formulas for use in a computer program. It is not a good formula for calculation on a desk or pocket calculator, where sums of squares and cross products are easily accumulated, but squares of differences must be obtained individually. The following formula may be more convenient:

(12)
$$\hat{\sigma}_{Y \cdot X} = \sqrt{\frac{(n-1)}{(n-2)}(s_Y^2 - b^2 s_X^2)},$$

where

$$s_X^2 = \frac{\sum X^2 - \frac{(\sum X)^2}{n}}{n-1}, \qquad s_Y^2 = \frac{\sum Y^2 - \frac{(\sum Y)^2}{n}}{n-1},$$

and b is the least squares estimate of B given by (2). Note that (12) can be rewritten as

(12')
$$\hat{\sigma}_{Y \cdot X} = \sqrt{\frac{\sum Y^2 - \frac{(\sum Y)^2}{n} - b^2 \left(\sum X^2 - \frac{(\sum X)^2}{n}\right)}{n-2}}.$$

Again, Comment 2 is relevant. The quantity s_X^2 is calculated by the familiar formula for the sample variance of the X values, n of them altogether, repeating any X according to the number of Y values at that level of X. Note that even though formula (12) uses the s_X, s_Y notation, the X values in the experiment are fixed by the experimenter, and s_X and s_Y themselves do not estimate the standard deviations of underlying populations. See Section 2, on correlation, for a different kind of study in which s_X and s_Y do have the usual interpretation as sample standard deviations that estimate population standard deviations.

4.   If Y does not have a straight-line relationship to X, then techniques for fitting a straight line to the data are inappropriate. But it may still be possible to transform either the Y variable or the X variable (or both) so that the new variables do provide a linear relationship. For example, suppose that theory or experience suggest that Y is exponentially related to X:

$$Y \cong C e^{DX}.$$

Then, taking the logarithm of Y, we have:

$$\log Y = \log C + DX.$$

Thus a transformed variable, $W = \log Y$, may be expected to be linearly related to X, with intercept A ($= \log C$) and slope B ($= D$). The statistical techniques for straight-line relationships can be applied to the W and X values, and the results can be translated back to the Y scale if necessary. It should be added that the transformation that is appropriate to attain approximate linearity often also achieves more constant variance, though this added dividend is not assured.

5.   Before fitting a straight line to a set of data, it is important to look at the scatterplot to be sure that the data are not clearly inconsistent with the linear model and the assumption of homogeneous variance. After fitting the straight line, the fit of the line to the data should be reexamined. This can be done by plotting the line on the scatterplot. It can be done a little more precisely by examining the deviation of each Y_i value from its best fitting value, $a + bX_i$, as they are displayed in the next to last column of Table 2. If the line does not fit, then the deviations (called **residuals**) usually show a systematic pattern of signs, indicating too many points above the line for low and high values of X, for example, and too many points below the line in the middle.

6.   In this section we described the two-sided tests of $A = A_0$ and $B = B_0$, respectively. The one-sided α level test of $A = A_0$ versus the alternative $A > A_0$ rejects when $(a - A_0)/\widehat{SD}(a)$ exceeds $t_{\alpha, n-2}$, and accepts otherwise. Similarly, the one-sided α level test of $A = A_0$ versus the alternative $A < A_0$ rejects when $(a - A_0)/\widehat{SD}(a)$ is less than $-t_{\alpha, n-2}$, and accepts otherwise.

Analogously, the one-sided α level test of $B = B_0$ versus the alternative $B > B_0$ rejects when $(b - B_0)/\widehat{SD}(b)$ exceeds $t_{\alpha, n-2}$, and accepts otherwise. The one-sided α level test of $B = B_0$ versus the alternative $B < B_0$ rejects when $(b - B_0)/\widehat{SD}(b)$ is less than $-t_{\alpha, n-2}$, and accepts otherwise.

7.   If tests (8) and (9) are each performed at individual α levels, then the probability of rejecting with at least one of the tests when $A = A_0$ and $B = B_0$ is greater than α (but less than 2α). To control for simultaneous testing, the hypothesis that $A = A_0$ and $B = B_0$ is rejected at the α level whenever

$$\mathscr{F}_{A_0, B_0} = \frac{n(a - A_0)^2 + 2n\bar{X}(a - A_0)(b - B_0) + (\sum X^2)(b - B_0)^2}{2(\hat{\sigma}_{Y \cdot X})^2} \geq F_{\alpha, 2, n-2}$$

where $F_{\alpha, 2, n-2}$ is the upper α percentile point of the F distribution with two degrees of freedom in the numerator and $n - 2$ degrees of freedom in the denominator. Values of the F percentiles are given in Table C9.

8.   In regression analysis it is common practice to call a variable that is controlled (set to desired values) by the experimenter an **independent variable**. The typical regression study is designed to observe (and later, describe mathematically) how changes in independent variables affect the values of **dependent**

variables. The dependent variables (often called the response variables) are not controlled by the experimenter. For example, in the problem that follows (Problem 1), age is an independent variable and the kidney-to-spine distance is a dependent variable. In Problem 4 the concentration of antibiotic is an independent variable and the zone diameter is a dependent variable.

We have limited our discussion in this chapter to the case of one independent variable. Frequently, an investigator wants to study the effects of several independent variables on a dependent variable. For such problems, model (1) can be generalized. For example, with k independent variables $W_1, W_2, \ldots, W_k$, the mean of the distribution of the dependent variable Y could be considered to be a linear function of the W's:

$$\mu_{Y \cdot W_1, W_2, \ldots, W_k} = A + B_1 W_1 + B_2 W_2 + \cdots + B_k W_k.$$

Again confidence intervals and tests of hypothesis concerning the parameters A, $B_1, \ldots, B_k$ could be developed. For treatment of these models and further generalizations the reader is referred to Draper and Smith (1966).

PROBLEMS

1. In a study by Friedland, Filly and Brown (1974), X-rays of the kidneys of normal children were used to measure the distance from the inside wall of the kidney to the spine, a distance easily visualized in the X-ray and sometimes useful in diagnosing kidney disease. Table 3, based on a subset of the Friedland *et al.* data, gives measurements obtained for the upper right kidney, together with the age of each child. It might be expected that the distance, as with most physical dimensions, would increase with age. Draw the scatterplot, calculate the least squares intercept and slope, and put the least squares line on the scatterplot. Test the hypothesis that the slope is zero.

2. The data in Table 4 are from a study by Wray and Aguirre (1969). They studied malnutrition in a preschool child population in the town of Candelaria, Columbia. Using a house-to-house survey, preschool children were weighed and measured and their mothers interviewed. The effect of family size on malnutrition was examined by grouping the children according to the number of living children in their families. Table 4 gives the proportion (number of malnourished children in a group divided by the total number of children in that group) of malnourished children for each group.

Is there sufficient evidence to claim that malnutrition tends to increase with the number of children in the family? (Your analysis should include a scatterplot.)

3. Illustrate with a set of data, or show directly, that $\sum [Y_i - (a + bX_i)] = 0$.

TABLE 3. Kidney to Spine Measurements from X-rays of 24 Normal Children

Age of Child (yr)	Distance of the Right Upper Pole Calyx to the Spine (in mm)
2	20
3	18
4	22, 25
5	17, 20, 20, 22
6	21, 22
7	20, 20, 22, 24
8	18, 25, 33
9	27, 31
10	18, 24, 34
11	25, 28

Source: G. Friedland, R. Filly, and B. Wm. Brown, Jr. (1974).

TABLE 4. Malnutrition in Preschool Children in Candelaria, Columbia in 1963

Number of Living Children in Family	Percentage of Malnourished Children
1	32.0
2	34.1
3	41.0
4	40.7
5	41.9
6	46.7
7	40.3
8 or more	46.2

Source: J. D. Wray and A. Aguirre (1969).

4. The basis for many techniques for measuring the potency of antibiotics is the agar diffusion method. Disks containing varying amounts of antibiotic are placed in a carpet of growing bacteria, and the diffusing antibiotic kills the bacteria in a circular area around the disk. The more antibiotic on the disk, the greater the diameter of the kill zone. The potency of a test preparation is compared with a standard preparation by testing both in the same study and estimating equivalent doses of the two. Here we present data only on a standard preparation to demonstrate the relationship between zone diameter and the concentration of antibiotic on the disk. The data were made available by Platt (1973) and are presented in Table 5. Plot the data using X = concentration and again using X = log concentration. Note that the relationship seems more linear on the log-dose scale, a very common phenomenon. Note that even on the log scale there is a slight tendency for curvature upward at the lower concentrations. At still lower doses this curvature is inevitable since the disks are 10 mm in diameter themselves, and thus zone diameters below 10 mm cannot be measured. This means that concentrations that are very low cannot be used in the assay, and the sensitivity of the assay is limited to the detection and measurement of strengths equivalent to .125 μ/ml of standard preparation, or greater.

TABLE 5. Zone Diameter Measurements (mm) for a Standard Preparation of Antibiotic

Concentration of Antibiotic (μ/ml)				
0.0125	0.025	0.050	0.100	0.200
12.3	14.1	17.6	19.9	22.7
12.1	15.0	16.8	19.8	22.7
10.8	14.8	17.0	20.0	23.0
11.7	13.8	17.2	20.5	24.1
11.9	14.1	17.4	20.8	24.3
12.0	14.3	17.5	20.7	23.8
12.2	13.7	17.7	20.9	23.5
12.0	14.2	16.9	20.6	23.7
12.1	13.6	16.7	20.5	23.1

Source: T. R. Platt (1973).

Calculate the intercept, slope and standard deviation about the line. Do this for both X = concentration and X = log concentration. It would be needless busywork to test the slope against a hypothesized value of zero, but other questions do pose statistical problems. Suppose a single disk had been dosed with 0.20 μ/ml of a test compound of unknown potency and had yielded a zone diameter of 16.3 mm. What would you estimate as the potency of the test preparation relative to the standard? Is it more potent or less, and by what factor? Can you attach any measure of statistical uncertainty to your answer? Statistical procedures for questions of this sort fall under the title of bioassay and are dealt with in a large and comprehensive book by Finney (1964).

5.   The data in Table 6 are from a study by Wardlaw and van Belle (1964). They discuss the mouse hemidiaphragm method for assaying insulin, which depends on the ability of the hormone to stimulate glycogen synthesis by the diaphragm tissue, *in vitro*. We briefly describe the quantitative assay and refer the interested reader to the Wardlaw–van Belle article for details. Hemidiaphragms are dissected from mice of uniform weight that have been starved for 18 hours. The tissues are incubated in tubes (usually three hemidiaphragms per tube), and after incubation the hemidiaphragms are washed with water and analyzed for glycogen content using anthrone reagent. The content is measured in terms of optical density. The procedure makes use of the fact that increasing the concentration of insulin in the incubation medium tends to increase glycogen synthesis by the hemidiaphragms. Specifically, for levels of insulin between 0.1 and 1.0 mu per milliliter there is an approximate linear relationship between glycogen content and log concentration of insulin.

Table 6 contains 12 pairs of observations corresponding to the Wardlaw–van Belle standard insulin data.

TABLE 6.   Glycogen Content of Hemidiaphragms   Measured   by   Optical Density

X (log dose)	Y (glycogen content)
log (0.3)	230
log (0.3)	290
log (0.3)	265
log (0.3)	225
log (0.3)	285
log (0.3)	280
log (1.5)	365
log (1.5)	325
log (1.5)	360
log (1.5)	300
log (1.5)	360
log (1.5)	385

Source: A. C. Wardlaw and G. van Belle (1964).

Draw the scatterplot for the 12 (X, Y) pairs. Determine the least squares estimates of A and B and draw the least squares line on the scatterplot.

2.	CORRELATION

Obesity and Blood Pressure

In the example described in Section 1, Ream and his colleagues were able to manipulate the intracranial pressure (X) and then, at various levels of pressure, they were able to obtain measurements (Y) by the new technique. In many studies directed at exploration of the relationship between two factors, neither factor can be manipulated by the investigator. Instead, a bivariate population is sampled, and measurements of the two factors are obtained for each of the n objects or subjects in the sample. Farquhar and his associates (1974) selected a random sample of people from several California communities in the course of their studies of heart disease prevention. We look here at a portion of their data, a sample of Mexican–American males between the ages of 35 and 60 from the town of Watsonville, and we focus on the question of whether men who are overweight also tend to have high blood pressure. Here neither factor is controlled or fixed in the study. The factors are simply measured as they are found among the men in the sample. The degree of overweight (X) is measured by the ratio of actual body weight to ideal body weight as given in certain standard tables; the blood pressure used is the systolic pressure (Y). The results for the 44 Mexican–American males are given in Table 7, and a scatterplot (and the least squares line) for the data is given in Figure 5. Note that since neither factor was controlled, the measurements of obesity cluster about a mean (of 1.19), with decreasing frequency toward the extremes, and the blood pressure measurements about their mean (of 127.95 mm Hg) with a similar pattern. There is no opportunity to purposely gather more data at extremes of the scales, as Ream *et al.* were able to do by manipulating the intracranial pressure in the dogs. Note further that the scatterplot does suggest that although almost all of the men were overweight to some extent (i.e., obesity greater than 1.00), the most obese men did tend to have higher blood pressures than the less obese men, though the tendency is not marked. We do need a way of quantitatively describing this relationship.

Fitting a Straight Line to the Data

A straight line can be fitted to the data using the criterion of least squares. It is important to note that in the intracranial pressure experiment described in Section

TABLE 7. Obesity and Blood Pressure Measurements on a Random Sample of Mexican–American Adult Males (aged 35 to 60) in a Small California Town

Obesity[1]	Blood Pressure[2]	Obesity[1]	Blood Pressure[2]
1.31	130	1.19	134
1.31	148	0.96	110
1.19	146	1.13	118
1.11	122	1.19	110
1.34	140	0.81	94
1.17	146	1.11	118
1.56	132	1.29	140
1.18	110	1.29	128
1.04	124	1.28	126
1.03	150	1.20	140
0.88	120	1.02	124
1.29	114	1.09	104
1.26	136	1.08	134
1.16	118	1.04	130
1.32	190	1.14	124
1.37	118	1.13	110
1.25	130	1.16	134
1.48	112	1.57	144
1.58	126	1.07	116
0.93	162	1.04	118
1.29	124	1.37	118
1.06	126	1.26	132

Source: J. W. Farquhar (1974).

[1] Ratio of actual weight to ideal weight from New York Metropolitan Life Tables.
[2] Systolic blood pressure in mm Hg.

1, one variable was manipulated, and the variables then measured. Thus it was natural to formulate the model in terms of the mean of the uncontrolled variable (Y) as a function of the controlled (X). Then the method of least squares was applied, minimizing the vertical distances of the points from the line.

In our survey sample of Mexican–Americans, neither variable is fixed. The two variables have been treated quite symmetrically thus far, and we have a choice as to which way to carry out the least squares analysis. We can think of estimating the

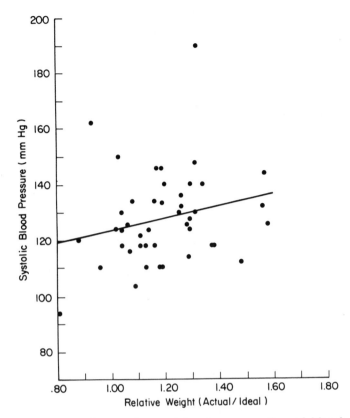

FIGURE 5. Scatterplot and least squares line of blood pressure versus relative weight.

regression of Y on X, that is, the mean of the Y populations for various values of X, by minimizing the vertical deviations of the points from the fitted line. We can also estimate the mean of the X variable for various levels of Y by minimizing the horizontal deviations. This is readily accomplished by simply reversing the roles of X and Y in the formulas given by (2). *These two approaches do not yield the same least squares line.* However, the two lines intersect at $(\bar{X}, \bar{Y})$.

At this point we choose to estimate the mean blood pressure for various levels of obesity, the natural approach if one is interested in estimating blood pressure from obesity. Table 8 shows the results of the basic calculations that yield b, a, $\hat{\sigma}_{Y \cdot X}$, and $\widehat{SD}(b)$. [Use the formulas given by (2) to obtain b and a, use (12) to obtain $\hat{\sigma}_{Y \cdot X}$, and use (7) to calculate $\widehat{SD}(b)$.] We find

$$b = 21.65, \qquad a = 102.10, \qquad \hat{\sigma}_{Y \cdot X} = 16.36, \qquad \widehat{SD}(b) = 14.50.$$

TABLE 8. Calculations for Least Squares
Estimates (to predict blood pressure from
obesity) Using the Data of Table 7

$\sum X = 52.53$	$\sum Y = 5,630$
$\sum X^2 = 63.9863$	$\sum Y^2 = 732,220$
$\sum XY = 6,749$	
$\bar{X} = 1.194$	$\bar{Y} = 127.95$
$s_X^2 = 0.0296$	$s_Y^2 = 275.25$
$s_X = 0.17$	$s_Y = 16.6$
$b = 21.65$	$a = 102.10$
$\hat{\sigma}_{Y \cdot X} = 16.36$	$\widehat{SD}(b) = 14.50$

Thus our estimate of the relationship between mean systolic blood pressure and relative weight is

$$Y = a + bX = 102.10 + (21.65)X.$$

This least squares line is imposed on the scatterplot of Figure 5. For men of relative weight $X = 1.00$ our estimate of mean blood pressure is

$$Y = 102.10 + (21.65)(1.00) = 123.8,$$

and for men 60% overweight the mean systolic blood pressure is estimated at

$$Y = 102.10 + (21.65)(1.60) = 136.7.$$

Since 120 mm Hg is considered normal systolic and 160 is considered to be high blood pressure, we can see that obese Mexican–American men can tend toward high blood pressure. However, even the very obese are not, on the average, in the hypertensive range (which is usually regarded as greater than 160 mm Hg). Note, however, that $\hat{\sigma}_{Y \cdot X} = 16.36$, suggesting that blood pressure generally ranges between plus and minus 33 units (roughly $= 2\hat{\sigma}_{Y \cdot X}$) from the mean, for men at any given level of obesity.

Inferences about the Slope

Although the slope is estimated at 21.65 mm Hg per unit increase in obesity ratio, this estimate is based on a sample of only 44 males and is subject to statistical variation. In fact, if we test the hypothesis that the slope is zero, using (9), we have:

$$\left| \frac{b}{\widehat{SD}(b)} \right| = 1.49.$$

We should refer the statistic $|b/\widehat{SD}(b)| = 1.49$ to the t distribution with $n - 2 = 42$ degrees of freedom. Table C5 contains entries for 40 and 60 degrees of freedom. We can either interpolate between 40 and 60 to get entries for 42 degrees of freedom, or use the entries for 40 degrees of freedom as a slightly conservative approximation, or simply use the fact that when the number of degrees of freedom is large, the t distribution is closely approximated by the normal distribution (Table C2). Referring the value 1.49 to Table C2, we obtain a one-sided P value of approximately .068 and a two-sided P value of .136. In particular, using a two-sided test, the null hypothesis $B = 0$ is not rejected at the .10 level. Thus although the slope is estimated at 21.65 mm Hg, the data are reasonably consistent with the null hypothesis that systolic blood pressure does not depend on relative weight. In other words, if these were the only data we had on this question, we would have to conclude that we do not have a persuasive case for rejecting the null hypothesis. Nevertheless, the data are suggestive, and 95% confidence limits for the slope, calculated according to expression (6) (with $z_{.025}$ approximating $t_{.025,42}$) are

$$21.65 \pm (1.96) \cdot (14.50) = 21.65 \pm 28.42$$

$$= -6.8 \text{ and } 50.1.$$

A Warning about "Causation"

The formulas for slope and intercept and their estimated standard deviations apply to both an experimental study in which the independent (X) variable is controlled and to a survey study in which neither variable, X nor Y, is controlled. However, the two types of studies differ in several important respects. In the experimental study, the independent variable is typically manipulated by the investigator, and if the dependent variable is shown to be related to the independent variable (i.e., $B \neq 0$), then it can be concluded that the data indicate a relationship between X and Y.

For example, if one could assign obesity to people randomly and then fatten them up or slim them down to their assigned levels, a valid conclusion might be drawn as to whether blood pressure could be affected by manipulating weight. In contrast, the survey being discussed here, in which neither variable is experimentally assigned or manipulated, cannot furnish definitive evidence concerning a causal relationship between the two variables. In the sample of Mexican–American males, if a relationship had been shown to exist between the two variables, it could be explained in several ways. One could maintain that obesity causes high blood pressure, or that high blood pressure causes obesity, or that both are caused by a third factor, for example, genetic predisposition toward both. In the third case it is clear that changing one of the variables, obesity or blood pressure (say by dieting or using drugs), need not change the other variable.

Pearson's Correlation Coefficient

The very symmetry of the two variables, X and Y, in the same survey underlines the dissatisfaction often felt in choosing one variable as the independent variable and one as the dependent variable. Often the investigator is not interested in predicting either variable from the other, and does not want to treat the data in this way. Fitting a straight line to the data seems even more unsatisfactory to him if he is aware that he will get a different least squares line, depending on which variable he chooses to call the independent variable.

For this symmetrical situation, a symmetrical measure of association between the two variables is needed. One possible measure is a **correlation coefficient**. There are a number of different correlation coefficients. Here we define and discuss Pearson's coefficient r, proposed by Karl Pearson in 1896. (In Section 3 we introduce a coefficient based on ranks that has certain advantages when compared with r.)

The Pearson correlation coefficient for n pairs of X, Y measurements is defined as follows:

(13)
$$r = \frac{\sum (X - \bar{X})(Y - \bar{Y})}{(n-1)s_X s_Y} = b\frac{s_X}{s_Y}$$

where $\bar{X}$ and $\bar{Y}$ are the sample means, s_X and s_Y are the sample standard deviations for the X and Y samples, respectively, and b is given by (2).

Formula (14) gives an equivalent definition of r, which tends to be easier to use for calculation:

(14)
$$r = \frac{n \sum XY - (\sum X)(\sum Y)}{\sqrt{[n \sum X^2 - (\sum X)^2] \cdot [n \sum Y^2 - (\sum Y)^2]}}.$$

Note that the coefficient r is symmetrical in X and Y. Furthermore, it is dimensionless.

Using the sums, sums of squares, and the sum of cross products given in Table 8 for the obesity–blood pressure data, we follow expression (14) to obtain

$$r = \frac{44(6,749) - (52.53)(5,630)}{\sqrt{[44(63.9863) - (52.53)^2] \cdot [44(732,220) - (5,630)^2]}}$$
$$= .224.$$

It can be shown that r always has a value between -1 and $+1$. Values close to zero indicate little relationship between X and Y; values close to -1 or $+1$ indicate that the points fall very nearly on a straight line with negative or positive slope, respectively. Thus for the blood pressure data, the correlation coefficient of .224 indicates a slight positive relationship, that is, a tendency for the data to fall on a line with positive slope but with much scatter.

The interpretation of the correlation coefficient r is greatly aided by knowing the following relationship:

$$(15) \qquad \hat{\sigma}_{Y \cdot X} = s_Y \sqrt{1 - r^2} \sqrt{\frac{n-1}{n-2}},$$

or equivalently,

$$(16) \qquad r^2 = 1 - \left\{ \frac{\hat{\sigma}_{Y \cdot X}^2}{s_Y^2} \cdot \frac{n-2}{n-1} \right\} = \frac{(n-1)s_Y^2 - (n-2)\hat{\sigma}_{Y \cdot X}^2}{(n-1)s_Y^2}.$$

Equation (15) [and the equivalent equation (16)] shows that the correlation coefficient r (or, more precisely, $1-r^2$) measures the relationship between the variance of the Y's for a fixed X ($\sigma_{Y \cdot X}^2$) and the variance of the Y's in the whole population disregarding X (i.e., σ_Y^2). Note that $\hat{\sigma}_{Y \cdot X}^2$ estimates $\sigma_{Y \cdot X}^2$ and s_Y^2 estimates σ_Y^2.

For the blood pressure data, $s_Y^2 = 275.25$. The estimated variance about the line, $\hat{\sigma}_{Y \cdot X}^2$, has already been calculated as $(16.36)^2$, but it could have been calculated by squaring the right-hand side of (15). Thus it is seen that a correlation coefficient of $r = .224$ indicates that nearly all $\{[1 - (.224)^2] \approx 95\%\}$ of the variation in the population is still present to get in the way of predicting blood pressure, even if the obesity of the individual is known. By symmetry, the same percentage variation remains in predicting obesity from blood pressure.

Test of Significance Based on r

The correlation coefficient computed from the sample is an estimate of the corresponding population correlation coefficient ρ defined for the underlying bivariate population. In this subsection we show how to use r to test the null hypothesis that the correlation in the population is zero:

$$H_0: \rho = 0.$$

Actually, we have already tested this hypothesis for the blood pressure data by testing the hypothesis that the slope B is zero. (See our earlier subsection

"Inferences about the Slope.") The hypothesis of zero correlation is equivalent (it can be shown that $B = \rho\sigma_Y/\sigma_X$) and the results are identical. The test statistic, which when $\rho = 0$ has a t distribution with $n - 2$ degrees of freedom, is

(17)
$$R = \frac{r\sqrt{n-2}}{\sqrt{1-r^2}}.$$

It can be shown that $R = b/\widehat{SD}(b)$, the latter being a statistic introduced earlier for testing that B is zero.

Significantly large values of R indicate the population correlation coefficient is positive; significantly small values indicate the population correlation coefficient is negative.

The value of $R = 1.49$, for the obesity–blood pressure data, agrees with our earlier calculation (as it must), which was in the context of the test of $B = 0$. The test shows that the sample correlation value $r = .224$ is not enough to put it beyond the possibility of a chance occurrence (the P value is $\approx .136$) in a sample from a population in which the true correlation is zero. Nevertheless, we estimate that there is a weak, positive association between obesity and blood pressure.

The significance test, based on referring the statistic R to the t distribution, is exact only when the X and Y variables are bivariate normally distributed. A precise definition of bivariate normalty is beyond the level of this text, but the reader should be aware of the need for the assumption of bivariate normality. In Section 3 we introduce a significance test that does not require an underlying bivariate normal population.

Discussion Question 3.   What are possible misuses of the Pearson correlation coefficient r? [The sample correlation coefficient r is often misused in the literature. See, e.g., the criticisms by Kruskal (1975) and Feinberg (1975) of a paper by Mazur and Rosa (1974).]

Discussion Question 4.   Do you think that the calculation of r is a suitable substitute for making a scatterplot? Explain.

Comments

9.   The distinction between the study with levels of X fixed by the investigator and the study with both X and Y random is an important one. The mathematics underlying the statistical inferences is different for the two situations, even though

the resulting statistical formulas for estimating the intercept, slope, and standard deviation about the line remain the same.

From the practical view, the distinction is important in that the fixed X situation forces the analysis to be carried out in one direction, and does test causality if the X values are imposed on the subjects as experimental units by the investigator. In the bivariate situation in which neither X nor Y is controlled, the analysis can be carried out in either direction, but the data shed no light directly on the question of causality.

For more discussion of regression and correlation methods, the reader is referred to Chapter 11 of Dixon and Massey (1969) and to the book by Draper and Smith (1966).

10. Frequently an investigator is tempted to calculate a correlation coefficient for data in which the X variable has been controlled. This is bad practice and leads to much misunderstanding. The best measure of the amount of variation of the individual values about the straight line is $\hat{\sigma}_{Y \cdot X}$; this measure is appropriate for data in which X has been controlled and also for bivariate survey data. For the latter situation only, the correlation coefficient can be used, but it serves another purpose. The statistic r^2 does not measure the amount of variation about the line, but, rather, the *proportion* of the total variation in Y that is "explained" by X [see expression (16)]. This is a useful concept in the bivariate survey situation, in which the population of Y measurements overall has meaning and s_Y estimates the standard deviation of the Y measurements in this population. However, in the study in which the X levels are fixed by the investigator, the populations of Y at the several levels of X are selected as well, and there is not an overall population of Y measurements in the background. In this case s_Y does not have its interpretation as an estimate of a population standard deviation. In fact, s_Y is indirectly under the control of the investigator, because if Y depends at all on X, the investigator could choose the X values in such a way (e.g., by taking them to be very spread out) that would tend to increase the variation in the Y observations. Thus s_Y could be greatly increased, with $\hat{\sigma}_{Y \cdot X}$ being relatively unchanged (the actual value of $\hat{\sigma}_{Y \cdot X}$ would change, but the change would tend to be small because $\hat{\sigma}_{Y \cdot X}$ estimates $\sigma_{Y \cdot X}$ and, when the homogeneous variance assumption is true, $\sigma_{Y \cdot X}$ is the same for all X). Now note, from equation (16), that when s_Y^2 is increased and $\hat{\sigma}_{Y \cdot X}^2$ is relatively unchanged, the net effect is a (meaningless) increase in r^2, and this could lead an investigator to erroneously conclude that there was a "stronger" dependence between the X and Y variables.

11. Sir Francis Galton used the concept of correlation, in a study published in 1889, to describe the relationship between the height of father and son . There is some correlation, but it is far from perfect. The scatterplot of results for a sample

of families shows rather marked variation, but an evident linear relationship between height of father and height of son. The problem can be looked at with regard to prediction in either direction, and there are two different regression lines. Now the average heights of fathers was about the same as the average heights of sons, so both lines went through the mean height point, say (5' 8", 5' 8"), but the two regression lines fell on opposite sides of the 45° line, and this presented an interesting paradox. Looked at it from the point of view of predicting son's height from father's height, unusually tall fathers had sons shorter than themselves, on the average, whereas unusually short fathers had sons taller than themselves, on the average. Thus there was an appearance of a *regression* toward the mean height, hence the term for all of these straight-line problems. One might feel that in a few generations all males would be the same height, 5' 8", but this is clearly erroneous, since one is led in quite the opposite direction by studying the regression of father's height on the son's height. Nevertheless, there are many instances of investigators who have been fooled by the phenomenon of regression toward the mean.

As an illustration, if one were to select all short bridegrooms, give each a vitamin A pill, and tell each his first son would be taller than he is as a consequence of this treatment, the prediction would work out well on the average, and with a little luck and a modest sample size, the significance test for the trial would look very good.

Now consider a real example. One might screen out patients with elevated blood pressure, operate on them, and see whether their blood pressures are significantly lowered. Again, the surgery would appear effective, even with a modest sample size, because of the phenomenon of regression toward the mean. Of course, inclusion of a control group, randomizing patients to surgery and to medical treatment, would serve to separate the regression effect from the effect that might be attributed to the surgical treatment itself. A paper by James (1973) describes this problem in more detail.

PROBLEMS

6. Table 9 presents the obesity and blood pressure measurements for Mexican–American females, paralleling the data for males used as our basic example in this section on correlation. Is there evidence of dependence between obesity and blood pressure in females? Test the null hypothesis that B is zero. Also calculate the correlation coefficient r.

7. Construct a set of data where $r = 0$. When will r equal -1? When will r equal $+1$?

TABLE 9. Obesity and Blood Pressure Measurements on a Random Sample of Mexican–American Adult Females (aged 35 to 60 years) in a Small California Town

Obesity[1]	Blood Pressure[2]	Obesity[1]	Blood Pressure[2]
1.50	140	1.54	130
1.59	150	1.43	128
1.43	130	1.74	128
1.63	132	1.51	118
2.39	150	1. 36	110
1.50	112	1.37	148
0.92	138	1.67	162
1.17	116	1.03	128
1.33	124	1.32	108
1.09	112	1.56	116
1.24	116	1.33	104
1.44	110	1.56	122
1.23	160	1.25	98
1.50	140	1.24	110
1.34	124	1.27	118
2.04	138	1.57	116
1.13	118	1.30	118
1.11	104	1.32	138
1.38	114	1.41	142
1.35	138	1.21	124
1.42	170	1.20	120
1.55	144	1.15	118
1.33	108	1.43	122
1.22	108	1.24	112
1.07	98	1.28	126
0.97	112	1.75	138
1.26	100	2.20	136
1.65	120	1.64	136
1.01	118	1.73	208

Source: J. W. Farquhar (1974).

[1] Obesity is actual weight divided by ideal weight from New York Metropolitan Life Tables.

[2] Blood pressure is systolic, in mm Hg.

8. Consider the heart transplant data of Table 4, Chapter 1. Use the 82 patients to compute the product moment correlation coefficient r between the variables "date of acceptance" and "age at acceptance date." Then use r to test if the population correlation coefficient is 0. (Take z_α as an approximation to the appropriate t distribution percentile point $t_{\alpha,80}$.)

9. Either verify directly, or illustrate by means of an example, that formula (14) (for r) is equivalent to formula (13).

3.	RANK CORRELATION

Pressure Reduction and Price in Wheelchair Cushions

The data in Table 10 relate to the Souther *et al.* (1974) wheelchair cushion study considered in Chapter 10. Column 2 of Table 10 contains ratings of 11 wheelchair cushions in terms of their ability to reduce pressure under the ischial tuberosities. Column 3 of Table 10 contains the 1973 prices of the 11 wheelchair cushions. Low values in column 2 of Table 10 are desirable in the sense that the cushions are

TABLE 10. Pressure Reduction and Price in Wheelchair Cushions

Cushion	Mean pressure (mm Hg)	Price ($)
RF	61.2	98
SF	67.4	298
SS	57.2	50
BB	53.9	14
CC	67.2	30
JP	48.1	49
DP	69.1	49
BF	58.9	129
LF	58.8	7
AP	67.4	65
JC	41.1	89

Source: S. G. Souther, S. D. Carr, and L. M. Vistnes (1974).

designed to reduce pressure. Does pressure reduction come at the expense of high prices? Is there any association between pressure reduction and price?

In Table 11 we have performed two separate rankings of the cushions. One ranking is based on price, the other on the cushions' ability to reduce pressure (as measured by the mean pressure values in column 2 of Table 10). For example, in the ranking of the cushions by prices, the lowest priced cushion (LF) gets **rank** 1, the second lowest (BB) gets rank 2, and so forth. Note that cushions JP and DP are tied for price ranks 4 and 5, and thus each receives the *average* of the ranks they are tied for, namely $\{(4+5)/2\} = 4.5$.

In the rankings of the cushions in terms of their performance in reducing pressure we have assigned rank 1 to the cushion having the smallest mean pressure (JC), rank 2 to the cushion that obtained the second smallest mean pressure (JP), and so forth. (Thus, as is the case with the ranking based on prices, low ranks are to be preferred to high ranks.) In the ranking based on mean pressures, cushions SF and AP are tied for ranks 9 and 10, and thus each is assigned the average rank 9.5.

We now need a measure of association between the two rankings. Spearman's **rank correlation coefficient** is one such measure.

TABLE 11.   Rankings of Cushions with Respect to Pressure Reduction and Price

Cushion	Pressure rank	Price rank
RF	7	9
SF	9.5	11
SS	4	6
BB	3	2
CC	8	3
JP	2	4.5
DP	11	4.5
BF	6	10
LF	5	1
AP	9.5	7
JC	1	8

Spearman's Rank Correlation Coefficient

A measure of association between two separate rankings on n items (subjects, patients, etc.) can be obtained as follows. For each item compute d, the difference between the two ranks that the item received, and then compute d^2. *Let $\sum d^2$* denote the sum of the squared d values, summed over all n items. Spearman's rank correlation coefficient is defined as

$$(18) \qquad r_s = 1 - \frac{6\sum d^2}{n^3 - n}.$$

It can be shown that the rank correlation coefficient r_s is simply the correlation coefficient r defined by (13) but where instead of the X and Y values being used in formula (13), we substitute the X and Y *ranks*. That is, r_s is simply the ordinary coefficient r computed by replacing in formula (13) each X by its X rank in the joint ranking of the n X's, and each Y by its Y rank in the separate joint ranking of the n Y's. As with r, r_s assumes that neither X nor Y is controlled by the investigator. However, r_s does not have the same interpretation as r in terms of the percentage of total variation in one variable "explained" by the other. [Recall equation (16).]

Table 12 illustrates the computation of r_s for the wheelchair cushion data of Table 10.

TABLE 12.   Computation of r_s for the Wheelchair Cushion Data

Cushion	Pressure rank	Price rank	d	d^2
RF	7	9	2	4
SF	9.5	11	1.5	2.25
SS	4	6	2	4
BB	3	2	−1	1
CC	8	3	−5	25
JP	2	4.5	2.5	6.25
DP	11	4.5	−6.5	42.25
BF	6	10	4	16
LF	5	1	−4	16
AP	9.5	7	−2.5	6.25
JC	1	8	7	49
				$\sum d^2 = 172$

From (18) we obtain, with $n = 11$,

$$r_s = 1 - \frac{6 \times 172}{(11)^3 - 11} = 1 - \frac{1032}{1320} = 1 - .782 = .218.$$

To test the significance of the value $r_s = .218$ exact tables are needed. Nevertheless, the reader should know that:

(i) r_s is always between -1 and $+1$

(ii) "Large" values of r_s indicate positive association; "small" values of r_s indicate negative association, and values of r_s close to zero indicate a lack of association. Table C10 enables us to say how large is "large" and how small is "small".

Before our discussion of the use of Table C10, we present two examples, one illustrating a situation in which r_s assumes its maximum value 1, the other being a case in which r_s assumes its minimum value -1.

Suppose, contrary to our Table 11 findings, that there had been perfect agreement between the pressure rank and the price rank for the 11 wheelchair cushions. For example, the results could have been like the data shown in Table 13.

TABLE 13. Situation of Perfect Agreement

Cushion	Pressure rank	Price rank	d	d^2
RF	7	7	0	0
SF	9	9	0	0
SS	5	5	0	0
BB	4	4	0	0
CC	8	8	0	0
JP	2	2	0	0
DP	11	11	0	0
BF	6	6	0	0
LF	3	3	0	0
AP	10	10	0	0
JC	1	1	0	0
				$\Sigma d^2 = 0$

The hypothetical rankings in Table 13 give (for this $n = 11$ case) the strongest possible indication that low pressures and low prices go hand in hand. From (18) we would then find

$$r_s = 1 - \frac{6 \times 0}{(11)^3 - 11} = 1 - 0 = 1.$$

Similarly, a hypothetical outcome in which there is perfect disagreement could be constructed as in Table 14.

TABLE 14.   Situation of Perfect Disagreement

Cushion	Pressure rank	Price rank	d	d^2
RF	7	5	−2	4
SF	9	3	−6	36
SS	5	7	2	4
BB	4	8	4	16
CC	8	4	−4	16
JP	2	10	8	64
DP	11	1	−10	100
BF	6	6	0	0
LF	3	9	6	36
AP	10	2	−8	64
JC	1	11	10	100
				$\Sigma d^2 = 440$

The hypothetical rankings in Table 14 give (for this $n = 11$ case) the strongest possible indication that low pressures go with high prices and high pressures go with low prices. From (18) we would find

$$r_s = 1 - \frac{6 \times 440}{(11)^3 - 11} = 1 - \frac{2640}{1320} = 1 - 2 = -1.$$

What can we say about the actual outcome of the rankings given in Table 12? For that outcome r_s is positive (.218). Is this value of .218 large enough to indicate that (roughly) low pressures go with low prices and high pressures go with high prices? We answer this question by referring the value $r_s = .218$ to Table C10.

Table C10 contains probabilities for the null distribution (the distribution under the hypothesis of independent rankings) of Spearman's rank correlation coefficient r_s for samples sizes $n = 1, 2, \ldots, 11$. The lower-tail values are for testing H_0, the null hypothesis of independence, against the alternative of negative association. The upper-tail values are for testing H_0 against the alternative of positive association. (For $n = 4$, 5, and 6 the table is arranged so that in each subtable the lower-tail values are presented first, with the upper-tail values beneath—and separated by a space from—the lower-tail values.) The lower-tail probability P, corresponding to a value c, is the probability, under the null hypothesis, that r_s will be less than or equal to c. The upper-tail probability P, corresponding to a value c, is the probability, under the null hypothesis, that r_s will be greater than or equal to c.

Consider the case $n = 11$. Then from Table C10 we find $Pr\{r_s \geq .527\} = Pr\{r_s \leq -.527\} = .050$. Thus in the wheelchair cushion example, if we want to test the null hypothesis

$$H_0: \text{ pressure and price are independent,}$$

against the alternative that pressure and price are positively associated (i.e., low pressures go with low prices and high pressures go with high prices), the $\alpha = .050$ test is reject H_0 if $r_s \geq .527$, accept H_0 if $r_s < .527$. If we wish to test H_0 against the alternative of negative association (i.e., low pressures go with high-priced cushions and high pressures go with low prices), the $\alpha = .050$ test is reject H_0 if $r_s \leq -.527$, accept H_0 if $r_s > -.527$. A two-tailed $\alpha = 2(.050) = .100$ test of H_0 versus negative and positive association alternatives is reject H_0 if $r_s \leq -.527$ or if $r_s \geq .527$, accept H_0 if $-.527 < r_s < .527$.

In Table 12, based on the wheelchair cushion data, we found $r_s = .218$. Thus, since $r_s = .218$ is between $-.527$ and $+.527$, we accept H_0 at the two-sided $\alpha = .100$ level. Thus there is not sufficient evidence to conclude that price and pressure are related. (From Table C10 the P value corresponding to $r_s = .218$ is .26. This P value is approximate rather than exact, because of the occurrence of ties and the subsequent use of average ranks. See Comment 12.)

Large Sample Approximation to the Test Based on Spearman's Rank Correlation Coefficient

For large n values (in particular, for values of n greater than 11, values not covered by Table C10) the statistic

$$(19) \qquad\qquad r_s^* = \sqrt{n-1} \cdot r_s$$

can be treated as having an approximate standard normal distribution. The test of independence can be performed by referring r_s^* to the standard normal table (Table C2). Significantly large values of r_s^* indicate positive association; significantly small values indicate negative association.

For our wheelchair cushion data we have $n = 11$, $r_s = .218$, and from (19) we find

$$r_s^* = \sqrt{10} \times (.218) = .69.$$

Referring $r_s^* = .69$ to the standard normal table, the one-sided P value is found to be approximately .25. Thus the large sample approximation leads us to the same conclusion we reached in the previous subsection, namely, the data indicate a lack of an association between pressure and price.

Advantages of r_s to r

Desirable characteristics of r_s, when compared with r, include:

1. To compute r_s we need only know the ranks, rather than the actual observations.
2. The statistic r_s is less sensitive than r to wildly outlying observations.
3. The significance test based on r is exact only when the X and Y variables have a "bivariate normal" distribution. As we stated earlier, a precise definition of this distributional restriction is beyond the level of our text, but it suffices that the reader be aware that the necessity for bivariate normality limits the applicability of r when n is small. (When n is large, the central limit theorem assures the validity of procedures based on r.) On the other hand, the significance test based on r_s is **nonparametric**, in that it requires only relatively mild assumptions about the underlying bivariate population governing the (X, Y) data—and in particular, it does not require bivariate normality. Thus procedures based on r_s are exact, even for small values of n. (See Comment 12.)

These advantages—wider applicability, relative insensitivity to outliers, and the removal of the restrictive assumption of underlying normal populations—are also to be found in the statistical procedures based on ranks covered in Chapters 12, 13, and 14.

Discussion Question 5. Consider the obesity-blood pressure data of Table 7. Let X denote the obesity variable and Y the blood pressure variable. Note that r, computed for the (X, Y) data, does not equal r, computed for the (log X, log Y) data. However, r_s computed for the (X, Y) data [by ranking the X's, then performing a separate ranking of the Y's, then computing $\sum d^2$ and finally r_s via (18)] is the same as r_s computed for the (log X, log Y) data [by ranking the logs of the X's, then performing a separate ranking of the logs of the Y's, then computing $\sum d^2$ and finally r_s via (18)]. Explain why r_s has this "invariance" property. Is such a property desirable?

Comments

12. The probabilities in Table C10, for the null distribution of r_s, are exact when the population of the (X, Y) variables is continuous, so that the chance of

ties among the observations is zero. If there are ties among the X's or the Y's, use average ranks (as we did in Table 11) to break ties. In the presence of ties, the P value obtained from Table C10 is approximate, not exact.

13.   The coefficient r_s was introduced by the eminent psychologist C. Spearman (1904), and it remains extremely popular today. As we have already noted, r_s is simply Pearson's r applied to the rankings of the X and Y observations (within their respective samples) rather than to the X and Y observations themselves. Kendall (1970), in the fourth edition of his book *Rank Correlation Methods*, has presented an excellent account of r_s and an alternative rank correlation coefficient called tau.

Summary of Statistical Concepts

In this chapter the reader should have learned that:
(i)   A regression problem considers the distribution of the dependent variable (Y) when the independent variable (X) is held fixed at several levels.
(ii)   There are many inferential procedures for the model specifying a linear regresssion of Y on X. These procedures include fitting a least squares line to the data, calculating confidence intervals and hypothesis tests concerning the slope and intercept, and obtaining a confidence band for the true regression line.
(iii)   A correlation problem considers the joint distribution of two variables, neither of which is restricted by the experimenter.
(iv)   Either Pearson's correlation coefficient or Spearman's rank correlation coefficient can be used to test for independence of two uncontrolled variables X and Y.

PROBLEMS

10.   A researcher is interested in the relationship between religiosity and coping behavior for patients undergoing a serious operation. Is the intensity of a patient's religious beliefs and feelings associated with the patient's ability to cope with the situation? On the basis of questionnaires and interviews the researcher could rank the patients with respect to religiosity (say, high ranks indicating high intensity of religious feelings) and also perform a *separate* ranking of these same patients with respect to coping ability (say, high ranks being given to good copers). Suppose the outcomes of the two separate rankings (one for religiosity and one for coping behavior) are as given in Table 15.

TABLE 15. Religiosity and Coping Behavior

Patient	Religiosity rank	Coping rank
1	2	1
2	3	5
3	4	4
4	6	3
5	1	2
6	5	6

Thus, for example, patient 5 had the lowest intensity of religious feelings, whereas patient 4 had the highest intensity of religious feelings. In terms of coping, patient 6 was the best coper, and patient 1 demonstrated the weakest coping ability among the six patients. Do the rankings in Table 15 indicate an association between religiosity and the ability to cope?

11. Consider the obesity–blood pressure data of Table 7. Rank the 44 obesity values from least to greatest. Then rank the 44 blood pressure values from least to greatest. Compute r_s. Then use the normal approximation to the distribution of r_s to test for independence of obesity and blood pressure.

12. Consider the obesity–blood pressure data of Table 9. Compute r_s and test for independence of obesity and blood pressure.

13. Take the case $n = 6$ and construct an example where r_s is approximately zero.

14. Consider the heart transplant data of Table 4, Chapter 1. Use the 82 patients to compute the rank correlation coefficient r_s between the variables "data of acceptance" and "age at acceptance date." Then use r_s to test for independence of the two variables. Compare your results with those based on r—see Problem 8. Discuss the relative difficulties of computing r and r_s for these data.

15. Construct an example that demonstrates the relative insensitivity of r_s to outliers.

16. Define regression curve, linear regression, scatterplot, the method of least squares, homogeneous variance, residuals, independent variable, dependent variable, Pearson's correlation coefficient, Spearman's rank correlation coefficient, ranks, nonparametric.

Supplementary Problems

1. McFadden and Pasanen (1975) have reported that when a subject hears two basic tones in one ear and the same two basic tones in the other ear but with one of the two tones modified very slightly (say 2 hertz difference in frequency), the subject will detect a beat that originates in the nervous system of the subject, because of some type of interaction. The beat is only heard when the two basic tones are fairly close in frequency themselves, say, separated in frequency by 200 hertz. Table 16 is based on a subset of the McFadden and Pasanen data. Table 16 contains data for two subjects, when one basic tone was 3550 hertz, the other lower by the amount given in the table, for one ear. In the other ear, the lower tone was either the same (no beat) or 2 hertz higher (possible beat). The subject was asked whether he could detect a beat, and the percentage of correct responses was recorded, where 50% represented chance performance.

TABLE 16. Percentage of Correct Discriminations between Beat and No-Beat Conditions

Frequency Separation of Basic Tones (in hertz)	Correct Discriminations (%)	
	Subject D.M.	Subject C.M.
25	94	96
75	85	80
125	92	84
175	74	64
225	63	68
275	65	60
325	66	51

Source: D. McFadden and E. G. Pasanen (1975).

Graph the results for each subject separately, showing the deterioration in ability to detect a beat as the basic tones are separated. Does the linear regression model seem to fit each set of data? Fit a straight line to each set of data.

2. Table 2 of Chapter 4 presents the results obtained in measuring the triglyceride levels in 30 plasma samples by two different methods. Taking differences and doing a t test on the mean of the sample of 30 differences, it can be shown that the mean difference is 2.5 and that this mean difference is significantly different from zero at the 1% level. Compute the Pearson correlation coefficient between the results obtained by the two methods. Is the correlation significantly different from zero? What does this correlation mean? How do you reconcile the conclusion that the two methods do not give the same results (one measures higher than the other) with the conclusion that the two methods are highly correlated?

3. In carefully controlled diet studies on hospitalized patients it has been reported by Keys, Anderson, and Grande (1965) that a change in the cholesterol content in the diet correlates .95 with a change in cholesterol level in the blood serum. However, in the Stanford Heart Disease Prevention Program, Farquhar (1974) found that the cholesterol changes in diet reported by subjects in the communities studied correlated with changes in serum only at the level of .2 to .3. Some of the attenuation must be due to the quality of the diet data obtainable by questionnaire, but the use of correlation for comparison should be examined. Are correlation coefficients a valid basis for comparing predictability of serum cholesterol from diet cholesterol in the two studies? Is correlation a valid measure of relationship in either or both studies?

4. Life insurance companies charge premiums on the basis of expected mortality rates among people taking out insurance. But the mortality rates they use as the basis for prediction of the number of deaths are generally somewhat out of current date and certainly out of date in the decades following the sale of the insurance. For this reason and for other reasons, (e.g., care in selecting healthy customers), insurance companies pay out less money than anticipated in premium setting. Some of this money is paid back to the insured persons in the form of dividends. Table 17 gives the ratio of dividends paid to premiums received over a 10-year period for one company.

TABLE 17. Ratio of Dividends Paid to Premiums Received

Year	New York Life
1965	31.7
1966	30.8
1967	31.0
1968	30.4
1969	30.4
1970	29.6
1971	29.5
1972	28.1
1973	27.0
1974	26.8

Source: Abstracted from information distributed by Northwestern Mutual (MDRT IR Index 1900.01, FO 29-2851 (11–75)).

Do the values seem to be changing linearly with time? Fit a straight line to the data. Estimate the 1975 ratio for New York Life. Can you criticize the use of the standard linear regression model for time series data such as those given in Table 17?

5. If an antibiotic is stored at high temperature, it eventually loses its potency, (i.e., its ability to kill bacteria). In an experiment to evaluate the stability of a particular antibiotic stored at 33° C, Nager (1972) obtained the data given in Table 18.

The measurements were made by removing aliquots from the bulk stored product at the specified times. Graph the data. Does the degradation or loss in potency appear to be linear? Fit a straight line to the data, and calculate 95% confidence limits for the rate of loss

TABLE 18.   Potency of an Antibiotic

Days of Storage at 33° C	Potency (in standard units)
0	623, 632, 628, 628
30	614, 607, 608, 613
60	596, 600, 605, 609
90	592, 582, 592, 589

Source: U. F. Nager (1972).

in potency per day. Estimate the loss in potency at 6 months, with 95% confidence limits, assuming that the loss will be linear over the whole 6-month period.

6. Suppose an investigator sets out to measure the relationship between muscular strength and income among teachers. In a random sample of teachers he obtains for the five males and six females in his sample the results given in Table 19. (Table 19 contains fictitious data, exaggerated to illustrate a point.)

TABLE 19.   Muscular Strength and Income of Teachers

Sex	Strength	Income
M	25	11.5
F	8	10.2
M	34	12.4
F	18	10.6
F	5	9.7
M	29	11.8
F	13	8.9
M	27	13.2
F	9	11.1
M	38	11.8
F	12	11.5

Compute both the Pearson and Spearman correlation coefficients for the sample of 11 individuals. Now calculate the correlations for males and for females separately. How do you explain your results? Draw a scatterplot with the points for females and males denoted by different symbols. This will explain the paradox.

7. In the light of the preceding problem, can you offer an explanation for the apparent paradox that within any given census tract there is a tendency for the foreign born to have a higher literacy rate than the native born; yet when a large group of census tracts is studied by examining the percent of foreign born and the literacy rate in each tract, there is a negative correlation between the two variables?

8. Describe some situations in which the regression curve of Y on X is not linear.

9. Verify directly that equation (16) and the square of the right-hand side of equation (14) are equivalent expressions for r^2.

10. Describe some data sets in which r_s could be computed, but in which r could not be computed.

11. The geographic variation in lung cancer mortality within the United States is striking, with many of the highest rates among men in Southern coastal areas, particularly in Louisiana, rather than in the urban, industrialized Northeast (Blot and Fraumeni [1976]). Table 20 lists lung cancer mortality rates among white males and females over the 20-year period 1950–1969 for Louisiana's 64 parishes, along with the percentage of the population in each parish living in urban areas in 1960. Do the mortality rates increase with urbanization? In both sexes? If so, what aspects of urban living may be associated with the increased lung cancer mortality?

TABLE 20. Lung Cancer Mortality and Urbanization in Louisiana—Lung Cancer Mortality Rate 1950–1969 (deaths/year/100,000 population)

Parish	Males	Females	Percentage Urban
Acadia	56.98	11.66	55.8
Allen	48.37	5.62	33.3
Ascension	55.39	4.04	33.3
Assumption	52.89	4.83	0
Avoyelles	47.72	4.64	25.1
Beauregard	36.96	9.87	37.5
Bienville	35.35	2.72	15.1
Bossier	48.14	6.86	66.0
Caddo	53.92	7.57	80.8
Calcasieu	54.10	8.24	73.8
Caldwell	34.62	8.94	0
Cameron	36.41	9.11	0
Catahoula	59.30	5.84	0
Claiborne	42.75	2.88	39.6
Concordia	66.85	5.04	43.3

TABLE 20—continued

Parish	Males	Females	Percentage Urban
De Soto	42.54	9.44	24.1
East Baton Rouge	44.43	6.26	85.1
East Carroll	51.72	8.21	40.1
East Feliciana	20.80	1.34	0
Evangeline	60.59	14.87	33.0
Franklin	38.84	7.09	17.0
Grant	40.65	10.51	0
Iberia	62.48	4.87	67.0
Iberville	46.02	5.73	25.6
Jackson	39.99	5.63	24.3
Jefferson	62.52	7.91	94.1
Jefferson Davis	52.60	10.10	62.8
Lafayette	51.60	9.52	55.6
Lafourche	58.00	5.45	41.5
La Salle	48.92	3.31	0
Lincoln	28.71	6.77	60.0
Livingston	52.11	5.72	22.1
Madison	54.41	3.34	57.1
Morehouse	38.64	3.02	45.1
Natchitoches	32.36	6.96	39.1
Orleans	64.10	7.73	100.0
Ouachita	49.01	5.33	79.1
Plaquemines	50.12	4.61	34.5
Pointe Coupee	38.62	6.46	17.6
Rapides	41.09	5.33	52.6
Red River	36.79	5.50	0
Richland	56.13	6.60	27.6
Sabine	23.34	4.01	17.0
St Bernard	71.80	5.44	66.0
St Charles	60.16	4.36	22.1
St Helena	34.63	4.60	0
St James	41.97	5.98	17.8
St John Baptist	36.06	2.60	47.8
St Landry	52.77	7.35	35.3
St Martin	61.03	8.42	33.6
St Mary	56.38	5.07	59.3
St Tammany	46.03	8.99	33.8

TABLE 20—continued

Parish	Males	Females	Percentage Urban
Tangipahoa	54.07	7.06	35.6
Tensas	68.52	2.43	0
Terrebonne	68.55	3.53	52.1
Union	29.47	5.79	15.5
Vermilion	44.92	5.70	40.3
Vernon	33.72	6.11	25.6
Washington	48.40	6.89	55.8
Webster	39.86	4.84	48.3
West Baton Rouge	54.27	5.71	39.1
West Carroll	36.08	6.24	0
West Feliciana	34.52	16.45	0
Winn	33.06	7.12	43.8

Sources: T. J. Mason and F. W. McKay (1974); U.S. Bureau of Census (1963).

REFERENCES

Blot, W. J. and Fraumeni, J. F., Jr. (1976). Geographic patterns of lung cancer: Industrial correlations. *Amer. J. Epidemiol.* **103**, 539–550.

Dixon, W. J. and Massey, F. J., Jr. (1969). *Introduction to Statistical Analysis*, 3rd edition. McGraw-Hill, New York.

Draper, N. R. and Smith, H. (1966). *Applied Regression Analysis*. Wiley, New York.

Farquhar, J. W. (1974). Personal communication.

Feinberg, S. E. (1975). Statistics, energy, and life-style. (Letter to *Science*.) *Science* **188**, 10.

Finney, D. J. (1964). *Statistical Method in Biological Assay*, 2nd edition. Hafner, New York.

Friedland, G. W., Filly, R., and Brown, B. Wm., Jr. (1974). Distance of upper pole calyx to spine and lower pole calyx to ureter as indicators of parenchymal loss in children. *Pediat. Radiol.* **2**, 29–37.

James, K. E. (1973). Regression toward the mean in uncontrolled clinical studies. *Biometrics* **29**, 121–130.

*Kendall, M. G. (1970). *Rank Correlation Methods*, 4th edition. Griffin, London.

Keys, A., Anderson, J. T., and Grande, F. (1965). Serum cholesterol response to changes in the diet II: The effect of cholesterol in the diet. *Metabolism* **14**, 759–765.

Kruskal, W. H. (1975). Statistics, energy, and life-style. (Letter to *Science*.) *Science* **188**, 10.

Mason, T. J. and McKay, F. W. (1974). U.S. Cancer Mortality by County: 1950–69. U.S. Government Printing Office, Washington, D.C.

Mazur, A. and Rosa, E. (1974). Energy and life-style. *Science* **186**, 607–610.

McFadden, D. and Pasanen, E. G. (1975). Binaural beats at high frequencies. *Science* **190,** 394–396.

Nager, U. F. (1972). Personal communication.

Platt, T. R. (1973). Personal communication.

Ream, A. K., Silverberg, G. D., Rolfe, J., Brown, B. Wm., Jr., and Halpren, J. (1975). Accurate measurement of intracranial cerebrospinal fluid pressure by an epidural technique. (Submitted for publication.)

Souther, S. G., Carr, S. D. and Vistnes, L. M. (1974). Wheelchair cushions to reduce pressure under bony prominences. *Arch. Phys. Med. Rehabil.* **55,** 460–464.

U.S. Bureau of Census (1963). U.S. Census of Population: 1960. U.S. Government Printing Office, Washington, D.C.

Wardlaw, A. C. and van Belle, G. (1964). Statistical aspects of the mouse diaphragm test for insulin. *Diabetes* **13,** 622–633.

Wray, J. D. and Aguirre, A. (1969). Protein-calorie malnutrition in Candelarria, Columbia. I. Prevalence; social and demographic factors. *J. Trop. Pediat.* **15,** 76–98.

CHAPTER 12

Analysis of Matched Pairs Using Ranks

Recall that Chapter 4 was devoted to the analysis of matched pairs data. The tests and confidence intervals of Chapter 4 are based on sample means, and these sample means have the undesirable property of being extremely sensitive to outlying observations. The nonparametric procedures of this chapter have the virtue of being relatively insensitive to outliers.

As in Chapter 4, let d be the amount by which the response of a treated subject in a matched pair exceeds the response of the control subject of the pair. (Note the excess may be negative; i.e., d may be a positive or a negative value.) With n matched pairs we view the d values as a random sample of size n from a population with mean μ. In Section 1 we present a nonparametric test of the hypothesis H_0: $\mu = 0$. Section 2 contains a nonparametric confidence interval for μ. The procedures of Sections 1 and 2 are called nonparametric because they are exact under very mild assumptions concerning the population of the d's. Recall that in Chapter 4 it was necessary to assume that the d's were from a normal population in order that the test and confidence interval, which used percentiles of the t distribution, would be exact.

1.	THE WILCOXON SIGNED RANK TEST

Decreasing Blood Pressure Through Operant Conditioning

Elevated arterial blood pressure increases the risk of coronary artery disease and cerebrovascular accidents. This increased risk may perhaps be lessened by lowering the blood pressure. The usual methods of lowering blood pressure are pharmacological or surgical. Benson, Shapiro, Tursky, and Schwartz (1971) conducted a study designed to lower arterial blood pressure through operant conditioning—feedback techniques. The seven patients in the study had moderate or severe hypertension, were ambulatory, and were attending the hypertension clinic of Boston City Hospital.

Systolic blood pressure was measured only in the laboratory; we refer the reader to Benson *et al.* for details of the apparatus. In conditioning sessions, when relatively lowered pressure occurred, this was fed back to the patient via a 100-msec flash of light and a simultaneous 100-msec tone of moderate intensity. The patients were informed that the tone and light were desirable, and that they should try to make them appear. Rewards for making them appear were incorporated into the system. Conditioning continued until no reductions in blood pressure occurred in five consecutive (separate) sessions.

Table 1 gives the median systolic blood pressure for the last five control sessions on each patient and the median systolic blood pressure for the last five conditioning sessions on each patient. In the control sessions pressure was recorded without feedback or reinforcement of lowered pressure.

TABLE 1. Median Systolic Blood Pressures (mm Hg)

Patient No.	Last Five Control Sessions	Last Five Conditioning Sessions	Conditioning Minus Control
1	139.6	136.1	−3.5
2	213.3	179.5	−33.8
3	162.3	133.1	−29.2
4	166.9	150.4	−16.5
5	157.8	141.7	−16.1
6	165.7	166.6	0.9
7	149.0	131.7	−17.3

Source: H. Benson, D. Shapiro, B. Tursky, and G. E. Schwartz (1971).

Signed Ranks

We now analyze the differences d between the control session and conditioning session values. Table 2 contains these seven differences, their absolute values (i.e., ignoring sign) $|d|$, and r, the rank of $|d|$ in the ranking from least to greatest of the seven $|d|$ values. Ranking from least to greatest means that the smallest $|d|$ value gets rank 1, the second smallest gets rank 2, and so forth.

TABLE 2.   Ranks of Absolute Differences

Patient	d	$|d|$	r
1	−3.5	3.5	2 (−)
2	−33.8	33.8	7 (−)
3	−29.2	29.2	6 (−)
4	−16.5	16.5	4 (−)
5	−16.1	16.1	3 (−)
6	0.9	0.9	1 (+)
7	−17.3	17.3	5 (−)

The parenthetical entry in the r column of Table 2 is the sign of the difference d. Thus for patient 3 the absolute difference is 29.2, and it receives the rank of 6 in the joint ranking of the seven $|d|$ values. The parenthetical $(-)$ to the right of the 6 indicates that the d difference (-29.2) for patient 3 is negative.

The rank values r corresponding to the negative d's are called the negative signed ranks. The rank values r corresponding to the positive d's are called the positive signed ranks. Thus in our example the negative signed ranks are 2, 3, 4, 5, 6, 7, and there is only one positive signed rank, namely, 1.

Suppose that conditioning does reduce pressure. Then the positive signed ranks would tend to be smaller than the negative signed ranks. The size of the positive signed ranks can be assessed via their sum. Call this statistic T^+. This is the Wilcoxon signed rank statistic. We have

$$T^+ = 1.$$

This value, $T^+ = 1$, certainly appears to be smaller than what we would expect if operant conditioning really has no effect on blood pressure. The question of whether $T^+ = 1$ is "small enough" to indicate a significant improvement (decrease) in blood pressure is considered in the next subsection.

Does Operant Conditioning Decrease Systolic Blood Pressure?

The null hypothesis is

H_0: operant conditioning has no effect on systolic blood pressure.

Let μ denote the true mean of the d population, that is, the (unknown) difference that might be obtained in a very large series. If H_0 is true, $\mu = 0$. If operant conditioning reduces pressure, then $\mu < 0$. If operant conditioning increases pressure, then $\mu > 0$. In view of our discussion in the preceding subsection, a reasonable procedure for testing H_0 is to refer the observed value of T^+ to percentiles of the null distribution of T^+. Significantly large values of T^+ indicate $\mu > 0$; significantly small values of T^+ indicate $\mu < 0$.

Table C11 gives percentiles of the null distribution of T^+ for the sample sizes $n = 3, 4, \ldots, 15$. The lower-tail values are for testing H_0, the hypothesis of no effect, against the alternative that the mean μ of the d population is negative. The upper-tail values are for testing H_0 against the alternative $\mu > 0$. (When H_0 is true, $\mu = 0$.) The lower-tail probability P, corresponding to a value c, is the probability, under the null hypothesis, that T^+ will be less than or equal to c. The upper-tail probability P, corresponding to a value c, is the probability, under the null hypothesis, that T^+ will be greater than or equal to c.

Consider the case $n = 7$. Then from Table C11 we find $Pr\{T^+ \le 4\} = Pr\{T^+ \ge 24\} = .055$. Thus if we wish to test H_0 versus $\mu < 0$, our $\alpha = .055$ test is reject H_0 if $T^+ \le 4$, accept H_0 if $T^+ > 4$. If we wish to test H_0 versus $\mu > 0$, our $\alpha = .055$ test is reject H_0 if $T^+ \ge 24$, accept H_0 if $T^+ < 24$. A two-tailed $\alpha = 2(.055) = .110$ test of H_0 versus the alternative $\mu \ne 0$ is reject H_0 if $T^+ \le 4$ or if $T^+ \ge 24$, accept H_0 if $4 < T^+ < 24$.

For the operant conditioning data, we have found $T^+ = 1$. Thus at $\alpha = .110$ (two-tailed) we reject H_0. More can be said. From Table C11 we find that when H_0 is true, $Pr\{T^+ \le 1\} = .016$. That is, the one-sided P value is .016, and the two-sided P value is $2(.016) = .032$. Thus there is strong evidence, on the basis of our limited data, to reject H_0 in favor of the alternative that operant conditioning reduces blood pressure. It should be noted, however, that in this uncontrolled study, there are other possible explanations for a decrease in blood pressure over the period of conditioning.

Large Sample Approximation to the Signed Rank Test

For large n values we can treat

$$(1) \qquad T^* = \frac{T^+ - [n(n+1)/4]}{\sqrt{\dfrac{n(n+1)(2n+1)}{24}}}$$

as an approximate standard normal variable. To test H_0 refer T^* to the standard normal table (Table C2). Significantly large values of T^* indicate $\mu > 0$; significantly small values of T^* indicate $\mu < 0$. This normal approximation works well when $n > 15$. Even for smaller sample sizes, it can yield P values that are very close to the true P value.

For our blood pressure data, we have $n = 7$, $T^+ = 1$, and from (1) we find

$$T^* = \frac{1 - [7(8)/4]}{\sqrt{\dfrac{7(8)(15)}{24}}} = -2.20.$$

Referring $T^* = -2.20$ to the standardized normal table, the one-sided P value is found to be approximately .014. This is in close agreement with the exact value .016 obtained in the previous subsection, and it suggests that the normal approximation is good even for relatively small values of n.

Discussion Question 1. What effect does a wildly outlying value (say, a very large treatment response in one matched pair) have on T^+?

Comments

1. If there are tied values amongst the $|d|$'s, the ranks should be assigned by giving each $|d|$ the average of the ranks for which it is tied. Consider a case in which $n = 6$ and in which, for example, the d values are as follows:

d	$|d|$	r	
-1.4	1.4	1.5	$(-)$
2.6	2.6	4	$(+)$
1.4	1.4	1.5	$(+)$
3.1	3.1	6	$(+)$
-2.6	2.6	4	$(-)$
2.6	2.6	4	$(+)$

The $|d|$ values 1.4 and 1.4 are tied for ranks 1 and 2, and hence they are each assigned the average rank $[(1+2)/2] = 1.5$. Similarly, the $|d|$ values 2.6, 2.6, 2.6 are tied for ranks 3, 4, 5, and thus each receives the average rank $[(3+4+5)/3] = 4$. The value of T^+, the sum of the positive signed ranks, is

$$T^+ = 4 + 1.5 + 6 + 4 = 15.5.$$

This statistic can be referred to Table C11 for a P value, but when ties occur, as in these data, the probabilities given in Table C11 are no longer exact—only approximate.

2.   If some d's are equal to zero, *before* ranking simply discard these values, reduce the sample size n accordingly, and proceed as described in this section but with a new (reduced) value of n.

3.   In the operant conditioning example, the null hypothesis is that the population mean μ of the d's is zero. If, for example, we wish to test the hypothesis that $\mu = 3$, we simply subtract 3 from each d, then rank the absolute values of $d - 3$, and add up the ranks of the absolute values corresponding to those patients for which $d - 3$ is positive. Call this sum of ranks T^+ and refer it to Table C11 for a P value. In general, to test the hypothesis $\mu = \mu_0$, where μ_0 is a specified nonzero number, obtain the modified observations $d' = d - \mu_0$ and compute T^+ using the d''s (rather than the d's).

4.   Consider a situation in which, rather than paired data, we have a single random sample from a population, and we wish to use the sample to make inferences about the unknown population mean μ. For example, skip ahead to Table 3, where tensile strengths are given for tape-closed and sutured wounds on the backs of rats. Our main interest in that problem is to compare tape-closed versus sutured wounds. However, cover up the suture column and suppose we were just interested in the mean tensile strength of tape-closed wounds. We could then consider the values in the tape column to be a sample of 10 d values from a common tape-closed population. To test, for example, the hypothesis that the mean tape-closed tensile strength is 1200, we subtract 1200 from each tape-closed value and compute T^+. The calculations are

d	$d-1200$	$|d-1200|$	r
1706	506	506	9 (+)
1627	427	427	8 (+)
1571	371	371	6 (+)
1984	784	784	10 (+)
1587	387	387	7 (+)

d	$d-1200$	$|d-1200|$	r
926	-274	274	3 $(-)$
931	-269	269	2 $(-)$
1282	82	82	1 $(+)$
1538	338	338	5 $(+)$
1516	316	316	4 $(+)$

$$T^+ = 9+8+6+10+7+1+5+4 = 50.$$

Referring $T^+ = 50$ to Table C11, we find the one-sided P value is .010; thus there is strong evidence to reject the hypothesis, that the mean tensile strength of the tape-closed wounds equals 1200, in favor of the alternative that the mean is greater than 1200.

5. To test the hypothesis that the control and treatment in matched pairs are equally effective, one can simply look at the sign of the d differences for the n pairs. If the null hypothesis is true, one would expect approximately half the pairs to show positive signs and the other half to show negative signs. Let B be the statistic that equals the number of pairs in which the difference d is positive. When H_0 is true, the statistic B has the binomial distribution (see Chapter 7) with parameters n and $p = \frac{1}{2}$. Thus Table C7, entered at $p = \frac{1}{2}$, can be used to find the P value corresponding to the observed value of B. Significantly large values of B indicate $\mu > 0$; significantly small values indicate $\mu < 0$.

Let us illustrate the test based on B using the operant conditioning data of Table 1. For those data $B = 1$, since (only) one of the seven d values is positive. From Table C7, we find (with $n = 7$)

$$Pr(B \le 1) = Pr(B = 0) + Pr(B = 1) = .0078 + .0547 = .0625.$$

This one-sided P value of .0625 suggests $\mu < 0$, although note that this P value is larger than the one-sided P value of .016 obtained using the signed rank statistic T^+.

The test based on B is called the sign test because it ignores all information in the data except the signs of the differences. It considers a sample in which all signs are positive but one to have the same amount of information against the null hypothesis, whether the odd sign corresponds to a numerical value nearest zero or furthest from zero. The signed rank test of this section is usually better in this regard, in that it uses information about the magnitude of d, not just its sign.

6. The signed rank test, and the sign test described in Comment 5, control the probability α of a Type-I error without requiring the assumption that the d's emanate from a normal population. It is enough to assume that the d's come from

a symmetric population (so that $Pr(d \geq \mu + a) = Pr(d \leq \mu - a)$ for all a) and that the d's are continuous variables (so that the probability of ties among the $|d|$'s is zero). If there are ties among the $|d|$'s we can use average ranks, but in that case the probabilities given in Table C11 are approximate, not exact.

7.   The signed rank and sign tests have, when compared to the normal theory test based on $\bar{d}$, the added advantage of being relatively insensitive to outlying observations.

8.   The hypothesis test based on T^+ was proposed by Wilcoxon (1945), a pioneer in the field of statistical methods based on ranks.

PROBLEMS

1.   Forrester and Ury (1969) illustrate the signed rank test on data from an investigation into the relative tensile strengths of tape-closed and sutured wounds. Four symmetrical incisions on the backs of rats were closed by suture or by surgical tape. The animals were killed in groups of 10 after 10, 20, 40, 60, 80, 100, and 150 days of healing, and the wounds were analyzed mechanically. Table 3 gives the results for the healing time of 100 days.

TABLE 3.   Tensile Strength (lb/in.²) of Taped and Sutured Wounds at 100 Days

Rat No.	Tape	Suture
1	1706	1540
2	1627	1571
3	1571	609
4	1984	1968
5	1587	881
6	926	883
7	931	1090
8	1282	1026
9	1538	1154
10	1516	851

Source: H. K. Ury and J. C. Forrester (1970).

Do the data indicate real differences in tensile strength for tape-closed versus sutured wounds? Use the signed rank test and compare with your analysis based on the statistic T [defined by equation (4) of Chapter 4]. See Problem 7 of Chapter 4.

2. Apply the signed rank test to the blood storage data of Table 1, Chapter 4, and compare your results with the results of Chapter 4 based on the statistic T [defined by equation (4) of Chapter 4].

3. Consider the data of Table 2, Chapter 4, and use the signed rank statistic to test the hypothesis that the two measurement methods are equivalent. Compare your results with those of Problem 5, Chapter 4.

4. Let T^- be the sum of the negative signed ranks. Verify directly, or via an example, that $T^+ + T^- = n(n+1)/2$. When H_0 is true, T^- has the same probability distribution as T^+. Can you give an intuitive explanation for this?

2.	**A NONPARAMETRIC CONFIDENCE INTERVAL BASED ON THE SIGNED RANK TEST**

What is the Estimated Reduction in Pressure Achieved via Operant Conditioning?

Let μ denote the mean of the d population. In the preceding section we found evidence that μ is less than 0, corresponding to the (alternative) hypothesis that operant condition does decrease blood pressure. It is then desired to get a measure of the amount of decrease. One such measure is a confidence interval for μ. We now describe how to obtain a confidence interval, associated with the T^+ statistic, for μ.

Consider the d values of Table 1, namely, $-3.5, -33.8, -29.2, -16.5, -16.1,$ $.9, -17.3$. Pair each d with every other d (including itself) and compute an average of the form $(d + d')/2$. With n differences, there are $n(n+1)/2$ such averages. These averages are called Walsh averages (named after John Walsh, who made many contributions to nonparametric methods, including techniques based on these averages). In our example $n = 7$, $\{n(n+1)/2\} = \{7(8)/2\} = 28$, and the 28 averages are

$$\frac{-3.5 + (-3.5)}{2} = -3.5, \qquad \frac{-3.5 + (-33.8)}{2} = -18.65,$$

$$\frac{-3.5 + (-29.2)}{2} = -16.35, \qquad \frac{-3.5 + (-16.5)}{2} = -10,$$

$$\frac{-3.5 + (-16.1)}{2} = -9.8, \qquad \frac{-3.5 + .9}{2} = -1.3,$$

$$\frac{-3.5+(-17.3)}{2}=-10.4, \qquad \frac{-33.8+(-33.8)}{2}=-33.8,$$

$$\frac{-33.8+(-29.2)}{2}=-31.5, \qquad \frac{-33.8+(-16.5)}{2}=-25.15,$$

$$\frac{-33.8+(-16.1)}{2}=-24.95, \qquad \frac{-33.8+.9}{2}=-16.45,$$

$$\frac{-33.8+(-17.3)}{2}=-25.55, \qquad \frac{-29.2+(-29.2)}{2}=-29.2,$$

$$\frac{-29.2+(-16.5)}{2}=-22.85, \qquad \frac{-29.2+(-16.1)}{2}=-22.65,$$

$$\frac{-29.2+.9}{2}=-14.15, \qquad \frac{-29.2+(-17.3)}{2}=-23.25,$$

$$\frac{-16.5+(-16.5)}{2}=-16.5, \qquad \frac{-16.5+(-16.1)}{2}=-16.3,$$

$$\frac{-16.5+.9}{2}=-7.8, \qquad \frac{-16.5+(-17.3)}{2}=-16.9,$$

$$\frac{-16.1+(-16.1)}{2}=-16.1, \qquad \frac{-16.1+.9}{2}=-7.6,$$

$$\frac{-16.1+(-17.3)}{2}=-16.7, \qquad \frac{.9+.9}{2}=.9,$$

$$\frac{.9+(-17.3)}{2}=-8.2, \qquad \frac{-17.3+(-17.3)}{2}=-17.3.$$

Next, order these 28 averages from least to greatest. The ordered values are -33.8, -31.5, -29.2, -25.55, -25.15, -24.95, -23.25, -22.85, -22.65, -18.65, -17.3, -16.9, -16.7, -16.5, -16.45, -16.35, -16.3, -16.1, -14.15, -10.4, -10, -9.8, -8.2, -7.8, -7.6, -3.5, -1.3, 0.9.

The lower end point of the confidence interval is the value, in the above list of *ordered* Walsh averages that occupies the bth position, where b is an integer. We explain how one determines b in a moment. The upper end point of the confidence interval is the value that occupies position $[n(n+1)/2]+1-b$ in the list of *ordered* Walsh averages.

The integer b depends on the desired confidence coefficient $1-\alpha$ and can be obtained from the equation:

(2) $b = \dfrac{n(n+1)}{2} + 1 - [\text{upper } \dfrac{\alpha}{2} \text{ percentile point of the null distribution of } T^+].$

For example, consider the operant conditioning data, and suppose we seek a confidence coefficient of 89%; then $1-\alpha = .89$, $\alpha = .11$, and $\alpha/2 = .055$. From Table C11, entered at $n = 7$, we find the upper .055 percentile point of the null distribution of T^+ to be 24. Thus, from (2),

$$b = \dfrac{7(8)}{2} + 1 - 24 = 5.$$

Then the 89% confidence interval for μ has as its lower end point the value occupying position 5 in the list of the ordered Walsh averages and as its upper end point the value occupying position 24. The interval is thus found to be $(-25.15, -7.8)$. In addition to providing an interval estimate of μ, this interval also has a hypothesis testing interpretation. Since the interval does not contain the point 0, we know that the two-tailed $\alpha = .11$ test based on T^+ leads to rejection of H_0.

For large n values (in particular, for values of n greater than 15, values not covered by Table C11), the integer b can be approximated by

(3) $$b \approx \dfrac{n(n+1)}{4} - z_{\alpha/2}\sqrt{\dfrac{n(n+1)(2n+1)}{24}},$$

where $z_{\alpha/2}$ is the upper $\alpha/2$ percentile point of the standard normal distribution. The value of b, found via (3), usually is not an integer, thus use the integer closest to b.

Discussion Question 2. For confidence intervals with very large (i.e., very close to 1) confidence coefficients, it is not necessary to compute all the Walsh averages to obtain the interval end points. Why?

Comments

9. Consider the $n(n+1)/2$ Walsh averages used to obtain a confidence interval for μ. When there are no tied observations and no d's equal to 0, the signed rank statistic T^+ is equal to the number of positive Walsh averages.

10. If one desires a point estimator $\hat{\mu}$ of the parameter μ, a good choice is the median of the $n(n+1)/2$ Walsh averages. In the blood pressure example, the

median of the 28 Walsh averages is the average of the fourteenth and fifteenth values in the list of the ordered Walsh averages. We find

$$\tilde{\mu} = \frac{-16.5 + (-16.45)}{2} = -16.48.$$

The standard deviation of the estimator $\tilde{\mu}$ can be approximated by $\widehat{SD}(\tilde{\mu})$ where

$$\widehat{SD}(\tilde{\mu}) = \left\{ \frac{\text{length of the } 1-\alpha \text{ confidence interval for } \mu}{2(z_{\alpha/2})} \right\}.$$

In the blood pressure example, where we set $1 - \alpha = .89$, we have

$$\widehat{SD}(\tilde{\mu}) = \frac{-7.8 - (-25.15)}{2(1.6)} = \frac{17.35}{3.2} = 5.4.$$

11.   The confidence interval of this section can also be applied to a single sample from a population to obtain an interval estimate of the population mean. Simply denote the values in the sample as d's, compute the Walsh averages, and obtain the interval end points as particular Walsh averages as described in this section. For example, refer to Table 3 and suppose we want a confidence interval for the population mean of taped-closed wounds. Then, letting the taped-closed values play the role of the d's, the $[10(11)/2] = 55$ Walsh averages are

$$\frac{1706 + 1706}{2} = 1706, \qquad \frac{1706 + 1627}{2} = 1666.5,$$

$$\frac{1706 + 1571}{2} = 1638.5, \qquad \frac{1706 + 1984}{2} = 1845,$$

$$\frac{1706 + 1587}{2} = 1646.5, \qquad \frac{1706 + 926}{2} = 1316,$$

$$\frac{1706 + 931}{2} = 1318.5, \qquad \frac{1706 + 1282}{2} = 1494,$$

$$\frac{1706 + 1538}{2} = 1622, \qquad \frac{1706 + 1516}{2} = 1611,$$

$$\frac{1627 + 1627}{2} = 1627, \qquad \frac{1627 + 1571}{2} = 1599,$$

$$\frac{1627 + 1984}{2} = 1805.5, \qquad \frac{1627 + 1587}{2} = 1607,$$

$$\frac{1627+926}{2}=1276.5; \qquad \frac{1627+931}{2}=1279,$$

$$\frac{1627+1282}{2}=1454.5, \qquad \frac{1627+1538}{2}=1582.5,$$

$$\frac{1627+1516}{2}=1571.5, \qquad \frac{1571+1571}{2}=1571,$$

$$\frac{1571+1984}{2}=1777.5, \qquad \frac{1571+1587}{2}=1579,$$

$$\frac{1571+926}{2}=1248.5, \qquad \frac{1571+931}{2}=1251,$$

$$\frac{1571+1282}{2}=1426.5, \qquad \frac{1571+1538}{2}=1554.5,$$

$$\frac{1571+1516}{2}=1543.5, \qquad \frac{1984+1984}{2}=1984,$$

$$\frac{1984+1587}{2}=1785.5, \qquad \frac{1984+926}{2}=1455,$$

$$\frac{1984+931}{2}=1457.5, \qquad \frac{1984+1282}{2}=1633,$$

$$\frac{1984+1538}{2}=1761, \qquad \frac{1984+1516}{2}=1750,$$

$$\frac{1587+1587}{2}=1587, \qquad \frac{1587+926}{2}=1256.5,$$

$$\frac{1587+931}{2}=1259, \qquad \frac{1587+1282}{2}=1434.5,$$

$$\frac{1587+1538}{2}=1562.5, \qquad \frac{1587+1516}{2}=1551.5,$$

$$\frac{926+926}{2}=926, \qquad \frac{926+931}{2}=928.5,$$

$$\frac{926+1282}{2}=1104, \qquad \frac{926+1538}{2}=1232,$$

$$\frac{926+1516}{2}=1221, \qquad \frac{931+931}{2}=931,$$

$$\frac{931+1282}{2}=1106.5, \qquad \frac{931+1538}{2}=1234.5,$$

$$\frac{931+1516}{2}=1223.5, \qquad \frac{1282+1282}{2}=1282,$$

$$\frac{1282+1538}{2}=1410, \qquad \frac{1282+1516}{2}=1399,$$

$$\frac{1538+1538}{2}=1538, \qquad \frac{1538+1516}{2}=1527,$$

$$\frac{1516+1516}{2}=1516.$$

The 55 ordered Walsh averages are then found to be: 926, 928.5, 931, 1104, 1106.5, 1221, 1223.5, 1232, 1234.5, 1248.5, 1251, 1256.5, 1259, 1276.5, 1279, 1282, 1316, 1318.5, 1399, 1410, 1426.5, 1434.5, 1454.5, 1455, 1457.5, 1494, 1516, 1527, 1538, 1543.5, 1551.5, 1554.5, 1562.5, 1571, 1571.5, 1579, 1582.5, 1587, 1599, 1607, 1611, 1622, 1627, 1633, 1638.5, 1646.5, 1666.5, 1706, 1750, 1761, 1777.5, 1785.5, 1805.5, 1845, 1984.

Suppose we desire a 98% confidence interval. Then $1-\alpha=.98$, $\alpha=.02$, $\alpha/2=.01$, and from equation (2)

$$b=\frac{10(11)}{2}+1-[\text{upper .01 percentile point of the null}$$

$$\text{distribution of } T^+]$$

$$=55+1-50=6,$$

where the upper .01 point of the null distribution of T^+ is found from Table C11, entered at $n=10$.

Thus the lower end point of the 98% confidence interval is the value that occupies the sixth position in the list of ordered Walsh averages, and the upper end point is the value that occupies position $[10(11)/2]+1-6=50$. From our list, we obtain the interval (1221, 1761).

If, instead of an interval estimator of the mean tensile strength of taped-closed wounds, we desire a point estimator, we can compute the estimator $\tilde{\mu}$ defined in Comment 10; namely, the median of the 55 Walsh averages. We find the twenty-eighth Walsh average in the list of ordered Walsh averages is

$$\tilde{\mu} = 1527.$$

The standard deviation of $\tilde{\mu}$ is estimated by (see Comment 10)

$$\widehat{SD}(\tilde{\mu}) = \frac{1761 - 1221}{2(z_{.01})} = \frac{540}{2(2.33)} = 115.9.$$

12.   In Comment 5 of Section 1 we described the sign test. This test also yields a confidence interval for μ. The interval based on the sign test is easier to compute than the signed rank interval of Section 2, but it tends to give longer intervals than the signed rank confidence interval. We illustrate the method using the Table 1 data.

First, order the n d values from least to greatest. The ordered values for the Table 1 data are $-33.8, -29.2, -17.3, -16.5, -16.1, -3.5, 0.9$. The lower end point of the confidence interval is the value in the list of *ordered d* values that occupies the vth position, where v is an integer. We explain how v is determined in a moment. The upper end point of the confidence interval is the value that occupies position $n+1-v$ in the list of *ordered d* values.

The integer v depends on the desired confidence coefficient $1-\alpha$ and can be obtained from the equation

(4) $\qquad v = n+1-[\text{upper } \alpha/2 \text{ percentile point of the}$
$\qquad\qquad\qquad \text{distribution of the statistic } B \text{ for the case } p = \tfrac{1}{2}].$

Recall, from Comment 5, that B is the number of pairs in which the difference d is positive. From Table C7, entered at $n = 7$, $p = .50$, we see that the upper .0547 percentile point of the distribution of B is 6. Thus to obtain a confidence interval with confidence coefficient $1 - 2(.0547) = .891$, we use (4) and find

$$v = 7 + 1 - 6 = 2.$$

Thus the lower end point of the 89% confidence interval based on the sign test is the value -29.2, which occupies position 2 in the list of the ordered d's. The upper end point is the value that occupies position 6 (note $n+1-v = 7+1-2 = 6$) in the list of the ordered d's. Thus the 89% confidence interval is $(-29.2, -3.5)$. Note that it is longer than the 89% confidence interval $(-25.15, -7.8)$ found using the signed rank test.

For n values greater than 10 (values not covered by Table C7), the integer v can be approximated by

(5) $\qquad\qquad\qquad\qquad\qquad v \approx \dfrac{n}{2} - z_{\alpha/2} \sqrt{\dfrac{n}{4}},$

where $z_{\alpha/2}$ is the upper $\alpha/2$ percentile point of the standard normal distribution. The value of v, found via (5), usually is not an integer, thus use the integer closest to v.

13.   The ranking and sorting tasks involved with applying the Wilcoxon signed rank test, and obtaining the associated confidence interval, are tedious when n is large. Such tasks are easily programmed for (and speedily performed by) a computer.

14.   The confidence interval for μ, based on the Walsh averages, was described graphically by Moses (1953). The point estimator $\hat{\mu}$, defined in Comment 10, was proposed by Hodges and Lehmann (1963). More details on these procedures can be found in Chapter 3 of Hollander and Wolfe (1973).

Summary of Statistical Concepts

In this chapter the reader should have learned:
(i)   Paired replicates data (and one sample data), analyzed in Chapter 4 using $\bar{d}$ (the sample mean of the n d-differences), can be analyzed using ranks.
(ii)   The Wilcoxon signed rank test and the corresponding confidence interval have desirable properties when compared with normal theory competitors. Advantages include availability of exact probabilities of Type-I errors and exact confidence coefficients under less restrictive assumptions concerning the underlying population, and relative insensitivity to outlying observations.

PROBLEMS

5.   For the data of Table 3, determine, for μ, a 95.2% nonparametric confidence interval based on the signed rank test. Compare your result with that of Problem 3, Chapter 4.

•6.   For the blood storage data of Table 1, Chapter 4, determine, for μ, a 99% nonparametric confidence interval based on the signed rank test and compare your result with the 99% interval found in Section 1 of Chapter 4.

7.   Suppose, for the data of Table 3, we require a 99.8% confidence interval. Obtain the nonparametric interval, based on the signed rank test, described in this section. Can you determine this interval *without* actually computing *all* of the $[10(11)/2] = 55$ Walsh averages?

8.   For the data of Table 2, Chapter 4, determine for μ a 95% confidence interval based on the sign test. Compare your result with that of Problem 1, Chapter 4.

9. Show directly, or verify in a special case, the assertion of Comment 9.

10. Construct an example that illustrates that the nonparametric confidence intervals based on the signed rank test and the sign test are relatively insensitive to outliers.

Supplementary Problems

1. Return to the data of Table 4, Chapter 4, and test for differences between the smokers and nonsmokers. Use the Wilcoxon signed rank test, and compare your results with the results you obtained for Supplementary Problem 2 of Chapter 4.

2. For the data of Table 4, Chapter 4, compute 99% confidence limits, based on the signed rank test, for the long-run average difference between smokers and controls. Compare your results with the results you obtained for Supplementary Problem 3 of Chapter 4.

3. For the data of Table 5, Chapter 4, use the Wilcoxon signed rank statistic to test the null hypothesis that stimulation of the vagus nerve has no effect on the blood level of immunoreactive insulin, against the one-sided alternative hypothesis that stimulation increases the blood level. Compare your results with those of Supplementary Problem 4 of Chapter 4.

4. For the data of Table 5, Chapter 4, use the confidence interval based on the signed rank test to determine 89% confidence limits for the mean of the "after minus before" population. Compare the interval with the 90% confidence interval obtained for Supplementary Problem 5 of Chapter 4.

5. For the data of Table 6, Chapter 4, use the Wilcoxon signed rank statistic to test the hypothesis that the mean oxidant content μ of dew water was .25 against the alternative that it was not equal to .25 (see Comment 4). Compare your results with the results you obtained for Supplementary Problem 9 of Chapter 4.

6. For the data of Table 6, Chapter 4, give a 94.8% confidence interval, based on the signed rank test, for the mean oxidant content μ (see Comment 11). Compare your results with the results you obtained for Supplementary Problem 10 of Chapter 4.

7. For the data of Table 4, Chapter 4, compute 99% confidence limits, based on the sign test (see Comment 12), for the long-run average difference between smokers and controls. Compare your results with the results you obtained for Supplementary Problem 2.

8. For the data of Table 5, Chapter 4, use the confidence interval based on the sign test (see Comment 12) to determine 89% confidence limits for the mean of the "after minus before" population. Compare this interval with the signed rank interval obtained in Supplementary Problem 4.

9. For the data of Table 4, Chapter 4, compute $\tilde{\mu}$ and $\widehat{SD}(\tilde{\mu})$ (see Comment 10). Compare these values with their parametric counterparts, namely, $\bar{d}$ and the estimated standard deviation of $\bar{d}$ [the latter is given by equation (2) of Chapter 4].

10. For the data of Table 5, Chapter 4, compute $\tilde{\mu}$ and $\widehat{SD}(\tilde{\mu})$ (see Comment 10). Compare these values with the values of $\bar{d}$ and the estimated standard deviation of $\bar{d}$ [the latter is given by equation (2) of Chapter 4].

REFERENCES

Benson, H., Shapiro, D., Tursky, B., and Schwartz, G. E. (1971). Decreased systolic blood pressure through operant conditoning techniques in patients with essential hypertension. *Science* **173**, 740–742.

Forrester, J. C. and Ury, H. K. (1969). The signed-rank (Wilcoxon) test in the rapid analysis of biological data. *Lancet*, February 1, 1969, 239–241.

*Hodges, J. L. Jr. and Lehmann, E. L. (1963). Estimates of location based on rank tests. *Ann. Math. Statist.* **34**, 598–611.

Hollander, M. and Wolfe, D. A. (1973). *Nonparametric Statistical Methods.* Wiley, New York.

Moses, L. E. (1953). Chapter 18 of Walker, H. M. and Lev, J. *Statistical Inference*, 1st edition. Holt, Rinehart and Winston, New York.

Ury, H. K. and Forrester, J. C. (1970). More on scale dependence of the Wilcoxon signed-rank test. *The American Statistician* **24(3)**, 25–26.

Wilcoxon, F. (1945). Individual comparisons by ranking methods. *Biometrics* **1**, 80–83.

Wilcoxon, F. and Wilcox, R. A. (1964). *Some Rapid Approximate Statistical Procedures*, 2nd edition. American Cyanamid Co., Lederle Laboratories, Pearl River, New York.

CHAPTER | 13

Analysis of the Two-Sample Location Problem Using Ranks

In this chapter we return to the two sample problem introduced in Chapter 5. The data consist of two independent random samples, sample 1 from population 1 and sample 2 from population 2. Let μ_1 denote the mean of population 1, and let μ_2 denote the mean of population 2. The null hypothesis asserts that the two populations are equivalent, and thus under this hypothesis $\mu_1 = \mu_2$. In Chapter 5 we tested for equality of population means using a test statistic based on the sample means. Also, the confidence interval for $\mu_2 - \mu_1$, introduced in Chapter 5, is based on the sample means. These sample means have the disadvantage of being extremely sensitive to wildly outlying observations. In this chapter we present nonparametric procedures that are relatively insensitive to outliers.

In Section 1 we present a rank test of the null hypothesis, and in Section 2, a corresponding confidence interval for $\mu_2 - \mu_1$. The test and interval do not assume the data are from normal populations, an assumption that was necessary in order for the test and interval of Chapter 5 (which used percentiles of the t distribution) to be exact.

| **1.** | **THE WILCOXON TWO-SAMPLE RANK SUM TEST** |

Measuring the Effects of Bacterial Infections on the Kidneys

In diagnosing the effects of bacterial infections on the kidneys, the diagnostic X-ray specialist looks for a thinner wall in the diseased kidney. Sometimes the outer edge of the kidney cannot be visualized well, but the inside wall can be seen easily, because the kidney is filled with radio-opaque fluid for the purpose of the radiography. In a well-known text by Williams (1958) it has been suggested that when the radiologist is faced with this difficulty he might measure the distance from the inside of the kidney wall to the edge of a well-defined point on the backbone. However, there were no data presented to substantiate the claim that this distance would tend to be decreased when the kidney is diseased. Friedland (1973) and his associates at Stanford University measured the distance (in mm) for a series of patients with diseased kidneys and for a series of patients X-rayed for kidney difficulties but proving to have normal kidneys. In Table 1 two such series of measurements are presented for 10-year-old children.

TABLE 1. Distance (in mm) from Right Upper Pole Calyx to Spine in Non-diseased and Diseased Kidneys of 10-year-olds

Nondiseased	Diseased
18	10
22	13
23	14
24	15
25	22
27	
27	
31	
34	

Source: G. Friedland, R. Filly, and B. Wm. Brown, Jr. (1974).

A Joint Ranking of the Observations

Let us denote the number of observations in the smaller sample by the letter n and the number of observations, in the sample with the larger sample size, by m. For the data in Table 1, $n = 5$ and $m = 9$. Furthermore, the population from which the sample of size m was obtained will be named population 1, and the population from which the sample of size n was obtained will be called population 2.

A comparison of the diseased and nondiseased groups can be obtained by ranking the 14 observations jointly, from least to greatest. Tied observations are assigned the average of the ranks tied for. Thus the two 22 values of Table 1, which are tied for ranks 6 and 7, each receive the average rank 6.5. The ranks of the observations of Table 1 are given in Table 2.

TABLE 2. Ranks of Distances

Nondiseased	Diseased
5	1
6.5	2
8	3
9	4
10	6.5
11.5	
11.5	
13	
14	

If the distance measured is generally smaller when the kidney is diseased, this should be reflected by the diseased group receiving smaller ranks than the nondiseased group. For our data the diseased group received ranks 1, 2, 3, 4, 6.5. The "smallness" of these ranks can be assessed via their sum. Call this statistic (the sum of the ranks obtained by the sample of size n) W. This is the Wilcoxon rank sum statistic. We have

$$W = 1 + 2 + 3 + 4 + 6.5 = 16.5.$$

This value, $W = 16.5$, appears to be smaller than one would expect if the diseased kidney and nondiseased kidney values were really from the same population. The question of whether $W = 16.5$ is "small enough" to indicate a real difference between the two populations is considered in the next subsection.

Does the Distance Tend to be Smaller when the Kidney is Diseased?

We can envision two underlying populations, namely, the population (population 1, say) of distances from nondiseased kidneys and the population (population 2) of distances associated with diseased kidneys. Then the data of Table 1 are viewed as independent random samples of sizes $m = 9$ and $n = 5$ from populations 1 and 2, respectively. Set $N = m + n = 14$.

The hypothesis H_0 of no differences between diseased and nondiseased distances can be formulated as

H_0: The diseased and nondiseased distance values can be viewed as one combined sample of size N from a single population.

That is, H_0 asserts that the two populations are identical and thus that the values in Table 1, instead of being thought of as samples of sizes 9 and 5 from populations 1 and 2, respectively, can be viewed as a single sample of size 14 from a common population.

Let μ_1 denote the mean of population 1, and let μ_2 denote the mean of population 2. Let $\Delta = \mu_2 - \mu_1$. When H_0 is true, $\mu_1 = \mu_2$, or equivalently, $\Delta = 0$. In view of our discussion in the preceding subsection, a reasonable procedure for testing H_0 is to refer the observed value of W to percentiles of the null distribution of W derived under the assumption that H_0 is true.

Table C12 gives percentiles of the null distribution of W for cases in which the sample sizes m and n are each no greater than 10. We let m denote the larger sample size and n the smaller sample size. The statistic W is the sum of the ranks obtained by the smaller sample. The lower-tail values are for testing H_0, the hypothesis of equivalent populations, against the alternative $\mu_2 < \mu_1$. The upper-tail values are for testing H_0 against the alternative $\mu_2 > \mu_1$. (For $m = 3, 4, \ldots, 7$, the table is arranged so that in each subtable the lower-tail values are presented first, with the upper-tail values beneath—and separated by a space from—the lower-tail values.) The lower-tail probability P, corresponding to a value c, is the probability, under the null hypothesis, that W will be less than or equal to c. The upper-tail probability P, corresponding to a value c, is the probability, under the null hypothesis, that W will be greater than or equal to c.

Consider the case $m = 4$, $n = 4$. Then from Table C12 we find $Pr\{W \le 13\} = Pr\{W \ge 23\} = .100$. Thus if we wish to test H_0 versus $\mu_2 < \mu_1$, our $\alpha = .100$ test is reject H_0 if $W \le 13$, accept H_0 if $W > 13$. If we wish to test H_0 versus $\mu_2 > \mu_1$, our $\alpha = .100$ test is reject H_0 if $W \ge 23$, accept H_0 if $W < 23$. A two-tailed $\alpha = 2(.100) = .200$ test of H_0 versus $\mu_1 \ne \mu_2$ is reject H_0 if $W \le 13$ or if $W \ge 23$, accept H_0 if $13 < W < 23$.

For the kidney data of Table 1, $m = 9$, $n = 5$, and we have found $W = 16.5$. From Table C12, we find that when H_0 is true, $Pr\{W \leq 17\} = .002$. Thus the one-sided P value for $W = 16.5$ is approximately .002; this can be taken as strong evidence that the distance does tend to be smaller when the kidney is diseased. (Our P value here is only approximate because when ties occur, as in the kidney data, the probabilities given in Table C12 are no longer exact.)

Large Sample Approximation to the Rank Sum Test

Set $N = m + n$. For large m and n values (in particular, for m, n values where either m or n (or both) exceeds 10, values that are not covered by Table C12) we can treat

(1)
$$W^* = \frac{W - [n(N+1)/2]}{\sqrt{\dfrac{mn(N+1)}{12}}}$$

as an approximate standard normal variable. To test H_0 refer W^* to the standard normal table (Table C2). Significantly large values of W^* indicate $\Delta > 0$; significantly small values of W^* indicate $\Delta < 0$.

For our kidney data we have $m = 9$, $n = 5$, $N = 14$, and from (1) we find

$$W^* = \frac{16.5 - [5(15)/2]}{\sqrt{\dfrac{9(5)(15)}{12}}} = -2.80.$$

Referring $W^* = -2.80$ to the standardized normal table (Table C2), we find the 'one-sided P value is .0026. This is in close agreement with the one-sided P value of .002, found (in the previous subsection) using Table C12. (Since Table C12 covers the case $m = 9$, $n = 5$, the reader may feel the use of the normal approximation here is unnecessary. The value of our calculation is that it illustrates how to use the normal approximation, and furthermore, it shows, in a specific case, how well the approximation works even for small m and n.)

Discussion Question 1.   What effect does a wildly outlying value (say a very large value in sample 2) have on W?

Comments

1. In the kidney example, the null hypothesis was that Δ, the difference between the two populations means, is zero. If, for example, we wish to test the

hypothesis that $\Delta = 4$, we simply subtract 4 from each observation in sample 2 to obtain a "modified" sample 2, and then rank all observations (those in sample 1 and those in the "modified" sample 2) jointly. Call the sum of ranks obtained by the modified sample 2 observations W and refer it to Table C12 for a P value. In general, to test the hypothesis $\Delta = \Delta_0$, where Δ_0 is a specified nonzero number, obtain the sample 2 modified observations by subtracting Δ_0 from each observation in sample 2.

2.   The Wilcoxon rank sum test controls the probability α of a Type-I error without requiring the assumption that the parent populations are normal. It is enough to assume that the two underlying populations are continuous so that the probability of obtaining tied observations is zero. (If there are ties we can use average ranks, but in such cases the probabilities given in Table C12 are approximate—not exact.) The null hypothesis H_0 asserts that the two populations are identical, but we emphasize that this common population need not be that of a normal distribution.

The Wilcoxon rank sum test, when compared with tests based on differences between sample means (see Chapter 5), has the added advantage of being relatively insensitive to outlying observations.

3.   The hypothesis test based on W was developed by Wilcoxon (1945) in the same pioneering article where he proposed the signed rank test (discussed in Chapter 12).

PROBLEMS

1.   Moorman, Chmiel, Stara, and Lewis (1973) have compared decomposition of ozone (O_3) in the nasopharynx of physiologically unaltered beagles under conditions of acute and chronic exposure. Table 3 gives tracheal O_3 concentrations for the group that received a daily concentration × time regimen of 1 ppm × 8 hours chronic (18 months) exposure and the group that received 1 ppm O_3 acute exposure. [The concentration of O_3 in the trachea was analyzed via a small Teflon catheter; see Moorman *et al.* (1973) for details.] The data of Table 3 do not appear in the Moorman *et al.* article but were kindly furnished by the authors.

One goal of the study was to compare the removal efficiency of chronically exposed dogs with that of previously unexposed dogs. Is there evidence that the mean of the underlying acute population is less than the corresponding mean of the chronic population (possibly indicating that acutely exposed subjects have better upper respiratory mechanisms for removing reactive gases than do chronically exposed subjects)? (Compare your results with those obtained in Problem 5, Chapter 5, where the analysis was based on sample means.)

TABLE 3. Tracheal Concentrations of Ozone in Beagles

Chronic Exposure	Acute Exposure
0.010	0.010
0.010	0.015
0.018	0.010
0.015	0.005
0.015	
0.010	
0.019	
0.015	

Source: W. J. Moorman, J. J. Chmiel, J. F. Stara, and T. R. Lewis (1973).

2. Consider the crude lipid composition data of Table 1, Chapter 5. Test the hypothesis that the mean of the control population equals the mean of the 1000 mg population. Use the Wilcoxon rank sum test, and compare your result with the analysis of Section 2, Chapter 5, based on the T' statistic.

3. Let W' denote the sum of the ranks, obtained by the observations in sample 1, when the N observations in sample 1 and sample 2 are ranked jointly. Show, or illustrate via an example, that $W + W' = [N(N+1)]/2$.

---

2.   A NONPARAMETRIC CONFIDENCE INTERVAL BASED ON THE RANK SUM TEST

What is the Estimated Difference between the Mean Distances for Non-diseased and Diseased Kidneys?

In Section 1 we found evidence that the population mean μ_2 corresponding to distances measured for diseased kidneys is significantly less than its counterpart μ_1, for the nondiseased population. We now describe how to obtain a confidence interval, associated with the W statistic, for the difference $\Delta = \mu_2 - \mu_1$.

For each of the $mn = 9(5) = 45$ ways we can pair a sample 1 observation with a sample 2 observation, compute a difference d of the form

$$d = [\text{sample 2 observation}] - [\text{sample 1 observation}].$$

For the data of Table 1, the 45 differences are

$10-18=-8$	$13-23=-10$	$14-25=-11$	$15-27=-12$
$10-22=-12$	$13-24=-11$	$14-27=-13$	$15-31=-16$
$10-23=-13$	$13-25=-12$	$14-27=-13$	$15-34=-19$
$10-24=-14$	$13-27=-14$	$14-31=-17$	$22-18=4$
$10-25=-15$	$13-27=-14$	$14-34=-20$	$22-22=0$
$10-27=-17$	$13-31=-18$	$15-18=-3$	$22-23=-1$
$10-27=-17$	$13-34=-21$	$15-22=-7$	$22-24=-2$
$10-31=-21$	$14-18=-4$	$15-23=-8$	$22-25=-3$
$10-34=-24$	$14-22=-8$	$15-24=-9$	$22-27=-5$
$13-18=-5$	$14-23=-9$	$15-25=-10$	$22-27=-5$
$13-22=-9$	$14-24=-10$	$15-27=-12$	$22-31=-9$
			$22-34=-12$

Next, order these 45 differences from least to greatest. The ordered values are $-24, -21, -21, -20, -19, -18, -17, -17, -17, -16, -15, -14, -14,$ $-14, -13, -13, -13, -12, -12, -12, -12, -12, -11, -11, -10, -10,$ $-10, -9, -9, -9, -9, -8, -8, -8, -7, -5, -5, -5, -4, -3, -3, -2, -1, 0,$ 4.

The lower end point of the confidence interval is the d value, in the foregoing list of *ordered d's*, that occupies the lth position, where l is an integer, the determination of which is about to be explained. The upper end point of the confidence interval is the d value that occupies position $mn+1-l$ in the list of the *ordered d's*.

The integer l depends on the desired confidence coefficient $1-\alpha$ and can be obtained from the equation:

(2) $$l=\frac{n(2m+n+1)}{2}+1-[\text{upper }\frac{\alpha}{2}\text{ percentile point of the}$$

null distribution of W].

For example, consider the kidney data, and suppose we seek a confidence coefficient of 88.8%; then $1-\alpha=.888$, $\alpha=.112$, and $\alpha/2=.056$. From Table

C12, entered at $m = 9$, $n = 5$ (upper tail), we find the upper .056 percentile point of the null distribution of W to be 50. Thus, from (2),

$$l = \frac{5(18+5+1)}{2} + 1 - 50 = 11.$$

Then the 88.8% confidence interval for Δ has as its lower end point the value occupying position 11 in the list of the ordered d's and as its upper end point the value occupying position 35. The interval is thus found to be $(-15, -7)$. In addition to providing an interval estimate of Δ, this interval also has a hypothesis-testing interpretation. Since the interval does not contain the point 0, we know that the two-tailed $\alpha = .112$ test based on W leads to rejection of H_0.

For large m and n values (in particular, for m, n values where either m or n is larger than 10, values that are not covered by Table C12), the integer l can be approximated by:

(3)
$$l \approx \frac{mn}{2} - z_{\alpha/2}\sqrt{\frac{mn(N+1)}{12}}$$

where $z_{\alpha/2}$ is the upper $\alpha/2$ percentile point of the standard normal distribution. The value of l, found via (3), is usually not an integer, thus use the integer closest to l.

Discussion Question 2. For confidence intervals with very large (i.e., very close to 1) confidence coefficients, it is not necessary to compute all of the mn d-differences to obtain the interval end points. Why?

Comments

4. If one desires a point estimator $\tilde{\Delta}$ of the parameter $\Delta = \mu_2 - \mu_1$, a good choice is the median of the mn d-differences. In our kidney example, the median of the 45 d-differences is the twenty-third value in the list of the ordered d's. We thus find

$$\tilde{\Delta} = -11.$$

The standard deviation of the estimator $\tilde{\Delta}$ can be approximated by $\widehat{SD}(\tilde{\Delta})$ where

$$\widehat{SD}(\tilde{\Delta}) = \frac{\{\text{length of the } 1-\alpha \text{ confidence interval for } \Delta\}}{2(z_{\alpha/2})}.$$

In our example, where we set $1-\alpha = .888$, we have

$$\widehat{SD}(\tilde{\Delta}) = \frac{-7-(-15)}{2(1.59)} = \frac{8}{3.18} = 2.52.$$

5. Consider the *mn* *d*-differences used to obtain the confidence interval for $\mu_2 - \mu_1$. Let U denote the number of these differences that are positive. Then (when there are no tied observations) it can be shown that $W = U + [n(n+1)/2]$. The statistic U is called the Mann–Whitney U statistic.

6. The reader has probably noted that many of the ranking and sorting tasks involved with applying the Wilcoxon rank sum test, and obtaining the associated confidence interval, are tedious when *m* and *n* are large. Such tasks are appropriate (and easily programmed) for a computer.

7. The nonparametric confidence interval based on the rank sum test is valid even if the data are not from normal populations. The confidence coefficient is exact when the underlying populations are continuous (so that the probability of tied observations is zero) and population 2 is of the same shape as population 1 but shifted (by the amount Δ) relative to population 1, as indicated in Figure 1.

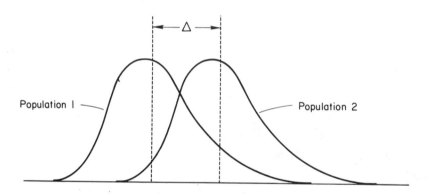

FIGURE 1. An example where population 2 has the same shape as population 1, but is shifted to the right by the amount Δ.

Furthermore, the nonparametric confidence interval is less sensitive to outliers than its competitor based on sample means (see Chapter 5). Recall also that the W test requires less information than the tests presented in Chapter 5; we need only know the ranks of the observations rather than their actual values. Note, however, that actual values are needed to obtain the confidence interval based on the W test.

8. The confidence interval for $\mu_2 - \mu_1$, based on the *mn* *d*-differences, was described graphically by Moses (1953). The point estimator $\tilde{\Delta}$, defined in Comment 4, was proposed by Hodges and Lehmann (1963). More details on these procedures can be found in Chapter 4 of Hollander and Wolfe (1973).

Summary of Statistical Concepts

In this chapter the reader should have learned:
(i)   The two sample problem, analyzed in Chapter 5 using the difference between sample means, can also be treated using ranks.
(ii)   The Wilcoxon rank sum test and the corresponding confidence interval have desirable properties when compared with the parametric tests discussed in Chapter 5. These properties include the availability of exact probabilities of Type-I errors and exact confidence coefficients under very mild assumptions (concerning the underlying population), and relative insensitivity to outlying observations.

PROBLEMS

4.   For the data of Table 3, determine a 95.2% nonparametric confidence interval for Δ. Compare your result with the 95% interval of Problem 2, Chapter 5.

5.   Consider the crude lipid composition data of Table 1, Chapter 5. Determine a 94.2% nonparametric confidence interval for Δ. Compare your result with the 95% interval obtained by formula (8) of Chapter 5.

6.   Table 4 is based on data of W. Bacon and can be found in van Belle (1971). The data are nuclear volumes of 8 cells of *H. surattensis* and 8 cells of the F_2 hybrid *H. surattensis sudanensis*. Do these data provide evidence that the nuclear volumes between strains differ?

TABLE 4.   Nuclear Volumes (in cubic microns) of *H. surattensis* and *H. surattensis sudanensis* F_2

H. sur	*H. sur-sud* F_2
144	361
91	216
216	728
307	421
216	1,156
91	258
339	421
263	144

Source: W. Bacon (1971).

7. Suppose, for the data of Table 1, we require a 99.8% confidence interval. Obtain the nonparametric interval described in this section. Can you determine this interval *without* actually computing *all* of the 45 d-differences?

8. Verify directly, or illustrate via an example, the assertion of Comment 5.

9. Construct an example that illustrates that the nonparametric confidence interval of this section is relatively insensitive to outliers.

10. For the data of Table 4, compute $\tilde{\Delta}$ and $\widehat{SD}(\tilde{\Delta})$ (see Comment 4).

Supplementary Problems

1. Return to the data of Table 2, Chapter 1, and use the rank sum statistic W to test for differences between the active-exercise group and the 8-week control group. Compare your results with those you obtained for Supplementary Problem 5 of Chapter 5.

2. For the data of Table 1, Chapter 5, compute $\tilde{\Delta}$ and $\widehat{SD}(\tilde{\Delta})$. Compare these values with $\hat{\Delta}$ and the estimated standard deviation of $\hat{\Delta}$ [defined by equations (1) and (5) respectively of Chapter 5, and calculated in Chapter 5].

3. Return to the data of Table 2, Chapter 5, and use the large sample approximation to the distribution of W to test for a possible effect of the different levels of protein on refractive characteristics of the right eye. Compare your results with those you obtained for Supplementary Problem 1 of Chapter 5.

4. Return to the data of Table 4, Chapter 5, and use W to test for differences between population means. Compare your results with those of Supplementary Problem 3, Chapter 5.

5. Return to the data of Table 4, Chapter 5, and obtain a 99.2% confidence interval for $\mu_2 - \mu_1$. Compare this interval with the 99% interval obtained for Supplementary Problem 3 of Chapter 5.

6. Return to the data of Table 5, Chapter 5, and use the large sample approximation to the distribution of W to test for differences in uptake rates for healthy volunteers and patients with rheumatoid arthritis. Compare your results with those obtained in Supplementary Problem 4 of Chapter 5.

7. For $m = 6$, $n = 6$, compare the one-sided $\alpha = .047$ test (of H_0 versus $\mu_2 > \mu_1$) based on W with the large sample approximation to the test based on W^*.

8. For $m = 10$, $n = 10$, compare the one-sided $\alpha = .045$ test (of H_0 versus $\mu_2 > \mu_1$) based on W with the large sample approximation to the test based on W^*. What do the results of this problem and Supplementary Problem 7 suggest about the accuracy of the large sample approximation as m and n increase?

9. For $m = 7$, $n = 6$, $\alpha = .022$, compare the value of l given by equation (2) with the approximate value given by equation (3).

10. For $m = 10$, $n = 9$, $\alpha = .022$, compare the value of l given by equation (2) with the approximate value given by equation (3). What do the results of this problem and Supplementary Problem 9 suggest about the accuracy, of the approximation given by (3), as m and n increase?

REFERENCES

Bacon, W. (1971). Personal communication.

Friedland, G. W., Filly, R., and Brown, B. Wm., Jr. (1974). Distance of upper pole calyx to spine and lower pole calyx to ureter, as indicators of parenchymal loss in children. *Pediat. Radiol.* **2**, 29–37.

*Hodges, J. L., Jr. and Lehmann, E. L. (1963). Estimates of location based on rank tests. *Ann. Math. Statist.* **34**, 598–611.

Hollander, M. and Wolfe, D. A. (1973). *Nonparametric Statistical Methods.* Wiley, New York.

Moorman, W. J., Chmiel, J. J., Stara, J. F., and Lewis, T. R. (1973). Comparative decomposition of ozone in the nasopharynx of beagles: Acute vs. chronic exposure. *Arch. Environ. Health* **26**, 153–155.

Moses, L. E. (1953). Chapter 18 of Walker, H. M. and Lev, J. *Statistical Inference*, 1st edition. Holt, Rinehart and Winston, New York.

Van Belle, G. (1971). Design and analysis of experiments in biology: Notes for a short course. Statistical Consulting Center Report. Florida State University.

Wilcoxon, F. (1945). Individual comparisons by ranking methods. *Biometrics* **1**, 80–83.

Wilcoxon, F. and Wilcox, R. A. (1964). *Some Rapid Approximate Statistical Procedures*, 2nd edition. American Cyanamid Co., Lederle Laboratories, Pearl River, New York.

Williams, D. I. (1958). *Urology in Childhood.* Springer-Verlag, Berlin.

CHAPTER 14

Methods for Censored Data

When a clinical trial is performed it is often necessary to admit the patients to the study one by one, as they come to the clinic for diagnosis and treatment. As each patient is diagnosed and admitted to the study, he is randomly assigned to one of the treatments that are to be compared. Then the patients are followed to the observation of some end point, for example, death. The treatments are compared to see which treatment resulted in longer survival. In nonfatal diseases, such as certain infections, the time required to cure the patient might be the point of major interest.

In such clinical trials, where observation is time to the occurrence of an end-point event, the data usually are analyzed before all patients have experienced the event. For example, in a study of the length of survival, many patients in the study may be alive at the time of analysis. For such a survivor, all that is known is that his survival time is at least as great as the survival time thus far. Incomplete information also occurs when a patient is lost to study (such as by moving or dying in an auto accident), having survived a known amount of time. These kinds of situations yield incomplete observations called **censored observations**.

This chapter introduces the **survival curve,** and shows how it is calculated in the presence of censored observations. Then a test procedure for comparing survival curves (formally, for testing the hypothesis of the equivalence of two treatments) is introduced. The test is an extension of the Wilcoxon rank sum test of Chapter 13 to samples containing censored observations.

TABLE 1.   Relapse-free Survival Time for Hodgkin's Disease Patients

Radiation of Affected Node		Total Nodal Radiation	
Relapse (Y) or not (N)	Days to Relapse or to Date of Analysis	Relapse (Y) or Not (N)	Days to Relapse or to Date of Analysis
Y	346	N	1699
Y	141	N	2177
Y	296	N	1968
N	1953	N	1889
Y	1375	Y	173
Y	822	N	2070
N	2052	N	1972
Y	836	N	1897
N	1910	N	2022
Y	419	N	1879
Y	107	N	1726
Y	570	N	1807
Y	312	Y	615
N	1818	Y	1408
Y	364	N	1763
Y	401	N	1684
N	1645	N	1576
Y	330	N	1572
N	1540	Y	498
Y	688	N	1585
N	1309	N	1493
Y	505	Y	950
N	1378	N	1242
N	1446	N	1190
Y	86		

Source: H. S. Kaplan and S. A. Rosenberg (1973).

A Clinical Trial in Hodgkin's Disease

Hodgkin's disease is a cancer of the lymph system. It starts in a lymph node, frequently in the neck, and if untreated, will spread to other nodes and to vital organs, eventually causing death. When diagnosed early, the affected node can be treated with radiation, and the signs and symptoms will disappear, for at least some months or years. The patient may eventually suffer a relapse, requiring further treatment, and if this happens his ultimate fate will be problematical.

At the Stanford Medical Center, Drs. H. S. Kaplan and S. A. Rosenberg initiated a clinical trial in early Hodgkin's disease (i.e., cases in which the disease was detected at an early stage), comparing the standard radiation treatment of the affected node with treatment of the affected node plus all nodes in the trunk of the body. Their theory was that this "total nodal" radiation would cure disease in the affected node and any undetected disease that had already spread from the node.

Patients were randomly assigned to the single node and the total node therapies. Over the course of 3 years (1967–1970), 49 patients were admitted to the study. These original 49 patients were subsequently followed and the data as of the fall of 1973 are presented in Table 1. (Starting in the spring of 1970, all further patients were assigned to the total nodal therapy, because at that time the relapse rate for this therapy appeared to be much better than for the standard therapy. Our analysis in this chapter sheds further light on whether the 1970 decision was well justified.)

From Table 1 it can be seen that by the fall of 1973 there had been 16 relapses among the 25 patients given the standard therapy, and only 5 relapses among the patients receiving the total nodal therapy. Thus the total nodal therapy does seem more effective in delaying or preventing relapse, but the varying lengths of observation for the patients confuses the picture and makes a judgment on statistical significance difficult.

Relapse-free Survival Curves

Censored observations in clinical trials can be displayed in terms of a function called the survival curve. If we think of a large group of patients followed from diagnosis and treatment to their dates of relapse, without censoring, we could plot the proportion surviving without relapse against time. We would get a curve that starts at 100% at time zero and decreases as various patients relapse. The survival curve has the following interpretation. Let X denote the relapse-free time of a patient selected at random from the population under consideration. Then for each t, the survival curve provides an estimate of $Pr(X > t)$. [For convenience, we will let $\bar{P}(t)$ denote $Pr(X > t)$.]

We would like to obtain relapse-free survival curves for the data in Table 1. For example, if X denotes the relapse-free time in days of a patient selected at random from the conceptual population of those early Hodgkin's disease patients to be treated by the affected node therapy, how should we estimate $Pr(X > 1800)$? A natural estimate of this probability would be obtained by counting the number of Table 1 patients (out of the total of 25 receiving the affected node therapy) that had relapse-free times that exceeded 1800, and then dividing this count by 25. Unfortunately, we cannot tell, for example, whether patient 21 will survive past 1800 days without a relapse. We only know that the actual days to relapse for patient 21 must exceed the censored value of 1309. Thus the censoring causes difficulties. However, these difficulties can be overcome. We describe shortly how to calculate a survival curve that reflects precisely the survival experience of the patients as they were observed, by first estimating conditional probabilities of relapse and then combining these to obtain estimates of the probability of escaping relapse.

In Table 2, the data of Table 1, for patients receiving radiation of affected nodes only, have been re-ordered, according to the number of days the patient survived or has been under observation since admission to the study. The second column shows whether the length of time was determined by the relapse of the patient or by censoring (i.e., whether or not the patient had relapsed by the time of analysis). The third column shows the number of patients still in the study and surviving free of relapse to the time indicated. The fourth column is the relapse rate at the indicated time: the number of relapses (0 or 1) divided by the number of patients in the study at the time (column three). Thus at 330 days, the relapse rate is $1/20$ or .05; at 1309 days the rate is $0/10$ or 0 since the time is a time of censoring for a patient and not a time of relapse. These ratios are estimates of conditional probabilities, the condition being survival to the specified time point. The last column gives the rate of relapse-free survival and is computed by multiplying the conditional probabilities of having no relapse, for all times from zero up to the time of interest. For example, the probability of having no relapse by the 296th day is calculated as $(1 - 1/25)(1 - 1/24)(1 - 1/23)(1 - 1/22)$ or .840. Notice that this method adjusts for losses of information due to censoring by estimating relapse rates at the specified times, based on those patients who have been followed long enough to contribute data. [This method is due to Kaplan and Meier (1958) and we often call the curve the Kaplan–Meier survival curve.]

Figure 1 shows the relapse-free survival curve calculated in Table 2 and also the curve for the patients receiving radiation of all nodes. The difference in the two

TABLE 2. Calculation of the Survival Curve from Censored Data for the Hodgkin's Patients Receiving Radiation of Affected Nodes Only

Time to Relapse or to Date of Analysis	Relapse (Y) or Not (N)	Patients at Risk	Estimated Conditional Probability of Relapse on Date, Given that the Patient has not Yet Relapsed	Estimated Probability of no Relapse to Date
86	Y	25	1/25	0.960
107	Y	24	1/24	0.920
141	Y	23	1/23	0.880
296	Y	22	1/22	0.840
312	Y	21	1/21	0.800
330	Y	20	1/20	0.760
346	Y	19	1/19	0.720
364	Y	18	1/18	0.680
401	Y	17	1/17	0.640
419	Y	16	1/16	0.600
505	Y	15	1/15	0.560
570	Y	14	1/14	0.520
688	Y	13	1/13	0.480
822	Y	12	1/12	0.440
836	Y	11	1/11	0.400
1309	N	10	0	0.400
1375	Y	9	1/9	0.356
1378	N	8	0	0.356
1446	N	7	0	0.356
1540	N	6	0	0.356
1645	N	5	0	0.356
1818	N	4	0	0.356
1910	N	3	0	0.356
1953	N	2	0	0.356
2052	N	1	0	0.356

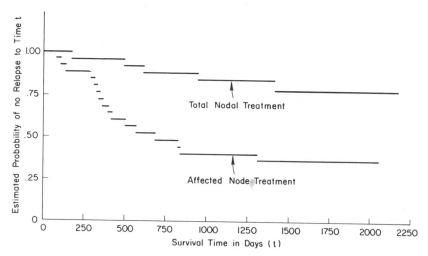

FIGURE 1. Kaplan–Meier curves for total nodal treatment and affected node treatment.

curves is quite striking. The estimated probability of relapse-free survival to any specified time is markedly higher for the group receiving total nodal radiation. However, these are estimated probabilities, based on rather small samples, and there is a question as to whether the two treatments might be equally effective, with the observed difference merely a result of the imbalances that can occur through random allocation of patients to treatments.

A Significance Test for Censored Data

Suppose our data consist of two independent random samples, sample 1 from population 1 and sample 2 from population 2. For cases where one or both samples contain censored observations, Gehan (1965) developed a simple significance test of the hypothesis that the two populations are the same. The test is performed by comparing every observation in the first sample with every observation in the second sample. For the data in Table 1 there are 25 observations in the first sample and 24 in the second, or $24 \times 25 = 600$ comparisons to make. For each comparison, score $+1$ if the second treatment (total nodal radiation) definitely "wins" (i.e., has a longer relapse-free survival), score -1 if the first treatment definitely wins, and score a 0 if censoring makes it impossible to say which treatment will win for a given pair. For example, the twenty-first patient in the affected node therapy group definitely wins against the fifth patient in the total nodal radiation group because the *actual* days to relapse for patient 21 in the affected node therapy group must exceed the given censored value of 1309, whereas the *actual* days to relapse for patient 5 in the total nodal radiation group is (the uncensored value) 173. Hence we score -1 for this comparison because the affected node therapy won. As another example, consider the comparison between patient 21 of the affected node therapy group and patient 6 of the total nodal radiation group. Here we observe two censored values, 1309 and 2070, and we cannot tell which patient will have the longer relapse-free survival. Thus for this comparison we score 0.

The scores for the 600 contests, computed using this scoring method, are presented in Table 3. The sum of the scores is 291. If the two treatments are

TABLE 3. Gehan Scores for Comparing the Two Censored Data Samples of Table 1

Patients with Total Nodal Radiation

	1	2	3	4	5	6	7	8	9	10	11	12	13	14	15	16	17	18	19	20	21	22	23	24
1	a				−1																			
2																								
3					−1																			
4	0	0	0	0	−1	0	0	0	0	0	0	0	−1	−1	0	0	0	0	−1	0	0	−1	0	0
5					−1								−1						−1			−1	0	0
6					−1								−1						−1					
7	0	0	0	0	−1	0	0	0	0	0	0	0	−1	−1	0	0	0	0	−1	0	0	−1	0	0
8					−1								−1						−1					
9	0	0	0	0	−1	0	0	0	0	0	0	0	−1	−1	0	0	0	0	−1	0	0	−1	0	0
10					−1																			
11																								
12					−1														−1					
13					−1																			
14	0	0	0	0	−1	0	0	0	0	0	0	0	−1	−1	0	0	0	0	−1	0	0	−1	0	0
15					−1																			
16					−1																			
17	0	0	0	0	−1	0	0	0	0	0	0	0	−1	−1	0	0	0	0	−1	0	0	−1	0	0
18					−1																			
19	0	0	0	0	−1	0	0	0	0	0	0	0	−1	−1	0	0	0	0	−1	0	0	−1	0	0
20					−1								−1						−1					
21	0	0	0	0	−1	0	0	0	0	0	0	0	−1	0	0	0	0	0	−1	0	0	−1	0	0
22					−1														−1					
23	0	0	0	0	−1	0	0	0	0	0	0	0	−1	0	0	0	0	0	−1	0	0	−1	0	0
24	0	0	0	0	−1	0	0	0	0	0	0	0	−1	−1	0	0	0	0	−1	0	0	−1	0	0
25																								
Sum	16	16	16	16	−19	16	16	16	16	16	16	16	−1	9	16	16	16	16	−5	16	16	5	15	15

Grand sum $= V = 291$.

[a] The 1 entries are deleted for easier reading.

equally effective, that is, if the null hypothesis is true, then this sum, denoted by V, has a distribution with mean 0 and standard deviation

$$(1) \qquad SD(V) = \sqrt{\frac{mn \sum h^2}{(m+n)(m+n-1)}},$$

where m and n are the sample sizes for treatments 1 and 2, respectively, and $\sum h^2$ is calculated as follows. First, rank all the observations together in one list, noting for each observation its sample membership and whether or not it is a censored observation. This has been done in Table 4 in the first three columns. Now, for each patient count the number of (other) patients whose days to relapse are *definitely* less than that patient's days to relapse. Call this count R_1. Similarly, for each patient count the number of (other) patients whose days to relapse are *definitely* greater than that patient's days to relapse. Call this count R_2. Then

TABLE 4. Calculation of the Gehan Sum of Scores and Its Standard Deviation

Treatment	Observation	Relapse (Y) or not (N)	R_1	R_2	h
1	86	Y	0	48	−48
1	107	Y	1	47	−46
1	141	Y	2	46	−44
2	173	Y	3	45	−42
1	296	Y	4	44	−40
1	312	Y	5	43	−38
1	330	Y	6	42	−36
1	346	Y	7	41	−34
1	364	Y	8	40	−32
1	401	Y	9	39	−30
1	419	Y	10	38	−28
2	498	Y	11	37	−26
1	505	Y	12	36	−24
1	570	Y	13	35	−22
2	615	Y	14	34	−20
1	688	Y	15	33	−18
1	822	Y	16	32	−16
1	836	Y	17	31	−14
2	950	Y	18	30	−12
2	1190	N	19	0	+19

TABLE 4—continued

Treatment	Observation	Relapse (Y) or not (N)	R_1	R_2	h
2	1242	N	19	0	+19
1	1309	N	19	0	+19
1	1375	Y	19	26	−7
1	1378	N	20	0	+20
2	1408	Y	20	24	−4
1	1446	N	21	0	+21
2	1493	N	21	0	+21
1	1540	N	21	0	+21
2	1572	N	21	0	+21
2	1576	N	21	0	+21
2	1585	N	21	0	+21
1	1645	N	21	0	+21
2	1684	N	21	0	+21
2	1699	N	21	0	+21
2	1726	N	21	0	+21
2	1763	N	21	0	+21
2	1807	N	21	0	+21
1	1818	N	21	0	+21
2	1879	N	21	0	+21
2	1889	N	21	0	+21
2	1897	N	21	0	+21
1	1910	N	21	0	+21
1	1953	N	21	0	+21
2	1968	N	21	0	+21
2	1972	N	21	0	+21
2	2022	N	21	0	+21
1	2052	N	21	0	+21
2	2070	N	21	0	+21
2	2177	N	21	0	+21

V = sum of the h values for sample 2 = 291.

$\sum h^2$ = sum of the squares of the h values for both samples = 31,512.

$$SD(V) = \sqrt{(24)(25)(31,512)/(49)(48)} = \sqrt{8038.78} = 89.66.$$

$$V^* = \frac{V}{SD(V)} = \frac{291}{89.66} = 3.25.$$

P=0.0012 (two-tailed; normal approximation.)

denote the difference by $h = R_1 - R_2$. The values R_1, R_2 and h are shown in Table 4 for each of the 49 patients. It can be shown that the sum of the h values *for sample* 2 is the Gehan sum of scores V. Now, $\sum h^2$ is obtained by summing the squares of *all* the h values, $m + n$ values altogether.

A formal (approximate) significance test is performed by referring

$$(2) \qquad\qquad V^* = \frac{V}{SD(V)}$$

to the standard normal table (Table C2). Significantly large values of V^* indicate μ_2, the mean of the treatment-2 time-to-relapse population, is greater than μ_1, the mean of the treatment-1 time-to-relapse population. Significantly small values of V^* indicate $\mu_2 < \mu_1$.

The calculations are shown in Table 4. Note that the two-tailed P value is .0012, indicating that the data are in strong disagreement with the null hypothesis. The conclusion that total nodal radiation is more effective than radiation of the affected nodes in preventing or at least delaying the recurrence of early stage Hodgkin's disease seems well justified.

Discussion Question 1.   Can you criticize the scoring system used by Gehan in terms of the manner in which it utilizes the information available in the data?

Comments

1.   Recall our Table 2 calculation of the Kaplan–Meier survival curve, for the Hodgkin's patients receiving radiation of affected nodes only. We did not have to account for ties in the "time to relapse or to date of analysis" variable, since there were no ties. The following, more general method for describing the Kaplan–Meier survival curve, also covers situations in which ties occur.

Suppose observations are available on n individuals. Let $Z_{(1)} \le Z_{(2)} \le \cdots \le Z_{(n)}$ denote the ordered values of the observations. That is, put the censored and uncensored observations in one list, in order, as in Table 2. Then $\hat{P}(t)$, the estimated probability of having no failure (in Table 2 "failure" meant relapse) by time t is

$$(3) \qquad\qquad \hat{P}(t) = \prod_{Z_{(i)} \le t} \left(\frac{n-i}{n-i+1} \right).$$

The symbol $\prod_{Z_{(i)} \le t}$ in (3) means that the fractions $(n-i)/(n-i+1)$, corresponding to all subscripts i that denote *uncensored* $Z_{(i)}$ values that are less than or equal to t, are to be multiplied together. If censored observations are tied with uncensored ones, the convention we use here when forming the list of the ordered $Z_{(i)}$'s is to treat the uncensored observations as being less than the censored observations. If the largest observation $Z_{(n)}$ is censored, then the product in (3) will not reach 0 as t gets arbitrarily large. In this case, that is, for those t values greater than $Z_{(n)}$, the estimate, of the probability of no failure by time t, is undefined.

The reader should find it instructive to return to the Hodgkin's Disease data and check that the formula given by (3) leads to the same estimated probabilities that we obtained in Table 2.

To illustrate the use of (3), and the ties convention, we consider the following example. Suppose in a clinical trial the survival times (in months) $Z_{(1)} \leq Z_{(2)} \leq \cdots \leq Z_{(7)}$ of seven patients are 4, 5, 5^c, 6^c, 7, 7^c, 9. Here 5^c denotes that the corresponding patient's survival time is incomplete but is known to be at least 5 months. Following our ties convention, we list the 5 before the 5^c, and for use in (3) we consider 5 to be less than 5^c. Then $\hat{P}(4)$, the estimated probability of surviving beyond 4 months, is found from (3) to be

$$\hat{P}(4) = \frac{7-1}{7-1+1} = \frac{6}{7} = .857.$$

From (3) we also find

$$\hat{P}(5) = \left(\frac{7-1}{7-1+1}\right) \cdot \left(\frac{7-2}{7-2+1}\right) = \left(\frac{6}{7}\right)\left(\frac{5}{6}\right) = .714.$$

Since the product $\prod_{Z_{(i)} \leq t}$ in (3) is over subscripts i corresponding to uncensored observations, we note

$$\hat{P}(6) = \hat{P}(5) = .714.$$

In fact, (3) shows that $\hat{P}(t)$ stays constant between consecutive censored observations and changes (decreases) only when t increases to an uncensored observation.
Continuing our calculation we find

$$\hat{P}(7) = \left(\frac{7-1}{7-1+1}\right) \cdot \left(\frac{7-2}{7-2+1}\right) \cdot \left(\frac{7-5}{7-5+1}\right) = \left(\frac{6}{7}\right)\left(\frac{5}{6}\right)\left(\frac{2}{3}\right) = .476.$$

$$\hat{P}(9) = \left(\frac{7-1}{7-1+1}\right) \cdot \left(\frac{7-2}{7-2+1}\right) \cdot \left(\frac{7-5}{7-5+1}\right) \cdot \left(\frac{7-7}{7-7+1}\right) = 0.$$

Figure 2 is a graph of $\hat{P}(t)$.

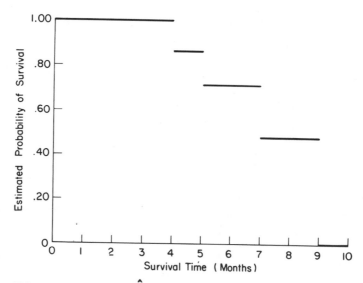

FIGURE 2. A plot of $\hat{\bar{P}}(t)$ for the example where the survival pattern is 4, 5, 5^c, 6^c, 7, 7^c, 9.

2. The standard deviation of $\hat{\bar{P}}(t)$ can be estimated from the data. The estimator, given by Kaplan and Meier (1958), is

(4)
$$\widehat{SD}(\hat{\bar{P}}(t)) = \hat{\bar{P}}(t)\sqrt{\sum_{Z_{(i)}\le t}\left[\frac{1}{(n-i)(n-i+1)}\right]}$$

where $Z_{(1)}\le Z_{(2)}\le\cdots\le Z_{(n)}$ denote the ordered observations, and the summation $\sum_{Z_{(i)}\le t}$ is over those positive integers i for which $Z_{(i)}\le t$ and $Z_{(i)}$ is an uncensored observation.

To illustrate the use of expression (4), we consider the clinical trial data of Comment 1. Then, for example, the estimated standard deviation of $\hat{\bar{P}}(7)$ is

$$\widehat{SD}(\hat{\bar{P}}(7)) = \hat{\bar{P}}(7)\sqrt{\frac{1}{(7-1)(7-1+1)} + \frac{1}{(7-2)(7-2+1)} + \frac{1}{(7-5)(7-5+1)}}$$

$$= (.476)\sqrt{\frac{1}{42} + \frac{1}{30} + \frac{1}{6}} = .23.$$

3. When there are no censored observations, Gehan's test reduces to the Wilcoxon two-sample test presented in Chapter 13. We do not prove this here but note that in the terminology of this chapter the number of $+1$ scores equals the statistic U defined in Comment 5 of Chapter 13.

4. Gehan's (1965) expression for $SD(V)$ is slightly more complicated than as given in equation (1). The form given in equation (1) is due to Mantel (1967).

Summary of Statistical Concepts

In this chapter the reader should have learned:

(i) In studies in which the variable being measured is time to a specific event, censored observations can arise because some subjects may be lost to the study and because some subjects may not yet have experienced the event when the data are analyzed.

(ii) In the presence of censored data, the Kaplan–Meier curve provides estimates, for times t, of the probability of "surviving" past t.

(iii) Gehan's test is a generalization of Wilcoxon's two-sample rank sum test and is appropriate when the data in one or both samples contain censored observations.

PROBLEMS

1. Early in the course of the Stanford Medical School heart transplantation program it became clear that the development of arteriosclerosis (accumulation of fatty deposits in the arteries) was significantly accelerated in the coronary arteries of patients receiving new hearts. As a result, a program of prevention of this complication was initiated, including diet, anticholesterolemic drugs, and other measures. Table 5, provided by Dr. R. B. Griepp (1974), shows the data on the first eight patients, transplanted before the initiation of the new program, and the first fourteen patients under the new program. The table contains only those patients surviving the first 3 months following transplant, since there is little time for a therapeutic program to manifest efficacy in the first 3 months. Note that the times shown are the times at which arteriosclerosis first manifested itself, through disease or death. Compute and graph the Kaplan–Meier survival curve for each group, showing the probability of surviving without manifest arteriosclerosis. Also compute the Gehan test statistic and obtain the P value for the hypothesis that the survival curves are really the same before and after initiation of the prevention program.

2. In a clinical trial designed to evaluate the efficacy of a new drug thought to be of value in the treatment of patients with a particular type of serious liver disease, Dr. P. B. Gregory (1974) of Stanford University obtained the data shown in Table 6. It can be seen that the group of patients randomly allocated to the new drug has experienced a higher death rate than the control group. Is the difference statistically significant by the Gehan significance

TABLE 5. Development of Arteriosclerosis Among Heart Transplant Patients, Before and After Initiation of an Arteriosclerosis Prevention Program

	Date of Operation	Date of Last Information; No Arteriosclerosis	Date Arteriosclerosis Evident
Group A:			
1	8/31/68		5/17/70
2	9/9/68		1/14/69
3	10/26/68		7/7/72
4	11/22/68		8/29/69
5	2/8/69		11/29/71
6	4/14/69		4/13/71
7	5/22/69	4/1/74	
8	7/16/69		11/29/69
Group B:			
1	1/3/70	4/1/74	
2	1/16/70	4/1/74	
3	7/4/70	4/1/74	
4	10/15/70	4/1/74	
5	1/11/71		10/1/73
6	3/22/71	4/1/74	
7	4/24/71		1/2/72
8	5/8/71		10/21/73
9	8/11/71		1/5/72
10	10/13/71		8/30/72
11	11/8/71	4/1/74	
12	12/15/71	4/1/74	
13	1/7/72	4/1/74	
14	3/4/72		9/6/73

Source: R. B. Griepp (1974).

test? Is there enough evidence here to justify terminating the allocation of additional patients to the new drug? What factors would be important in making this decision?

*3.   Consider a sample of n survival times and suppose all observations are observed to failure (i.e., there are no censored observations in the sample). Let $Z_{(1)} < Z_{(2)} < \cdots < Z_{(n)}$ denote the ordered survival times. Show that the Kaplan–Meier survival curve in this uncensored situation estimates the probability of surviving past time $Z_{(1)}$ as $(n-1)/n$, the

TABLE 6.   Severe Viral Hepatitis Study

Patient Number	Treatment[a]	Length of Observation[b]	Status[c]
1	D	6	D
2	P	16	A
3	P	2	D
4	D	1	D
5	D	1	D
6	P	4	A
7	D	16	A
8	D	3	A
9	P	16	A
10	D	8	D
11	P	16	A
12	P	2	D
13	D	4	D
14	P	16	A
15	P	5	A
16	D	1	D
17	D	10	A
18	P	2	A
19	P	2	A
20	D	10	D
21	P	16	A
22	D	1	A
23	P	16	A
24	D	15	A

Source: P. B. Gregory (1974).
[a] D = drug; P = placebo.
[b] In weeks.
[c] A = alive; D = dead.

probability of surviving past time $Z_{(2)}$ as $(n-2)/n$, and, in general, estimates the probability of surviving past time $Z_{(i)}$ as $(n-i)/n$. [Note that these estimates are quite natural since, in our observed uncensored sample, there are $n-i$, out of a total of n, that survive past time $Z_{(i)}$.] Show also that in this case of no censored survival times, the estimated probability of surviving past time t is given by {number of observations that are greater than t}/n.

•4.   Return to the heart transplant data of Table 4, Chapter 1. Consider the variable that indicates the waiting time for a donor. This is the variable T_2 for the 52 transplant patients and (ignoring the 3 deselected patients in the nontransplant group) it is given by the censored value T_1 for the nontransplant patients. That is, in the nontransplant group, the waiting time variable is censored either because the patient died before a donor was available or because the patient was still waiting for a donor at the closing date. Thus using 52 uncensored observations and 27 censored observations, compute the Kaplan–Meier curve for the waiting time for a donor.

5. Return to the heart transplant data of Table 4, Chapter 1. For the 52 transplant patients, consider the variable T_3 = days from transplant to death or closing date. Using this variable, compute the Kaplan–Meier curve for the posttransplant survival time.

6. Show either directly or via an example that the Gehan statistic V, initially defined as the sum of $m \times n$ scores (where each score is either $-1, 0,$ or $+1$), is also equal to the sum of the h values for sample 2.

7. Return to Comment 1 and show that if $Z_{(n)}$ is a censored observation, the probability estimator defined by (3) does not reach 0 as t gets arbitrarily large.

8. The data in Table 7 are from a study discussed by Siddiqui and Gehan (1966) and also considered by Bryson and Siddiqui (1969). The data are survival times (measured from the date of diagnosis) of 43 patients suffering from chronic granulocytic leukemia. For the data of Table 7 compute the Kaplan–Meier survival curve. (See Problem 3, which points out how the Kaplan–Meier calculations are simplified when all the observations are uncensored.)

TABLE 7.   Ordered Survival Times (days from diagnosis)

7	429	579	968	1877
47	440	581	1077	1886
58	445	650	1109	2045
74	455	702	1314	2056
177	468	715	1334	2260
232	495	779	1367	2429
273	497	881	1534	2509
285	532	900	1712	
317	571	930	1784	

Source: M. M. Siddiqui and E. A. Gehan (1966).

9. Use the data of Table 7 to estimate the probability of a patient surviving at least 2056 days, given that the patient has survived 881 days.

10. Define censored observation and survival curve.

Supplementary Problems

1. Refer again to the data of Table 4, Chapter 1 and consider the 52 patients who received new hearts. Analyze the data for these patients to determine whether the waiting time for a new heart (T_2) is associated with survival time after getting the heart (T_3). One way to do this is to divide the 52 patients into the 26 patients with waiting times of 21 days or less, and the 26 patients who had to wait longer. Then the two groups can be compared, first obtaining a Kaplan–Meier curve for each, and then calculating the Gehan test for the null hypothesis that survival time distributions are the same for the two groups.

*2. In the first papers describing survival experience of fairly large groups of heart transplant patients, Messmer, *et al.* (1969) and Clark, *et al.* (1971) compared the survival experience of patients who received new hearts with the survival experience of those chosen as heart transplant candidates who died before hearts were found for them. Can you see the several sources of bias favoring the transplant patient in this approach? What approach would you suggest?

3. Table 8 presents data on a sample of people who entered a retirement home at varying times since it opened in 1964. The data were supplied by Bortz (1975). The table gives the month and year each person entered and the month and year of death if the person had not survived to the time of the survey in July 1975. Calculate and graph the Kaplan–Meier survival curve for this sample of 46 persons, measuring survival from time of entry into the retirement home.

TABLE 8. Entry and Survival Dates for a Sample of People Entering a Retirement Home

Case #	Sex	Birth Date	Entry Date	Date of Death
1	F	9/96	1/64	5/74
2	F	5/79	1/64	11/72
3	F	10/79	3/64	9/72
4	F	7/89	1/64	5/73
5	F	6/90	1/64	6/73
6	F	4/94	1/64	4/73
7	F	3/91	1/64	1/73
8	F	4/97	4/67	
9	M	1/92	10/72	
10	M	12/91	1/64	
11	F	8/98	2/75	
12	F	3/90	1/68	
13	F	6/92	2/68	
14	F	11/92	1/64	
15	F	9/87	6/71	

TABLE 8—continued

Case #	Sex	Birth Date	Entry Date	Date of Death
16	F	2/89	1/64	
17	F	7/90	1/64	
18	M	7/88	6/72	
19	F	1/88	1/64	
20	F	10/96	3/75	
21	F	5/83	1/64	
22	F	7/95	1/64	
23	F	1/87	9/66	
24	M	10/86	12/64	
25	F	12/93	1/64	
26	F	4/93	6/72	
27	M	6/95	1/64	
28	F	3/01	7/67	
29	M	5/88	9/70	
30	F	7/01	9/66	
31	F	2/96	1/64	
32	F	11/86	7/72	
33	F	10/84	3/65	
34	F	10/94	6/73	7/73
35	F	7/83	1/64	2/69
36	M	6/91	1/64	2/64
37	M	7/83	1/64	7/65
38	F	6/91	1/64	2/70
39	F	2/80	1/64	1/65
40	F	7/89	1/64	10/66
41	F	7/85	1/64	8/67
42	F	4/92	1/64	10/69
43	F	8/83	6/66	1/68
44	F	6/02	4/69	11/70
45	F	6/90	3/67	8/67
46	F	4/92	1/64	7/68

Source: W. M. Bortz (1975).

4. Refer to the data of Table 8 and calculate the Gehan test for the null hypothesis that the survival experience of people entering the home in 1964 is the same as the experience of those entering later.

5. Referring again to the data of Table 8, compare survival times for the males and females by calculating a Gehan test of the null hypothesis of equivalent survival distributions for males and females.

*6. Explain why Gehan's test can be viewed as a generalization of the Wilcoxon two sample test presented in Chapter 13.

7. Is the Gehan test relatively insensitive to outlying observations? Explain.

8. In the Stanford heart transplant program, the queue of candidates waiting for a heart usually consists of at most one patient, and patients receive hearts in order of queue, matched with donor on blood type. Under these circumstances, is there a chance for selection bias in assigning a donor heart to one candidate or another? Explain.

9. Transplantations in the Stanford heart transplant program are not done on a randomized basis. Thus a suitable control group, to be used as a basis for comparison with the survival experience among transplanted patients, is not readily available. How might one design a randomized study?

10. The survival data of Table 4, Chapter 1 give no information as to the possibly improved quality of life for the patient receiving a new heart. How could a study be performed to determine whether or not the quality of life is improved?

REFERENCES

Bortz, W. M. (1975). Personal communication.

*Bryson, M. C. and Siddiqui, M. M. (1969). Some criteria for aging. *J. Amer. Statist. Ass.* **64,** 1472–1483.

Clark, D. A., Stinson, E. B., Griepp, R. B., Scroeder, J. S., Shumway, N. E., and Harrison, D. C. (1971). Cardiac transplantation in man. VI. Prognosis of patients selected for cardiac transplantation. *Ann. Int. Med.* **75,** 15–21.

*Gehan, E. A. (1965). A generalized Wilcoxon test for comparing arbitrarily singly-censored samples. *Biometrika* **52,** 203–223.

Griepp, R. B. (1974). Personal communication.

Gregory, P. B. (1974). Personal communication.

*Kaplan, E. L. and Meier, P. (1958). Nonparametric estimation from incomplete observations. *J. Amer. Statist. Ass.* **53,** 457–481.

Kaplan, H. S. and Rosenberg, S. (1973). Personal communication.

*Mantel, N. (1967). Ranking procedures for arbitrarily restricted observations. *Biometrics* **23,** 65–78.

Messmer, B. J., Nora, J. J., Leachman, R. D., and Cooley, D. A. (1969). Survival times after cardiac allografts. *Lancet* **1,** 954–56.

Siddiqui, M. M. and Gehan, E. A. (1966). *Statistical Methodology for Survival Time Studies.* Communication of National Cancer Institute.

Glossary

$=$	equals
$\neq$	unequal
$<$	less than
$\leq$	less than or equals
$>$	greater than
$\geq$	greater than or equals
$13 < W < 23$	shorthand for saying $13 < W$ and $W < 23$
$\approx$	approximately equals
$|x|$	absolute value of x
$Pr(A)$	probability of the event A
$Pr(A|B)$	probability of the event A given that the event B has occurred
H_0	null hypothesis
P	significance probability
z_α	upper α percentile point of the standard normal distribution

$t_{\alpha,\nu}$	upper α percentile point of student's t distribution with ν degrees of freedom
$\chi^2_{\alpha,\nu}$	upper α percentile point of the chi-squared distribution with ν degrees of freedom
F_{α,ν_1,ν_2}	upper α percentile point of the F distribution with ν_1 degrees of freedom in the numerator and ν_2 degrees of freedom in the denominator
$\binom{n}{x}$	binomial coefficient—the number of combinations of n things taken x at a time

Alternative hypothesis. Any admissible hypothesis alternative to the null hypothesis.

Area sampling. Sampling used when no complete frame of reference is available. The total area under investigation is divided into small subareas that are sampled at random or by some restricted random process. Each of the chosen subareas is then fully inspected and enumerated, and may form a frame for further sampling if desired.

Average rank. Assume that X is one of a set of n observations and that X has the same value as (is tied with) some of the other observations. The average rank of X, in the ranking of the n observations, is the numerical average of those ranks that would be assigned to X and assigned to the observations having the same value as X, if these tied observations could be distinguished.

Binomial coefficient. The binomial coefficient $\binom{n}{x}$ denotes the number of patterns with x successes and $n - x$ failures. For example, with $n = 5$ and $x = 2$, $\binom{5}{2} = 10$. The 10 patterns (letting S denote success and F failure) are:

$$SSFFF \quad SFSFF \quad SFFSF \quad SFFFS \quad FSSFF$$

$$FSFSF \quad FSFFS \quad FFSSF \quad FFSFS \quad FFFSS$$

Binomial distribution. The probability distribution of the number of successes B in n independent trials, where each trial has two outcomes (conveniently labeled success and failure), and the probability of success p is the same for each trial.

Block. The name given, mainly in experimental design, to a group of items under treatment or observation. For example, a block may comprise a group of contiguous plots of land, all the animals in a litter, the results obtained by a single operator, or meteorological observations on a series of days at a given place. The purpose for grouping the items in blocks is to isolate sources of heterogeneity, the items in a block being as homogeneous as possible.

Censored observation.   An observation is said to be censored when its value is not known although its existence is known. The true value may, for example, be known to be greater than some number or less than some number, but its exact value is not determined. For example, in a posttransplant survival study, we may be interested in the number of days from the transplant to death. When the data are analyzed, patient A may have survived 342 days and still be alive. Patient A's observation is said to be censored on the right. We only know that the actual number of posttransplant survival days for patient A is some number that exceeds 342. As another example, in a study in walking in the newborn a researcher may be interested in the number of days (from birth) it takes for the infant to learn to walk unassisted. The researcher tells the parents of the children in the study to return with the children for continual testing beginning at 240 days of age. When child A appears for the first time, it is determined that he already can walk unassisted. Here the time to walking alone observation is censored on the left. The true value is some unknown number less than 240.

Chi-squared distribution.   The distribution may be regarded as that of the sum of squares of k independent normal variates, where each normal variate emanates from a (normal) population with mean 0 and standard deviation 1. The parameter k is then the number of degrees of freedom.

Clinical trial.   A controlled experiment on human beings, designed to investigate the efficacy of some form of medical treatment.

Coefficient of variation.   The standard deviation of a distribution divided by the mean of the distribution, sometimes multiplied by 100.

Comparative experiment.   An experiment designed to make comparisons between a control and one or more treatments.

Completely randomized experiment.   A very simple form of experimental design in which the treatments are allocated to the experimental units purely on a chance basis.

Conditional probability (of an event). The probability of occurrence of an event B computed under the assumption that another event A has occurred. Denoted by

$Pr(B|A)$, the conditional probability can be found via the formula

$$Pr(B|A) = \frac{Pr(A \text{ and } B)}{Pr(A)}.$$

This formula assumes that $Pr(A) \neq 0$.

Confidence coefficient. The confidence coefficient of a confidence interval for a parameter μ is the probability that the random interval will contain the value μ.

Confidence interval. A $1 - \alpha$ confidence interval for an unknown parameter μ is an interval computed from the sample data. It has the property that, in repeated sampling, $100(1 - \alpha)$ percent of the intervals obtained will contain the value μ.

Contingency table.   The members of an aggregate may be classified according to qualitative or quantitative characteristics. Where the characteristics are qualitative, a classification according to two of them may be given in a two-way table known as a contingency table. If, for example, the characteristic A is r-fold and the characteristic B is k-fold, the contingency table will have r rows and k columns. It is then often called an $r \times k$ contingency table, or simply an $r \times k$ table.

Continuous variable.   A variable that can assume values in a continuous range. When recording an observation on a continuous variable, a refinement of the measuring instrument yields a more precise observation.

Contrast (in population means).   Any linear combination of the population means which has the property that the coefficients, which multiply the population means, sum to zero. Thus,

$$\sum a\mu = a_1\mu_1 + a_2\mu_2 + \cdots + a_k\mu_k$$

is a contrast in the k populations means $\mu_1, \mu_2, \ldots, \mu_k$ if the a's sum to zero; that is, if

$$\sum a = a_1 + a_2 + \cdots + a_k = 0.$$

Contrast (in sample means).   Any linear combination of the sample means which has the property that the coefficients, which multiply the sample means, sum to zero. Thus

$$\sum a\bar{X} = a_1\bar{X}_1 + a_2\bar{X}_2 + \cdots + a_k\bar{X}_k$$

is a contrast in the k sample means $\bar{X}_1, \bar{X}_2, \ldots, \bar{X}_k$ if the a's sum to zero; that is, if

$$\sum a = a_1 + a_2 + \cdots + a_k = 0.$$

Correction for continuity.   When a statistic is discrete, but its distribution is being approximated by a continuous function (such as the normal function), probabilities can sometimes be more accurately obtained by entering the tables of the continuous function, not with the actual values of the statistic, but with slightly corrected values. These values are then said to be corrected for continuity.

Correlation coefficient. A correlation coefficient is a measure of the interdependence between two variables. It is a measure that takes on values between -1 and $+1$ with the intermediate value of 0 indicating the absence of association, and with the -1 and $+1$ values indicating perfect negative association and perfect positive association, respectively.

Critical region. A critical region of a test statistic is the set of values of the test statistic that lead to rejection of the null hypothesis.

Cumulative frequency (of a class interval). The number of measurement values falling in the interval plus the number falling in lower intervals.

Cumulative relative frequency (of a class interval). The cumulative frequency expressed as a proportion of the total number of measurement values.

Cumulative relative frequency polygon. A plot, in percentages, of cumulative relative frequency. At the right-hand end point of each class interval, at a height equal to 100 times the cumulative relative frequency of that interval, a dot is placed on a graph. Then the successive dots are connected by straight line segments to form the polygon.

Degrees of freedom.   In this text, the term degrees of freedom can be taken to be simply a name attached to a parameter indexing a family of distributions. For the reader seeking a precise answer to the question, what are degrees of freedom? see the definition in Kendall and Buckland (1971) and the article by Good (1973).

Density.   A curve describing a population. The curve must be nonnegative and include a finite area between itself and the measurement axis. The probability that a randomly selected value X from the population will be between the points a and b is given by

$$Pr(a \leq X \leq b) = \frac{\text{area under the curve between } a \text{ and } b}{\text{total area under the curve}}.$$

It is convenient to scale the curve so that the total area under the curve is 1. Then $Pr(a \leq X \leq b)$ reduces to

$$Pr(a \leq X \leq b) = \text{area under the curve between } a \text{ and } b.$$

Dependent variable. This term is used in contradistinction to independent variable in regression analysis. When a random variable Y is expressed as a function of a controlled variable X, Y is known as the dependent variable.

Discrete variable. A variable that assumes a finite number or, at most, a countable number of possible values.

Distribution. Description of a set of measurement values along with frequencies of occurrence. A distribution of measurements is portrayed by a histogram or by a cumulative relative frequency polygon.

Double blind (clinical trial). A clinical trial in which neither physician nor patient are aware of which treatment the patient has received.

Experimentwise error rate. Suppose that several inferences are to be made from the same set of data. The experimentwise error rate is the probability that at least one (i.e., one or more) of the inferences will be wrong.

_F_ Distribution. A distribution, due to R. A. Fisher, which can be described as the distribution of the statistic $\mathscr{F} = s_2^2/s_1^2$ where s_1^2 is the sample variance of sample of size m from a normal population with variance σ^2 and s_2^2 is the sample variance of an independent sample of size n from a normal population with variance σ^2. The statistic $\mathscr{F}$ is said to have an F distribution with $n-1$ degrees of freedom in the numerator and $m-1$ degrees of freedom in the denominator. For the mathematical form of F distributions see, for example, Kendall and Buckland (1971).

False-negative result. An error a diagnostic test makes when it gives a disease-free indication to a person who really has the disease.

False-positive result. An error a diagnostic test makes when it gives a disease indication to a person who does not have the disease.

Finite population. A finite collection of individuals.

Finite population correction. If a sample of n values is drawn without replacement from a finite population of size N, the standard deviation of the sample mean $\bar{X}$ can be written as

$$\sigma_{\bar{X}} = \sqrt{\frac{N-n}{N-1}} \cdot \frac{\sigma}{\sqrt{n}},$$

where σ is the population standard deviation. The factor $(N-n)/(N-1)$ is sometimes called the finite population correction.

Frame.   A list, map, or other specification of the units that constitute the available information relating to the population designated for a particular sampling scheme.

Frequency.   The number of occurrences of a given type of event, or the number of members of a population falling into a specified class.

Frequency polygon.   A polygon obtained from the histogram as follows. For each bar top of the histogram, determine the midpoint. Then connect adjacent midpoints of bar tops by straight lines.

Frequency table.   A table that shows the distribution of the frequency of occurrence of a given characteristic according to some specified set of class intervals.

Grouped data.   Data listed in terms of class frequencies. With grouped data we may not know the exact values of the measurements falling within the class intervals.

Histogram.   A frequency diagram in which rectangles proportional in area to the class frequencies are erected on sections of the horizontal axis, the width of each section representing the corresponding class interval of the measurement.

Homogeneous variance. (See **Homoscedasticity**)

Homoscedasticity.   In a bivariate situation, if the variance of one variate is the same for all fixed values of the other, the distribution is said to be homoscedastic in the first variate; if not, it is heteroscedastic.

Hypothesis.   (See **Null Hypothesis** and **Alternative Hypothesis**)

Independent events.   Two events are independent if the probability of one is the same whether or not the other event is given. That is,

$$Pr(B|A) = Pr(B) \quad \text{and} \quad Pr(A|B) = Pr(A).$$

From either one of these equations, and the multiplication rule for obtaining conditional probabilities, it is seen that when A and B are independent,

$$Pr(A \text{ and } B) = Pr(A) \cdot Pr(B).$$

It is often preferable to use the last equation as the definition of independence since it avoids difficulties arising when $Pr(A)$ or $Pr(B)$ is zero.

Independent variable.   This term is used in contradistinction to "dependent variable" in regression analysis. When a random variable Y is expressed as a function of a controlled variable X, X is known as the independent variable. Many authors prefer to call X the regressor, since use of the word independent here has no connection with mathematical or statistical independence.

Least squares.   A method of estimation by which the estimates, of the parameters to be estimated, are determined by minimizing a certain sum of squares involving the observations and the parameters.

Linear regression (of Y on X).   When the regression curve of Y on X is a straight line, that is, when $\mu_{Y \cdot X} = A + BX$, then we say there is a linear regression of Y on X.

Mean.   (See **Population Mean, Sample Mean**)

Median.   (See **Population Median, Sample Median**)

Multiple comparison procedure.   A statistical procedure that on the basis of the sample observations makes a number (more than one) of statements (decisions) concerning the various parameters of interest.

Multiple stage sampling. Sampling by stages, where the sampling units at each stage are subsampled from the larger units chosen at the previous stage.

Nonparametric method.   A method, for example, of testing a hypothesis or obtaining a confidence interval, that does not require knowledge of the form of the underlying parent population.

Normal distribution.   A symmetric, bell-shaped probability distribution represented by the mathematical equation

$$f(x) = \frac{1}{\sigma\sqrt{2\pi}} e^{-(x-\mu)^2/2\sigma^2}, \quad -\infty < x < \infty.$$

The value $f(x)$ is the height of the normal curve at the point x. The parameters μ and σ are the mean and standard deviation, respectively. The mean μ can be any number and the standard deviation σ can be any positive number. Thus we have really defined a "family" of normal distributions, different members of the family corresponding to different choices of the pair (μ, σ).

Null hypothesis.   In general, this term relates to a particular hypothesis under test, as distinct from the alternative hypotheses that are under consideration.

One-sided alternative.   An alternative hypothesis that only allows the deviation from the null hypothesis to be in one direction. For example, if the null hypothesis asserts that the parameter of interest μ is equal to some specified value μ_0, the alternative $\mu > \mu_0$ is a one-sided alternative.

One-tailed test.   A test of hypothesis for which the critical region is wholly located at one end of the distribution of the test statistic.

Outlier.   An observation that is found to lie an abnormally long way from its fellow observations.

P value.   (See **Significance Probability**)

Parameter.   In typical mathematics usage it is an unknown quantity that may vary over a certain set of values. In statistics it usually occurs as a quantity associated with a conceptual population. For example, the mean μ of a normal population is a population parameter.

*p*th percentile (of a population).   A value such that p percent of the objects in the population have measurements less than this value. For example, if there are eight objects in the population and their corresponding measurements are 4, 6, 19, 34, 37, 41, 45, 67, then each number that is greater than 41 and less than 45 is a 75th percentile because each has the property that 75% of the measurements in the population are below it.

Pooled estimate (of variance).   If two samples are from the same population or from different populations having equal variances σ^2, then the sample variances can be pooled or averaged to give an estimate of σ^2. If the sample sizes are m and n, with corresponding sample variances s_1^2 and s_2^2, the formula for the pooled estimate is

$$s_p^2 = \frac{(m-1)s_1^2 + (n-1)s_2^2}{m+n-2}.$$

Population.   A term used to describe any finite or infinite collection of individuals.

Population mean.   The most widely used measure of location of the population. For a finite population with measurement values $X_1, X_2, \ldots, X_N$, it is defined as

$$\mu = \frac{X_1 + X_2 + \cdots + X_N}{N}.$$

Population median.   That value which divides the total population into two halves. For a continuous variate X it may be defined by the equation $Pr(X > M) = Pr(X < M) = \frac{1}{2}$, M being the median value. For a discontinuous variate, ambiguity may arise which can only be removed by some convention. For a population of

$2N+1$ objects, the median is the value of the $(N+1)$th (ordered) object. For a population of $2N$ objects, it is customary to make the median unique by defining it to be the average of the Nth and $(N+1)$th ordered objects.

Population standard deviation.   The most widely used measure of dispersion of the population. For a finite population with measurement values $X_1, X_2, \ldots, X_N$, it is defined as

$$\sigma = \sqrt{\frac{\Sigma(X-\mu)^2}{N}},$$

where μ is the population mean.

Population variance.   A population parameter that measures dispersion. It is equal to the square of the population standard deviation.

Power.   The power of a test against a specified alternative is the probability of (correctly) rejecting the null hypothesis; that is, the probability of rejecting the null hypothesis when in fact the alternative is true.

Prevalence rate (of a disease). The proportion of people in the population who have the disease in question.

Probability (of an event).   A number between 0 and 1 indicating how likely the event is to occur.

Randomized blocks experiment.   An experiment in which each block consists of a set of experimental units, and treatments are allocated to the various units within the blocks in a random manner.

Randomized clinical trial.   A clinical trial in which the treatments are assigned to the subjects by some random allocation. The random allocation eliminates bias in the assignment of treatments to patients and establishes the basis for statistical analysis.

Random sample (of n objects). A sample of n objects (each with an associated measurement) such that for any given interval, each object's measurement has an equal and independent chance of falling in the interval and the chance is given by the area of the histogram or density over that interval.

Rank.   We rank observations when we order them according to their size and assign to them numbers (called ranks) that correspond to their positions in the ordering. We usually rank from least to greatest and use, when ranking n observations, the integers $1, 2, \ldots, n$ as ranks. Then the rank of X among a set of n observations is equal to $1 +$ (the number of other observations in the set that are less than X). This definition needs to be modified when there are tied observations. In such cases average ranks are used.

Regression curve (of Y on X).   A curve that represents the regression equation. For a particular point on the curve, the abscissa is the value X and the ordinate is $\mu_{Y \cdot X}$, the mean of the distribution of Y for that specific fixed value X.

Relative frequency.   The frequency in an individual group of a frequency distribution expressed as a fraction of the total frequency.

Residual.   In the context of linear regression, the difference, between an observed value Y_i and the predicted value $a + bX_i$, is called a residual.

Sample.   A subset of a population obtained to investigate properties of the parent population.

Sample mean.   The most widely used estimator of the population mean. For a sample of size n with measurement values $X_1, X_2, \ldots, X_n$, the sample mean is

$$\bar{X} = \frac{X_1 + X_2 + \cdots + X_n}{n}.$$

Sample median.   That value that divides the sample into two halves. For an odd sample size, say a sample of $2n + 1$ objects, denote the ordered values of the measurements by $X_{(1)} \leq X_{(2)} \leq \cdots \leq X_{(2n+1)}$. Then the sample median is $X_{(n+1)}$, the measurement that occupies position $n + 1$ in the list. For an even sample size, say a sample of $2n$ objects, with measurements listed in order as $X_{(1)} \leq X_{(2)} \leq \cdots \leq X_{(2n)}$, it is customary to make the sample median unique by defining it as the average of $X_{(n)}$ and $X_{(n+1)}$; that is,

$$\frac{X_{(n)} + X_{(n+1)}}{2}.$$

Sample standard deviation.   The most widely used estimator of the population standard deviation. For a sample of size n with measurement values $X_1, X_2, \ldots, X_n$, the sample standard deviation is

$$s = \sqrt{\frac{\sum (X - \bar{X})^2}{n - 1}},$$

where $\bar{X}$ is the sample mean.

Scatterplot.   A diagram for examining the relationship between two variables X and Y. Each (X, Y) observation pair is represented by a point whose coordinates on ordinary rectangular axes are the X and Y values. A set of n (X, Y) observations thus provides n points on the plot and the scatter or clustering of the points exhibits the relationship between X and Y.

Sensitivity. The probability that the diagnostic test will yield a positive result when given to a person who has the disease.

Significance probability. For a given set of data and test statistic, the significance probability is the lowest value of α (the probability of a Type-I error) that we can use and still reject the null hypothesis. This "lowest value" is often called the P value.

Simple random sampling. Simple random sampling, of n objects from a population of N objects, is any procedure that assures each possible sample of size n an equal chance or probability of being the sample selected; that is, any procedure that assures each possible sample has probability $1\Big/\binom{N}{n}$ of being the sample selected.

Specificity. The probability that the diagnostic test will yield a negative result when given to a person who does not have the disease.

Standard deviation. (See **Population standard deviation, Sample standard deviation**)

Standard error. Consider a statistic T. A standard error of T is an estimator of the standard deviation of T. In practice, the true standard deviation $SD(T)$ is usually unknown, and hence must be estimated from the data. Such an estimate is called a standard error and we denote it by $\widehat{SD}(T)$. When the statistic under consideration is clear from context, we sometimes shorten our notation to $\widehat{SD}$.

Standard normal distribution. A normal distribution with mean $\mu = 0$ and standard deviation $\sigma = 1$.

Standardized deviate. The value of a deviate that is reduced to standardized form (zero mean, unit variance) by subtracting the parent mean and then dividing by the parent standard deviation.

Statistics. Numerical data relating to an aggregate of individuals; the science of collecting, analyzing, and interpreting such data.

Stratified random sampling. Sampling in which the population is first divided into parts, known as strata, then a simple random sample is obtained from each of the strata.

Survival curve. In the context of survivorship studies, if X denotes the number of days a patient must wait for a specified event (such as death or relapse), then the survival curve gives, for each number t, an estimate of $Pr\{X > t\}$.

t **Distribution.** This distribution, originally due to "Student", is, among other things, the distribution of $T = (\bar{d} - \mu)/(s/\sqrt{n})$ where $\bar{d}$ is the sample mean of a random sample of size n from a normal population with mean μ, and s is the sample standard deviation. In this case the distribution has $n-1$ degrees of freedom. The distribution is used for making inferences about the unknown population mean μ.

Test statistic. A function of a sample of observations that provides a basis for testing a statistical hypothesis.

Two-sided alternative. An alternative hypothesis that permits the deviation from the null hypothesis to be in either direction. For example, if the null hypothesis asserts that the parameter of interest μ is equal to some specified value μ_0, the alternative "$\mu > \mu_0$ or $\mu < \mu_0$" is a two-sided alternative.

Two-tailed test. A test for which the critical region consists of both extremes of the sampling distribution of the test statistic. It is customary, but not essential, to assign one-half of the probability of rejection to each extreme, giving a symmetrical test.

Type-I error. A false rejection of the null hypothesis; that is, a rejection of the null hypothesis when in fact it is true.

Type-II error. A false acceptance of the null hypothesis; that is, an acceptance of the null hypothesis when in fact an alternative hypothesis is true.

Upper *p*th percentile (of a population). A value such that p percent of the objects in the population have measurements greater than or equal to this value. For example, if there are eight objects in the population and their corresponding measurements are 4, 6, 19, 34, 37, 41, 45, 67, then each number that is greater than 41 and less than or equal to 45 is an upper 25th percentile because each has the property that 25% of the measurements in the population are greater than or equal to it.

REFERENCES

Good, I. J. (1973). What are degrees of freedom? *The American Statistician* **27**, 227–228.
Kendall, M. G. and Buckland, W. R. (1971). *A Dictionary of Statistical Terms*, 3rd ed. Oliver & Boyd, Edinburgh.

Answers to Selected Questions and Solutions to Selected Problems

Discussion Questions

1. One way would be to compare $Pr(D_{\text{Jan.}}|B_{\text{Dec.}})$ with $Pr(D_{\text{Jan.}})$, compare $Pr(D_{\text{Feb.}}|B_{\text{Jan.}})$ with $Pr(D_{\text{Feb.}})$, and so forth.

Problems

1. Let D_0 be the event "death in interval 0–10" and let D_1 be the event "death in interval 10–20." Then

$$Pr(\text{death by age } 20) = Pr(D_0 \text{ or } D_1)$$

$$= Pr(D_0) + Pr(D_1)$$

(by Rule III of probability)

$$= \frac{11}{275} + \frac{10}{275} = \frac{21}{275} = .076.$$

3. Let A be the event "digit is a 0." Let B be the event "digit is in row 1."

$$Pr(A|B) = \frac{Pr(A \text{ and } B)}{Pr(B)} \qquad \text{(by the multiplication rule)}$$

Now,

$$Pr(A \text{ and } B) = \frac{2}{100},$$

$$Pr(B) = \frac{10}{100},$$

and thus

$$Pr(A|B) = \frac{2/100}{10/100} = \frac{2}{10} = .2.$$

5. Define events as follows:

$$B: \text{person has diabetes}$$

$$A: \text{person's test is positive.}$$

Then sensitivity $= 52.9\%$ implies

$$Pr(A|B) = .529,$$

and specificity $= 99.4\%$ implies

$$Pr(\text{not } A | \text{not } B) = .994.$$

Also we are told the prevalence rate is .008, that is,

$$Pr(B) = .008.$$

Then, by formula (3) of Chapter 2, we find

$$Pr(B|A) = \frac{(.529)(.008)}{(.529)(.008) + (.006)(.992)} = .416.$$

$$Pr(\text{not } B | \text{not } A) = 1 - Pr(B|\text{not } A)$$

$$= 1 - \left\{ \frac{(.471)(.008)}{(.471)(.008) + (.994)(.992)} \right\}$$

[by formula (4) of Chapter 2]

$$= 1 - .004 = .996.$$

CHAPTER 3	SOLUTIONS

Discussion Questions

1. Many of the intervals may then contain frequencies that are very small—possibly zero. This will make it difficult to detect regularity in the shape of the distribution.

3. Certainly not. In particular, each candidate does not have an equal chance of surviving until a donor is found. (Some candidates are less hardy than others. Also, for some it is harder to find a suitable donor.)

Problems

1.

Interval	Frequency	Relative Frequency	Cumulative Relative Frequency
75–99	3	.006	.006
100–124	6	.012	.018
125–149	14	.028	.046
150–174	26	.052	.098
175–199	36	.072	.170
200–224	49	.098	.268
225–249	63	.126	.394
250–274	65	.130	.524
275–299	57	.114	.638
300–324	48	.096	.734
325–349	46	.092	.826
350–374	33	.066	.892
375–399	24	.048	.940
400–424	16	.032	.972
425–449	4	.008	.980
450–549	10	.020	1.000

3.

	X	X^2	$X-\mu$	$(X-\mu)^2$
	1	1	−6	36
	2	4	−5	25
	2	4	−5	25
	3	9	−4	16
	3	9	−4	16
	4	16	−3	9
	6	36	−1	1
	10	100	3	9
	16	256	9	81
	23	529	16	256
Totals:	70	964	0	474

$$\mu = \frac{\sum X}{N} = \frac{70}{10} = 7.$$

$$\text{median} = \frac{3+4}{2} = 3.5.$$

$$\sigma = \sqrt{\frac{\sum (X-\mu)^2}{N}} = \sqrt{\frac{474}{10}}$$

$$= \sqrt{47.4} = 6.88.$$

Using the alternative formula for σ we have:

$$\sigma = \sqrt{\frac{N(\sum X^2) - (\sum X)^2}{N^2}} = \sqrt{\frac{10(964) - (70)^2}{(10)^2}} = \sqrt{\frac{4740}{100}} = \sqrt{47.4} = 6.88.$$

5. For the data in Table 3 of Chapter 3:

$$x_{.05} = 149.5 + \frac{25-23}{26}(25) = 149.5 + 1.9 = 151.4$$

$$x_{.25} = 199.5 + \frac{125-85}{49}(25) = 199.5 + 20.4 = 219.9$$

$$x_{.50} = 249.5 + \frac{250-197}{65}(25) = 249.5 + 20.4 = 269.9$$

$$x_{.75} = 324.5 + \frac{375-367}{46}(25) = 324.5 + 4.3 = 328.8$$

$$x_{.95} = 399.5 + \frac{475-470}{16}(25) = 399.5 + 7.8 = 407.3$$

7. Let $Z = (X - 100)/15$, where $X = $ I.Q. score. Then

$$Pr(X < 90) = Pr\left(Z < \frac{90 - 100}{15}\right) = Pr(Z < -.67) = Pr(Z > .67) = .2514,$$

the latter value being read from Table C2. Similarly,

$$Pr(X > 145) = Pr\left(Z > \frac{145 - 100}{15}\right) = Pr(Z > 3) = .0013,$$

and

$$Pr(120 < X < 140) = Pr\left(\frac{120 - 100}{15} < Z < \frac{140 - 100}{15}\right)$$

$$= Pr(1.33 < Z < 2.67) = Pr(Z > 1.33) - Pr(Z > 2.67)$$

$$= .0918 - .0038 = .0880.$$

9. Let

$$Z_1 = \frac{81 - \mu}{\sigma_{\bar{x}}} = \frac{81 - 78}{(9/\sqrt{16})} = 1.33.$$

Then

$$Pr(\bar{X} > 81) = Pr(Z > Z_1) = Pr(Z > 1.33),$$

where Z is standard normal. From Table C2 the probability is found to be .0918.

CHAPTER 4	SOLUTIONS

Discussion Questions

1. For 95% confidence, $\alpha = .05$, $\alpha/2 = .025$, and from Table C5 we find $t_{.025,29} = 2.045$. Thus the 95% confidence interval for μ is

$$\bar{d} \pm (t_{.025,29})(\widehat{SD}) = 1.10 \pm 2.045(1.05)$$

$$= 1.10 \pm 2.15$$

$$= -1.05 \text{ and } 3.25.$$

The 95% interval is narrower than the 99% interval derived in Chapter 4. This is to be expected. Here we are satisfied with a lower probability (.95 versus .99) that the interval will contain the true value μ, and by sacrificing some confidence we can obtain a narrower interval.

Problems

1.
$$\bar{d} = \frac{74}{30} = 2.467$$

$$s = \sqrt{\frac{\sum d^2 - \frac{(\sum d)^2}{n}}{n-1}} = \sqrt{\frac{384 - \frac{(74)^2}{30}}{29}} = 2.636$$

$$\widehat{SD} = \frac{s}{\sqrt{n}} = \frac{2.636}{\sqrt{30}} = .481$$

$$t_{.025,29} = 2.045.$$

Thus the 95% confidence interval for μ is

$$\bar{d} \pm (t_{.025,29})(\widehat{SD}) = 2.467 \pm (2.045)(.481) = 2.467 \pm .984$$
$$= 1.48 \text{ and } 3.45.$$

3.
$$\bar{d} = \frac{-3095}{10} = -309.5$$

$$s = \sqrt{\frac{2,137,175 - \frac{(-3095)^2}{10}}{9}} = 361.98$$

$$\widehat{SD} = \frac{361.98}{\sqrt{10}} = 114.47$$

$$t_{.025,9} = 2.262.$$

Thus the 95% confidence interval for μ is:

$$\bar{d} \pm (t_{.025,9})(\widehat{SD}) = -309.5 \pm (2.262)(114.47)$$
$$= -309.5 \pm (258.93)$$
$$= -568.4 \text{ and } -50.6.$$

5. $\bar{d} = 2.467$, $\widehat{SD} = .481$, so

$$T = \frac{\bar{d}}{\widehat{SD}} = \frac{2.467}{.481} = 5.13.$$

From Table C5 we find $t_{.0005,29} = 3.659$. Thus since $T = 5.13$ is greater than 3.659, the one-sided P value is seen to be less than .0005, and the two-sided P value is less than .001. Thus there is very strong evidence that the two measurement methods are not equivalent.

7. $\bar{d} = -309.5$, $\widehat{SD} = 114.47$, so

$$T = \frac{\bar{d}}{\widehat{SD}} = \frac{-309.5}{114.47} = -2.70.$$

From Table C5 we find $t_{.025,9} = 2.262$ and $t_{.01,9} = 2.821$. Since $|T| = 2.70$ is between 2.262 and 2.821, the one-sided P value is between .01 and .025. The two-sided P value is thus between .02 and .05. Thus there is strong evidence that tensile strengths for sutured wounds are less than tensile strengths for tape-closed wounds.

CHAPTER 5	SOLUTIONS

Discussion Questions

1. The ascorbic acid study could have been planned as a paired replicates study. Using male guinea pigs of the same strain, the pigs could be paired on the basis of factors—such as initial body weight—that might affect the lipid fraction.

Problems

1. $m = 9$, $n = 5$

$$\bar{X} = \frac{231}{9} = 25.67, \qquad \bar{Y} = \frac{74}{5} = 14.8$$

$$\hat{\Delta} = \bar{Y} - \bar{X} = -10.87$$

$$\sum X^2 = 6113 \qquad \sum Y^2 = 1174$$

$$S_p = \sqrt{\frac{6113 - \frac{(231)^2}{9} + 1174 - \frac{(74)^2}{5}}{12}} = \sqrt{21.9} = 4.68.$$

At $\alpha = .10$, $\alpha/2 = .05$, and from Table C5, $t_{.05,12} = 1.782$. From display (8) of Chapter 5, the desired interval is

$$-10.87 \pm (1.782)(4.68)\sqrt{\tfrac{1}{9} + \tfrac{1}{5}} = -10.87 \pm 4.65 = -15.52 \text{ and } -6.22.$$

3. If we change the Y value 9.3, in Table 1 of Chapter 5, to 33, then, of course, $\bar{X}$ remains 19.72, but $\bar{Y}$ is now given by

$$\bar{Y} = \frac{(13.8 + 33 + 17.2 + 15.1)}{4} = \frac{79.1}{4} = 19.78.$$

We then obtain

$$\hat{\Delta} = \bar{Y} - \bar{X} = 19.78 - 19.72 = .06.$$

The outlying value 33 changes the point estimator $\hat{\Delta}$ dramatically (from -5.88 to .06). It has a similar undesirable effect on the confidence interval.

5. $H_0: \Delta = 0$ versus $H_a: \Delta < 0$. Compute

$$T' = \frac{\bar{Y} - \bar{X}}{s_p \sqrt{\dfrac{1}{m} + \dfrac{1}{n}}},$$

and if $T' < -t_{\alpha, m+n-2}$, the null hypothesis can be rejected at the α level. Here $m = 8$, $n = 4$, so we have

$$T' = \frac{-.004}{(.00377)\sqrt{\frac{1}{8} + \frac{1}{4}}} = -1.73.$$

Suppose we set $\alpha = .05$. From Table C5, we find $t_{.05,10} = 1.812$. Since $T' = -1.73$ exceeds -1.812, the null hypothesis cannot be rejected at the $\alpha = .05$ level. However, the one-sided P value is only slightly greater than .05, and the investigator might well take the position that the data suggest that the acute exposure population mean is less than that of chronic exposure.

CHAPTER 6	SOLUTIONS

Problems

1.   The statistical analysis of the Gunther and Bauman data deals most directly with the question of whether the anesthetic difference observed among the women included in the study (67% of the women delivering babies at Stanford, 88% of all women receiving caudal anesthetics during the study period) could be attributed solely to imbalances between the two treatment groups because of the random assignment, or whether it is due in part to a difference in the efficacies of the two methods of caudal anesthesia. The latter was concluded, and the larger question then is whether the conclusion could be generalized. Since there seems to be nothing special in the characteristics of the study women, nor in the techniques of the investigators, it is concluded that the difference in effectiveness of the two anesthetic methods would be observed by other investigators at Stanford and elsewhere if they replicated the study. However, nothing in the study itself and the statistical analysis of the results directly supports this generalization. At the time of the study Stanford Hospital was unusual in the high proportion of caudal anesthetics used, and it is conceivable, though not likely, that the difference in efficacy for the two methods in a general obstetrical population might be much smaller, or even reversed, in some very select and unusual subset of patients.

3. Planning a clinical trial is a difficult problem, requiring the cooperative effort of physicians, statisticians, nurses, pharmacists, and others, depending on the area of medicine, the treatments under study, the measurements to be taken, and other factors. The following protocol reminder list, adapted from "Some problems in the clinical trials of anti-cancer agents," by M. Schneiderman, R. Taylor, and S. Fand, joint meeting of the Institute of Mathematical Statistics and Biometrics Society, March 9, 1957, furnishes some comprehensive and detailed guidance for the development of a detailed plan or *protocol* for a clinical trial.

Protocol Reminder Sheet

I   Why is the study being made?
    A.   What is the medical rationale?
    B.   Is the question to be answered precisely stated so that the cooperators agree on its specific as well as its general objectives?
    C.   How small a change is worth finding?

II   Who will be studied?
    A.   Are *all* patients seen with the disease(s) being studied recorded?
    B.   If some don't go into this study, why?
        1.   Will they go into another? How does one choose which patients go into which study? Random assignment? If not, why not?
        2.   Medical or ethical reasons? What are they, specifically?
    C.   How will patients with diagnosis in doubt be handled? Will there be review? By whom?
    D.   Will categories be set up for similar patients? Will this be a basis for randomization?

III   How will they be studied?
    A.   How will different patients be assigned to the different treatments? Is the randomization scheme an adequate bias suppressor?
    B.   Will the study be double blind? If not, why not?
    C.   What doses? For how long? What route?
    D.   How do patients get taken off the drug? For toxicity? Is toxicity objectively defined?
    E.   What other treatments may be given that won't confuse the results?
    F.   What other treatments may *not* be given?

IV   How big a study will this be?
    A.   How many patients will be needed to achieve the specific aims set out in I?
    B.   How long will it take to get them?
    C.   What is the breakdown at the separate institutions?

     D. Is the study planned to require as small a number of patients and as short a length of time as is feasible?

     E. Is there enough drug on hand for the whole study? Any problems of deterioration? Of generalization?

V Definition of treatment effects

     A. Rigor of criteria used

         1. Are there built-in checks on the comparability of data collected by the different investigators?

     B. Objectivity of measures

         1. Is there awareness that a double-blind study allows for the objective use of less objective criteria?

         2. Objective evaluation when double-blind is not possible?

VI Clinical and laboratory observations

     A. Are date of onset of symptoms and date of diagnosis recorded?

     B. Pretreatment evaluation—specific plan and pattern?

     C. Periodic evaluations—specific plan and pattern?

     D. Will the data be collected and recorded in a uniform fashion? On standard tabular forms, and not as a narrative?

VII Evaluation of results

     A. When is patient evaluation done?

         1. Is there defined a cutoff point? In time? In patient response?

     B. How is evaluation done?

         1. Are the evaluations blind?

     C. What are the provisions for follow-up?

VIII Analysis of trial (Note: We refer to *evaluation* of a case, but *analysis* of a clinical trial.)

     A. Is there a statement as to when to analyze the results of the trial?

         1. If there is a preliminary analysis is there provision for protection of the validity of the final analysis?

     B. Is provision made to use proper techniques to allow inclusion of patients with long survival, if survival is a criterion?

     C. Preplanning of final tabulations?

     D. Precision of estimates? How big a difference could one have missed, in view of the size of the experiments?

Some of the most important points concern the specifications of the patients to be studied, the details of treatment, and the nature of the measurements on outcome that are to be made. A beginning in the development of a plan for the aspirin trial might be to decide to recruit volunteers from some large pool, for example, the employees of a large corporation. Only persons using aspirin occasionally for headache would be chosen. Each would be given a bottle of similar-appearing tablets. The tablets would all have equal amounts of aspirin, but some persons would get tablets with antiacid additive, some without, the assignment being random. Each person would receive a form on which to record each episode requiring aspirin, the number of tablets taken, and the degree and rapidity of relief, each on a scale from zero to 10. Of course, they would be asked to refrain from taking other headache remedies.

This is just a sketch of an approach. The details would have to be spelled out very carefully and the whole study carried out under careful medical monitoring, with special attention to the informed consent of the patients and the protection of the health of the patients. Statistical questions yet to be resolved would be the nature of the final analyses and the numbers of patients needed to provide a reliable comparison.

5. There is a current and lively debate in the medical literature on the necessity for randomized clinical trials, in general, and with specific reference to certain therapies that are of questionable efficacy or that are prescribed more widely than might be appropriate. The putative benefits of tonsillectomy are to be found in an improved quality of life, with fewer episodes of minor illness. The costs are largely financial, although several hundred children in the United States die each year as a result of the surgery itself, and there are some claims of reduced resistance to certain diseases, such as Hodgkin's disease and mononucleosis. Clearly, it would be useful to both physician and parents to have definitive data on the balance among these benefits and risks, and to know the indications that would dictate surgery. Just as clearly, there would be difficulties in carrying out a clinical trial to obtain such information. Such a trial can be randomized only with great difficulty, because of patient and physician reluctance; further, it would be almost impossible to do the study as a blinded trial. Additionally, there are almost insurmountable measurement problems because of the difficulty of gaging improvements in the quality of life, and because of the very low incidence of undesirable events, such as post operative death and later serious disease.

The arguments pro and con for carrying out a clinical trial to measure costs and benefits of appendectomy would parallel those for tonsillectomy. There are a certain number of acute cases that would find universal agreement on the need for surgery to avert imminent death. Most appendectomies, however, are elective operations, and the decision to operate or not is not based on well-specified data and rules. The cost of the operation and the risk involved are not trivial, and the benefits, compared with medical therapy, including antibiotic treatment when needed, are not known with any precision. Therefore, as with tonsillectomy, there is a clear need for the kind of firm, unbiased evidence that a clinical trial could furnish. Yet, the difficulties of carrying out such a trial are even greater, since any surgeon electing to do the operation will be convinced of the need and most reluctant to randomize. Ignoring the ethical difficulties and the problems of patient consent, there is always the difficulty of blinding surgical trials and the problem of measuring the success in terms of long-range quality of life.

The question of randomizing in a heart transplant study is different from that for tonsillectomy and appendectomy in that the latter procedures are already in widespread use. If, indeed, they are being used too broadly, the cost is measured in hundreds of young lives and millions of dollars. Heart transplantation will never be adopted as common therapy, so there is not the need for persuasive, reproducible and public evidence of its net benefit. The investigators and their patients need some evidence, but they may well be satisfied with evidence based on careful study of the clinical data for patients receiving new hearts and comparable, though nonrandomized, patients who didn't receive a new heart. See, for example, the study of the early Stanford heart transplant survival data by Turnbull, Brown, and Hu (1974). (The Turnbull–Brown–Hu reference is given at the end of Chapter 6.)

7. Each part of this question would take a committee of experts several months to accomplish. The cost estimates would be critical to the planning and might even require pilot studies in order to secure good data on travel costs, medical examination costs, interview time, and so forth. Thus here we give the briefest of study sketches and cost estimates that are first approximations based on other experiences.

Strategy I.   We would seek out some cancer registry system already in existence, and there are several good ones in the United States. Then stomach cancer cases would be culled from the registry records. For each case found, we would select one or more controls, matching on certain variables such as age, race, sex, and socioeconomic status. For example, we might go to the neighborhood in which the cancer case lives (or last lived) and obtain controls from the neighborhood according to well-specified rules. Then we would set about measuring the aspirin consumption of stomach cancer cases and controls, using care to keep the method of measurement the same for case and control if possible.

Strategy II.   Prospective study of past data would require a file on a very large group of people who were asked about their aspirin consumption many years ago, since it is generally agreed that an agent that causes cancer will take some 10 or 20 years to become manifest. It is possible that certain insurance companies have routinely asked specifically about aspirin usage when issuing life insurance policies. If so, their policy holders could be traced and deaths could be examined as to cause. Thus, stomach cancer rates could be compared for aspirin users and nonusers.

Strategy III.   If the prospective study were to be done prospectively, one might begin by surveying a large population to find out about their aspirin usage. Then a large group of apparently healthy users and a matched group of nonusers would be selected and followed (i.e., checked for health intermittently) through the years, until a comparison of stomach cancer rates could be calculated, with small standard errors.

Strategy IV.   A randomized clinical trial would require subjects relatively young with minimal history of aspirin consumption, all willing to follow an assignment of aspirin abstinence or aspirin consumption, as prescribed, for a period of some 10 to 20 years. Each subject would be followed, and each death checked as to cause, until reliable stomach cancer rates could be determined on the two groups.

   With regard to costs, it is clear that cost depends rather directly on the number of cases studied, though certain basic costs of planning, central salary, rent, and so forth may not depend much on the number of cases studied. In the retrospective study, strategy I, one might expect the percent of aspirin takers to be 50% in the control group. If aspirin does cause stomach cancer the percent of aspirin takers should be higher among the cancer cases. It needn't be 100%, unless we hypothesize that aspirin is the only cause of stomach cancer and that our data on both stomach cancer and aspirin are perfect. If we had 100 cancer cases and 100 controls we could detect an increase in the percentage of aspirin users from 50% to 70% with good statistical power. If such an aim is acceptable, the cost could be calculated, based on culling the records for 100 cancer cases (say $5.00 per case), visiting neighborhoods to choose controls, and measuring aspirin usage by interview of subjects and/or relatives ($100.00 a subject, 200 subjects). Thus the study might cost in the order of $25,000 to $50,000, including planning, analysis and reporting.

On the other hand, the prospective study, strategy III, would require the accumulation of a total of 200 cases of stomach cancer among users and nonusers of aspirin to be relatively sure of detecting an increased risk of 50% among users. This would imply roughly 100 cases among the controls and 100 to 150 cases among the aspirin users, depending on whether or not aspirin was a cause of cancer. Given an incidence of 100 cases of stomach cancer per 100,000 persons per year, it is clear that the samples required would be in the order of tens of thousands of aspirin users and like numbers of nonusers. The actual numbers would depend on the number of years each person was observed: 100,000 persons followed for 1 year, or 10,000 persons observed for 10 years, during the period when the aspirin might be expected to cause cancer, might be equally acceptable. The costs in either case could be astronomical. Figuring the cost of recruiting and observing a case at a minimum of $10.00, the cost would be several millions of dollars.

The cost of using a ready-made file for a prospective study (strategy II) would be substantially less than the several millions necessary for strategy III, and might be competitive with the costs of the retrospective study (strategy I). The cost of a randomized trial (strategy IV) would certainly be several fold more than the nonrandomized prospective study (strategy III) because of the problems of recruitment, patient protection, treatment monitoring, and so forth.

| CHAPTER 7 | SOLUTIONS |

Discussion Questions

1. Not without further investigation. For example, there may be some factor which causes many deaths over a 3-year period that does not operate over the shorter 1-year period.

Problems

1. A 95% confidence interval for the population sensitivity rate is

$$\hat{p} \pm (1.96)\widehat{SD}(\hat{p}) = \frac{22}{30} \pm (1.96)\sqrt{\frac{\left(\frac{22}{30}\right)\left(\frac{8}{30}\right)}{30}} = .575 \text{ and } .892.$$

3. From Table C6, entered at $n = 5$, $B = 4$, and $\alpha = .10$, the interval with confidence coefficient (at least) .90 is found to be $p_L = .3426$, $p_U = .9898$.

5. For 95% confidence limits, the large sample approximation yields:

$$\hat{p} \pm (1.96)\widehat{SD}(\hat{p}) = .8 \pm (1.96)\sqrt{\frac{(.8)(.2)}{10}} = .55 \text{ and } 1.$$

From Table C6, entered at $n = 10$, $B = 8$, and $\alpha = .05$, the interval with confidence coefficient (at least) .95 is found to be $p_L = .4439$ and $p_U = .9748$. The large sample approximation interval and the interval obtained from Table C6 are in reasonable agreement.

7. If student A is simply guessing, the probability p that he will obtain the correct answer on a given question is 1/5 or .2. To test $p = .2$ compute

$$B^* = \frac{B - n(.2)}{\sqrt{n(.2)(.8)}} = \frac{8 - 25(.2)}{\sqrt{25(.2)(.8)}} = 1.5.$$

Referring $B^* = 1.5$ to Table C2, the one-sided P value (for testing $p = .2$ versus the alternative $p > .2$) is found to be .0668. Thus, for example, we cannot, at the $\alpha = .05$ level, reject the hypothesis of guessing.

CHAPTER 8	SOLUTIONS

Discussion Questions

1. Select a new graduating class of nurses and randomly assign them to the operating room and general duty groups.

Problems

1. Let p_1 denote the probability that a pregnant anesthetist will have a miscarriage. Let p_2 denote the probability that a pregnant member of the other-specialists group will have a miscarriage.

(a) $\hat{p}_1 = \frac{14}{37} = .378$, $\hat{p}_2 = \frac{6}{58} = .103$.

$$\widehat{SD}(\hat{p}_1) = \sqrt{\frac{\hat{p}_1(1-\hat{p}_1)}{m}} = \sqrt{\frac{(.378)(.622)}{37}} = .080.$$

$$\widehat{SD}(\hat{p}_2) = \sqrt{\frac{\hat{p}_2(1-\hat{p}_2)}{n}} = \sqrt{\frac{(.103)(.897)}{58}} = .040.$$

(b) $\widehat{SD}(\hat{p}_1 - \hat{p}_2) = \sqrt{[\widehat{SD}(\hat{p}_1)]^2 + [\widehat{SD}(\hat{p}_2)]^2} = \sqrt{(.08)^2 + (.04)^2} = .089.$

An approximate 95% confidence interval for $p_1 - p_2$ is

$$(\hat{p}_1 - \hat{p}_2) \pm (z_{\alpha/2})\widehat{SD}(\hat{p}_1 - \hat{p}_2) = (.378 - .103) \pm (1.96)(.089) = .275 \pm .174$$

$$= .10 \text{ and } .45.$$

3. Let p_1 denote the population rate of tonsillectomy for the students without M.D. parents and let p_2 denote the population rate of tonsillectomy for the students with M.D. parents.

$$\hat{p}_1 = \frac{134}{241} = .556, \qquad \hat{p}_2 = \frac{21}{54} = .389$$

$$\widehat{SD}(\hat{p}_1) = .0320, \qquad \widehat{SD}(\hat{p}_2) = .0663$$

$$\widehat{SD}(\hat{p}_1 - \hat{p}_2) = \sqrt{[\widehat{SD}(\hat{p}_1)]^2 + [\widehat{SD}(\hat{p}_2)]^2} = .074.$$

An approximate 95% confidence interval for $p_1 - p_2$ is

$$(\hat{p}_1 - \hat{p}_2) \pm (z_{\alpha/2})\widehat{SD}(\hat{p}_1 - \hat{p}_2) = (.556 - .389) \pm (1.96)(.074) = .167 \pm .145$$

$$= .02 \text{ and } .31.$$

5.
$$\chi_c^2 = \frac{112\{|(30)(40) - (26)(16)| - 56\}^2}{(56)(56)(46)(66)}$$

$$= 6.23.$$

From row 1 of Table C8 we find $\chi_{.05,1}^2 = 3.84$ and $\chi_{.01,1}^2 = 6.63$. Hence referring the observed value $\chi_c^2 = 6.23$ to Table C8 yields a P value between .01 and .05. Thus we conclude that the data are not consistent with the null hypothesis.

CHAPTER 9	SOLUTIONS

Discussion Questions

1. The use of the 1-year survival rates can be criticized. It is possible that as the transplant program gains experience, the chance of early posttransplant death could decrease without an increase in the chances of long-term posttransplant survival (e.g., 1-year survival rates might increase but the expected lengths of posttransplant survival times might even decrease). A more complete data set (such as the one given in Table 4 of Chapter 1) can be more informative.

Problems

1. $x^2 = \dfrac{(315-312.75)^2}{312.75} + \dfrac{(101-104.25)^2}{104.25} + \dfrac{(108-104.25)^2}{104.25} + \dfrac{(32-34.75)^2}{34.75} = .47.$

From Table C8 (entered at 3 degrees of freedom) we see that the P value is approximately between .90 and .95. Thus the data support the theoretical probabilities.

3. $x^2 = \dfrac{(19-20)^2}{20} + \dfrac{(21-20)^2}{20} + \dfrac{(26-20)^2}{20} + \dfrac{(18-20)^2}{20} + \dfrac{(23-20)^2}{20} + \dfrac{(13-20)^2}{20}$

$= 5.00.$

From Table C8, entered at five degrees of freedom, we find the P value is between .40 and .50. The data support the null hypothesis. Conclude that M. Hollander was tossing a fair die.

5. $x^2 = \dfrac{(17)^2}{12} + \dfrac{(16)^2}{12} + \dfrac{(3)^2}{12} + \dfrac{(3)^2}{8} + \dfrac{(4)^2}{8} + \dfrac{(17)^2}{8} - 60$

$= 25.4.$

From Table C8, entered at $(2-1) \times (3-1) = 2$ degrees of freedom, we see that the P value is less than .001. Thus there is strong evidence that the three conditions are not equivalent.

7. $x^2 = \dfrac{(14)^2}{12.121} + \dfrac{(6)^2}{7.879} + \dfrac{(17)^2}{10.303} + \dfrac{(0)^2}{6.697} + \dfrac{(9)^2}{17.576} + \dfrac{(20)^2}{11.424} - 66$

$= 22.4.$

From Table C8, entered at $(3-1) \times (2-1) = 2$ degrees of freedom, we find that the P value is less than .001. Thus there is strong evidence that the latecomers and earlycomers do not estimate their position with the same degree of optimism.

9. Note that, under H_0, $E_{11} = n_{..} \times Pr(A_1) \times Pr(C_1)$. To get an estimate of E_{11} simply estimate $Pr(A_1)$ by $n_{.1}/n_{..}$ and estimate $Pr(C_1)$ by $n_{1.}/n_{..}$.

11. (a) $C = 497(29+24) + 560(24) = 39{,}781$

$D = 19(560+269) + 29(269) = 23{,}552$

$S = C - D = 16{,}229$

$$n_{..}^3 = (1398)^3 = 2{,}732{,}256{,}792$$

$$\sum n_{.i}^3 = (516)^3 + (589)^3 + (293)^3 = 366{,}878{,}322$$

$$V = \frac{72(1{,}326)(2{,}732{,}256{,}792 - 366{,}878{,}322)}{3(1{,}398)(1{,}397)} = 38{,}543{,}560.2$$

$$S^* = \frac{S}{\sqrt{V}} = \frac{16{,}229}{6{,}208.3} = 2.61; \; P = .0045 \text{ (one-sided).}$$

Thus there is strong evidence of an increasing trend in the carrier rates as the tonsil sizes increase.

(b)
$$\chi^2 = \frac{(19)^2}{26.575} + \frac{(29)^2}{30.335} + \frac{(24)^2}{15.090} + \frac{(497)^2}{489.425}$$

$$+ \frac{(560)^2}{558.665} + \frac{(269)^2}{277.910} - 1{,}398 = 7.88.$$

Referring $\chi^2 = 7.88$ to the chi-squared distribution with $k - 1 = 2$ degrees of freedom, the P value is found to be between .01 and .05. Thus the χ^2 test also indicates that H_0 is not true. (Note that the χ^2 P value is larger than the two-sided P value of .009 obtained with the S^* test. This is consistent with the theory of the two tests. The S^* test concentrates its power on trend alternatives, whereas the χ^2 test guards against a larger class of alternatives, and thus, in the presence of a trend, the χ^2 test usually yields a higher P value than the S^* test.)

13. $C = 1{,}610{,}358$ $D = 943{,}034$

$$S = C - D = 667{,}324$$

$$n^3 = (3581)_.^3 = 45{,}921{,}171{,}941$$

$$\sum n_{.i}^3 = (534)^3 + (746)^3 + (784)^3 + (705)^3 + (443)^3 + (299)^3 + (70)^3$$

$$= 1{,}513{,}739{,}375$$

$$V = \frac{1416(2165)(45{,}921{,}171{,}941 - 1{,}513{,}739{,}375)}{3(3581)(3580)}$$

$$\sqrt{V} = 59{,}495.5$$

$$S^* = \frac{667{,}324}{59{,}495.5} = 11.2.$$

From Table C2, the P value corresponding to $S^* = 11.2$ is seen to be (much) less than .0002, and we conclude there is very strong evidence of an increasing trend.

CHAPTER 10	SOLUTIONS

Discussion Questions

1. In a randomized clinical trial, the null hypothesis is that the several treatments have the same effects, that is, they are interchangeable. In particular, this implies that they will achieve the same mean effects, but more generally, this statement of the null hypothesis also implies that the treatments would also manifest the same variability in response among patients, and in fact display the same distributional form. Thus in a randomized clinical trial, it may make very good sense to anticipate that under the null hypothesis both means and standard deviations of responses will be alike for the several treatments under test.

Problems

1. $n_1 = 20$, $n_2 = 18$, $n_3 = 19$, $k = 3$, $N - k = 54$.

Sample	Sample Size	Mean	Standard Deviation
Normal	20	186.0255	158.9505
All. Diabetic	18	181.8178	144.7238
All. Diab/Insulin	19	112.7200	105.8308

pooled standard deviation = 138.625

Since three intervals are to be computed, and overall confidence of 95% is desired, for each individual t interval we use a confidence coefficient of 98.33%. (See Comment 2 of Chapter 10.)

Thus we use equation (2) of Chapter 10 with $\alpha = .0167$ and $\alpha/2 = .0083$. With 54 degrees of freedom we can approximate $t_{.0083,54}$ by $z_{.0083} \approx 2.4$. The Scheffé interval is analogous to (2) but uses [see equation (3) of Chapter 10] $\sqrt{2F_{.05,2,54}}$ in place of $t_{.0083,54}$. We use $F_{.05,2,60} = 3.15$ as an approximation to $F_{.05,2,54}$ so that $\sqrt{2F_{.05,2,54}} \approx \sqrt{2(3.15)} = 2.51$.

Thus the t interval for $\mu_2 - \mu_1$ is

$$181.8178 - 186.0255 \pm (2.4)(138.625)\sqrt{\frac{1}{20} + \frac{1}{18}}$$

$$= -4.21 \pm 108.1.$$

The Scheffé interval for $\mu_2 - \mu_1$ is

$$-4.21 \pm (2.51)(138.625)\sqrt{\frac{1}{20} + \frac{1}{18}} = -4.21 \pm 113.0.$$

Similarly, the t interval for $\mu_3 - \mu_1$ is

$$-73.3 \pm (2.4)(138.625)\sqrt{\frac{1}{20} + \frac{1}{19}} = -73.3 \pm 106.6.$$

The Scheffé interval for $\mu_3 - \mu_1$ is

$$-73.3 \pm (2.51)(138.625)\sqrt{\frac{1}{20} + \frac{1}{19}} = -73.3 \pm 111.5.$$

The t interval for $[(\mu_2 + \mu_3)/2] - \mu_1$ is

$$\left(\frac{181.8178 + 112.72}{2} - 186.0255\right) \pm (2.4)(138.625)\sqrt{\frac{1}{20} + \left(\frac{1}{4}\right)\left(\frac{1}{18}\right) + \left(\frac{1}{4}\right)\left(\frac{1}{19}\right)}$$

$$= -38.8 \pm 92.3.$$

The Scheffé interval for $[(\mu_2 + \mu_3)/2] - \mu_1$ is

$$-38.8 \pm (2.51)(138.625)\sqrt{\frac{1}{20} + \left(\frac{1}{4}\right)\left(\frac{1}{18}\right) + \left(\frac{1}{4}\right)\left(\frac{1}{19}\right)}$$

$$= -38.8 \pm 96.6.$$

For this example, where there are only three contrasts of interest, the Scheffé procedure, developed for all possible contrasts, is too conservative, and the t intervals should be preferred.

Note that at an overall 95% level of confidence all three t intervals contain the point zero, and thus the claim that the two treated groups are different from the normal group is not as strong as a first impression of the data might have suggested.

3. A first approach is to calculate the analysis of variance table (see Table 2 of Chapter 10). We obtain

Source	SS	DF	MS	$\mathscr{F}$
Among samples	14.4950	2	7.2475	6.71
Within samples	9.7275	9	1.0808	
Total	24.2225	11		

Referring to Table C9 we see that 6.71 is well above the upper .025 percentile point of the F distribution with two degrees of freedom in the numerator and nine degrees of freedom in the denominator ($F_{.025,2,9} = 5.71$) so that the omnibus null hypothesis can be rejected at the $\alpha = 2.5\%$ level.

5. Multiply the MS_A in (7) by $k - 1$ to obtain the sum of squares among samples. Multiply s_p^2 in (8) by $N - k$ to obtain the sum of squares within samples. Add these two expressions together, and the result simplifies to $\sum X^2 - (T^2/N)$, the expression in Table 2 for the total sum of squares.

7. The analysis of variance table for this set of data is (see Table 9 of Chapter 10):

Source	SS	DF	MS	$\mathcal{F}$
Treatments	10.9460	2	5.4730	$\mathcal{F}_T = .40$
Subjects	108.3253	9	12.0361	
Residuals	246.6807	18	13.7045	
Total	365.9520	29		

Note that almost all the measurements in Table 12 of Chapter 10 are positive, suggesting that all drugs are efficacious in preventing motion sickness, compared to placebo. The analysis of variance table indicates that the variations among the three treatments (drugs) might well have come about by chance, the $\mathcal{F}_T$ statistic being well below its expected value of approximately unity under the null hypothesis of equivalence among the drugs. Specifically, the $\alpha = .05$ test rejects H_0 if $\mathcal{F}_T$ exceeds $F_{.05,2,18}$. From Table C9 we find $F_{.05,2,15} = 3.68$ and $F_{.05,2,20} = 3.49$. Thus $F_{.05,2,18}$ is between 3.49 and 3.68. We could interpolate to determine an approximation to $F_{.05,2,18}$, but it is already clear that our observed value $\mathcal{F}_T = .40$ is less than $F_{.05,2,18}$, and hence we accept H_0 at $\alpha = .05$.

9. The conservative approach inherent in application of the Scheffé procedure carefully controls the probability of rejecting the null hypothesis erroneously (i.e., when it is true). This is often the most important error, since rejection of the null hypothesis usually leads to some positive claim, for example, a manufacturer's assertion that his product is better than competitors. The consumer should enjoy the conservatism of the Scheffé procedure, because the procedure makes it hard to assert claims of superiority.

CHAPTER 11 | SOLUTIONS

Discussion Questions

1. Whereas the scatterplot gives a visual idea of the relationship between Y and X, the least squares line provides a quantitative measure of the relationship, valid when the true regression curve of Y on X is a linear regression.

Problems

1.
$$\sum X = 167 \qquad a = 16.48$$
$$\sum X^2 = 1,309 \qquad b = .96$$

$$\Sigma Y = 556 \qquad \hat{\sigma}_{Y \cdot X} = 4.09$$

$$\Sigma Y^2 = 13{,}384 \qquad \widehat{SD}(b) = .34$$

$$n = 24$$

$$\Sigma XY = 4{,}010.$$

To test $B = 0$ we compute

$$\left| \frac{b - 0}{\widehat{SD}(b)} \right| = \left| \frac{.96}{.34} \right| = 2.82.$$

Since $n = 24$ we enter Table C5 at $n - 2 = 22$ degrees of freedom. From Table C5 we find $t_{.005,22} = 2.819$, and thus the two-sided P value is approximately .01. Thus the X-ray measurement is clearly related to age, and averages about 1 mm more ($b = .96$ mm/yr) for each year of age.

3.
$$\Sigma[Y_i - (a + bX_i)] = \Sigma Y_i - na - b \Sigma X_i$$

$$= \Sigma Y_i - n(\bar{Y} - b\bar{X}) - bn\bar{X}$$

$$= n\bar{Y} - n\bar{Y} + bn\bar{X} - bn\bar{X} = 0.$$

5.
$$\log(.3) = -.5229, \qquad \log(1.5) = .1761,$$

$$\Sigma X = -2.0808, \qquad \Sigma X^2 = 1.8266$$

$$\Sigma Y = 3{,}670, \qquad \Sigma Y^2 = 1{,}153{,}750$$

$$\Sigma XY = -454.638 \qquad b = 123.99$$

$$a = 327.33.$$

7. Consider the following set of three data points:

$$X: 2, 4, 6$$

$$Y: 7, 7, 7$$

Since the value of Y is the same for all values X, the least-squares line has slope $b = 0$, and thus r is 0 [see equation (13) of Chapter 11].

There are other ways in which the sample correlation coefficient r can be 0. For example, the data points

$$X: 2, 4, 6$$

$$Y: 7, 9, 7$$

also yield $r = 0$.

The following data points all lie on a straight line with a negative slope, yielding a correlation of $r = -1$.

$$X: 2, 4, 6$$

$$Y: 9, 8, 7$$

Similarly, $r = +1$ for the following data points that lie on a straight line with positive slope:

$$X: 2, 4, 6$$

$$Y: 7, 8, 9$$

9. For the numerator of (13) we have:

$$\Sigma (X - \bar{X})(Y - \bar{Y}) = \Sigma XY - \bar{X} \Sigma Y - \bar{Y} \Sigma X + n\bar{X}\bar{Y}$$

$$= \Sigma XY - (\Sigma X)(\Sigma Y)/n.$$

For the denominator of (13) we have:

$$(n-1)s_X s_Y = \sqrt{(n-1)s_X^2(n-1)s_Y^2}$$

$$= \sqrt{\Sigma (X - \bar{X})^2 \Sigma (Y - \bar{Y})^2}$$

$$= \sqrt{[\Sigma X^2 - (\Sigma X)^2/n] \cdot [\Sigma Y^2 - (\Sigma Y)^2/n]}.$$

Then, writing numerator over denominator and multiplying each by n we obtain (14).

11. As a partial check on your clerical accuracy in ranking the data in Table 7 of Chapter 11, compare your rankings for the individual at the top of the first column and the next-to-last individual in the first column:

> obesity value 1.31 has obesity rank 35.5, blood pressure value 130 has blood pressure rank 27,

> obesity value 1.29 has obesity rank 32.5, blood pressure value 124 has blood pressure rank 19.5.

From the paired ranks we have:

$$\Sigma d^2 = 10,630.5$$

$$r_s = .2508.$$

From equation (19) of Chapter 11 we then find

$$r_s^* = \sqrt{43} \cdot r_s = 1.645,$$

corresponding to a two-sided P value of approximately .10. Thus the rank correlation coefficient between obesity and blood pressure is of borderline statistical significance, when tested against the null hypothesis of independence of obesity and blood pressure.

13. In the case $n = 6$, there are many paired rankings of six in which r_s is approximately 0 but none (with untied data) for which r_s is exactly 0. For example, for the following pairs,

1	6
2	1
3	3

4	4
5	2
6	5

$\sum d^2 = 36$ and $r_s = -.029$. (See Table C10.)

15. For example, in Table 7 if obesity value 1.58 were actually 1.98, the obesity ranks would remain the same and r_s would be unaltered. If the 1.57 obesity value had been measured with large error and should really have been 1.21, its corresponding obesity rank would change from 43 to 27, and the obesity values 1.31, 1.31, 1.34, 1.56, 1.29, 1.26, 1.32, 1.37, 1.25, 1.48, 1.29, 1.29, 1.29, 1.28, 1.37, 1.26 would each have their obesity rank increased by one, with all other obesity ranks remaining unchanged. This would change r_s only slightly (from .2508 to .2232).

CHAPTER 12	SOLUTIONS

Discussion Questions

1. The statistic T^+ is the sum of the positive signed ranks. Thus a wildly outlying value, say a very large treatment response, does not seriously affect the T^+ statistic. The very large observation receives the largest rank, but, beyond attaining this largest rank, the fact that this observation is extremely "far" from the other observations is not taken into account. Contrast this relative insensitivity with the more extreme sensitivity of the T test of Chapter 4.

Problems

1.

Rat Number	d	$\lvert d \rvert$	r
1	−166	166	5 (−)
2	−56	56	3 (−)
3	−962	962	10 (−)
4	−16	16	1 (−)
5	−706	706	9 (−)
6	−43	43	2 (−)
7	159	159	4 (+)
8	−256	256	6 (−)
9	−384	384	7 (−)
10	−665	665	8 (−)

Summing the positive signed ranks (there is only one positive signed rank in this example), we find $T^+ = 4$. From Table C11 (entered at $n = 10$) we find that when H_0 is true, $Pr\{T^+ \le 4\} = .007$. Thus the one-sided P value is .007, and the two-sided P value is .014. Thus there is strong evidence that the tape-closed wounds are stronger than the sutured wounds. (This signed rank analysis is in agreement with the results obtained using the T test of Chapter 4. See the solution to Problem 7 of Chapter 4.)

3.

Plasma Sample	d	$\|d\|$	r
1	5	5	23 (+)
2	3	3	10.5 (+)
3	4	4	17.5 (+)
4	1	1	3.5 (+)
5	1	1	3.5 (+)
6	5	5	23 (+)
7	0^a		
8	0^a		
9	3	3	10.5 (+)
10	4	4	17.5 (+)
11	0^a		
12	−1	1	3.5 (−)
13	0^a		
14	4	4	17.5 (+)
15	5	5	23 (+)
16	−6	6	25 (−)
17	1	1	3.5 (+)
18	3	3	10.5 (+)
19	2	2	7 (+)
20	1	1	3.5 (+)
21	3	3	10.5 (+)
22	1	1	3.5 (+)
23	3	3	10.5 (+)
24	4	4	17.5 (+)
25	4	4	17.5 (+)
26	9	9	26 (+)
27	4	4	17.5 (+)
28	4	4	17.5 (+)
29	4	4	17.5 (+)
30	3	3	10.5 (+)

[a] Note that, following the recommendation of Comment 2 of Chapter 11, we discard the d values of 0 and reduce the sample size so that here n is reduced to 26. We then rank the (remaining) 26 nonzero values of $|d|$. Ties are broken using average ranks.

Summing the positive signed ranks we find $T^+ = 322.5$. We then use the normal approxima-tion. We find

$$T^* = \frac{T^+ - \dfrac{n(n+1)}{4}}{\sqrt{\dfrac{n(n+1)(2n+1)}{24}}} = \frac{322.5 - \dfrac{26(27)}{4}}{\sqrt{\dfrac{26(27)(53)}{24}}} = 3.73.$$

Referring $T^* = 3.73$ to the standard normal table (Table C2) the one-sided P value is found to be less than .0002, and thus the two-sided P value is less than .0004. Thus there is very strong evidence that the two measurement methods are not equivalent. (This signed rank analysis is in agreement with the results obtained using the T test of Chapter 4. See the solution to Problem 5 of Chapter 4.)

5. The 55 ordered Walsh averages are $-962, -834, -813.5, -706, -685.5, -673, -665,$ $-609, -564, -545, -524.5, -509, -502.5, -489, -481, -460.5, -436, -415.5, -401.5,$ $-384, -381, -374.5, -361, -360.5, -354, -340.5, -320, -275, -273.5, -256, -253,$ $-220, -213.5, -211, -200, -166, -156, -149.5, -136, -112.5, -111, -104.5, -91,$ $-56, -49.5, -48.5, -43, -36, -29.5, -16, -3.5, 51.5, 58, 71.5, 159$. For a confidence coefficient of 95.2%, $1 - \alpha = .952$, $\alpha = .048$, and $\alpha/2 = .024$. From Table C11, entered at $n = 10$, we find the upper .024 percentile point of the null distribution of T^+ to be 47. Thus from equation (2) of Chapter 11,

$$b = \frac{10(11)}{2} + 1 - 47 = 9.$$

Thus the 95.2% confidence interval for μ has as its lower end point the value occupying position 9 in the list of the ordered Walsh averages and as its upper end point the value occupying position 47. The interval is thus found to be $(-564, -43)$. (Compare with the interval found in the solution to Problem 3 of Chapter 4.)

7. For a 99.8% confidence interval, $1 - \alpha = .998$, $\alpha = .002$, and $\alpha/2 = .001$. From Table C11, entered at $n = 10$, the upper .001 percentile point of the null distribution of T^+ is found to be 55. Then, from equation (2) of Chapter 11,

$$b = \frac{10(11)}{2} + 1 - 55 = 1.$$

Thus the 99.8% confidence interval for μ has as its lower end point the value occupying position 1 in the list of the ordered Walsh averages; that is, the lower end point is the *smallest* Walsh average. Similarly, the upper end point is the *largest* Walsh average. Now note that the smallest Walsh average equals the smallest d value and the largest Walsh average equals the largest d value. Thus, *without computing any Walsh averages in this case*, the interval is found to be $(-962, 159)$.

CHAPTER 13	SOLUTIONS

Discussion Questions

1. A very large value in sample 2 does not seriously affect the W statistic. The very large observation receives the largest rank but, beyond attaining this largest rank, the fact that this observation is extremely "far" from the other observations is not taken into account. This is a desirable property of the W statistic, since the outlying observation may simply represent a gross error in obtaining, or recording, the measured response. Contrast this relative insensitivity of the W statistic with the more extreme sensitivity of the two-sample test of Chapter 5, based on the sample means.

Problems

1. The ranks obtained in the joint ranking of the 12 observations are as follows:

Chronic exposure	Acute exposure
4	4
4	8.5
11	4
8.5	1
8.5	
4	
12	
8.5	

Hence we have $W = 4 + 8.5 + 4 + 1 = 17.5$. From Table C12 (entered at $m = 8$, $n = 4$, lower tail) we see the P value for $W = 17.5$ is between .077 and .107. Thus although we cannot reject at the $\alpha = .05$ level, the investigator might well take the position that the data suggest that the acute exposure population mean is less than the mean of the chronic exposure population. (Recall the solution to Problem 5 of Chapter 5, and note that the results obtained here are in agreement.)

3. The statistic $W + W'$ equals the sum of the sample 1 and sample 2 ranks. Thus $W + W'$ equals the sum of the first N integers. That is

$$W + W' = 1 + 2 + \cdots + N.$$

But it can be shown that the sum of the first N integers is equal to $N(N+1)/2$. Thus $W + W' = N(N+1)/2$.

5. The $4 \times 4 = 16$ ordered d-differences are $-14.5, -12.4, -10, -8.7, -8.7, -7.9, -6.6$, $-6.6, -6.1, -4.5, -4.2, -2.9, -1.6, -.8, -.3, 1.8$. For a 94.2% confidence interval, $1 - \alpha = .942$, $\alpha = .058$, and $\alpha/2 = .029$. From Table C12 (entered at $m = n = 4$, upper tail) we find the upper .029 percentile point of the null distribution of W to be 25. Thus from equation (2) of Chapter 13,

$$l = \frac{4(8+4+1)}{2} + 1 - 25 = 2.$$

Then the 94.2% confidence interval for Δ has as its lower end point the value occupying position 2 in the list of the ordered d's and as its upper end point the value occupying position 15. The interval is thus $(-12.4, -.3)$. [Recall that by using formula (8) of Chapter 5 we obtained a 95% confidence interval for Δ of $(-12.02, .26)$. This interval was based on the assumption that the two parent populations are normal with equal variances. The confidence interval obtained here does not require the assumption of normal parent populations.]

7. For a 99.8% confidence interval, $1 - \alpha = .998$, $\alpha = .002$, and $\alpha/2 = .001$. From Table C12 (entered at $m = 9$, $n = 5$, upper tail), we find the upper .001 percentile point of the null distribution of W to be 59. Thus, from equation (2) of Chapter 13,

$$l = \frac{5(18+5+1)}{2} + 1 - 59 = 2.$$

Then the 99.8% confidence interval for Δ has as its lower end point the value occupying position 2 in the list of the ordered d's and as its upper end point the value occupying position 44. The interval is found to be $(-21, 0)$. To find this confidence interval we need to compute only the two smallest and two largest d values. This can be done quickly as follows. Let $X_{(1)} \le X_{(2)} \le \cdots \le X_{(m)}$ denote the *ordered* sample 1 values. Similarly, let $Y_{(1)} \le Y_{(2)} \le \cdots \le Y_{(n)}$ denote the *ordered* sample 2 values. Finally, let $d_{(1)} \le d_{(2)} \le \cdots \le d_{(mn)}$ denote the $m \times n$ ordered d-differences. A little thought will convince you that

$$d_{(1)} = Y_{(1)} - X_{(m)} = [\text{smallest sample 2 value}] - [\text{largest sample 1 value}],$$

$$d_{(2)} = \text{the smaller of } Y_{(1)} - X_{(m-1)} \text{ and } Y_{(2)} - X_{(m)},$$

$$d_{(mn-1)} = \text{the larger of } Y_{(n)} - X_{(2)} \text{ and } Y_{(n-1)} - X_{(1)},$$

$$d_{(mn)} = Y_{(n)} - X_{(1)} = [\text{largest sample 2 value}] - [\text{smallest sample 1 value}].$$

For the kidney data of Table 1, Chapter 13, we immediately find the two smallest sample 1 values are 18 and 22, and the two largest sample 1 values are 31 and 34. In the notation just introduced, $X_{(1)} = 18$, $X_{(2)} = 22$, $X_{(8)} = 31$, $X_{(9)} = 34$. Similarly, for the sample 2 values, $Y_{(1)} = 10$, $Y_{(2)} = 13$, $Y_{(4)} = 15$, $Y_{(5)} = 22$. Thus

$$d_{(1)} = 10 - 34 = -24,$$

$$d_{(2)} = \text{the smaller of } 10 - 31 \text{ and } 13 - 34,$$

$$= \text{the smaller of } -21 \text{ and } -21 = -21,$$

$$d_{(44)} = \text{the larger of } 22 - 22 \text{ and } 15 - 18,$$

$$= \text{the larger of } 0 \text{ and } -3 = 0,$$

$$d_{(45)} = 22 - 18 = 4.$$

Thus our lower end point is directly (without computing the 45 d-differences) found to be $d_{(2)} = -21$, and the upper end point is directly found to be $d_{(44)} = 0$.

CHAPTER 14 | **SOLUTIONS**

Discussion Questions

1. An obvious weakness in the Gehan scoring system occurs for those cases where we score zero. Recall that we score zero when it is impossible to say which of the treatments will win in a particular comparison. This score ignores the values of the observations, when in fact these values provide information about which treatment will win. For example, the twenty-first patient in the affected node therapy group (Table 1 of Chapter 14) had a censored value of 1309, and the second patient in the total nodal radiation group had a censored value of 2177. Since 2177 is much larger than 1309, one would guess that the actual unknown days to relapse for the patient with censored observation 2177 will exceed the actual unknown days to relapse for the patient with censored observation 1309. Thus the data suggest that scoring zero (a value that, in this context, represents indifference) is inappropriate. We should score some value that is closer to 1 (than it is to -1) to reflect the view that the total nodal radiation treatment will probably win in this particular comparison. In this spirit, B. Efron ("The two sample problem with censored data," *Proceedings of the Fifth Berkeley Symposium on Mathematical Statistics and Probability, Vol. IV*, 831–854, 1967, University of California Press, Berkeley, Calif.) has designed a test with a scoring system that improves upon the Gehan scoring system.

Problems

1. In the Group A and Group B tables, for brevity we use the word relapse to signify manifestation of arteriosclerosis.

GROUP A

Time to Relapse or to Date of Analysis	Relapse (Y) or Not (N)	Patients at Risk	Estimated Conditional Probability of Relapse on Date, Given that the Patient has Not Yet Relapsed	Estimated Probability of No Relapse to Date
128	Y	8	1/8	0.875
137	Y	7	1/7	0.750
281	Y	6	1/6	0.625
625	Y	5	1/5	0.500
730	Y	4	1/4	0.375
1025	Y	3	1/3	0.250
1351	Y	2	1/2	0.125
1776	N	1	0	0.125

GROUP B

148	Y	14	1/14	0.929
254	Y	13	1/13	0.857
323	Y	12	1/12	0.786
552	Y	11	1/11	0.714
816	N	10	0	0.714
839	N	9	0	0.714
876	N	8	0	0.714
898	Y	7	1/7	0.612
995	Y	6	1/6	0.510
1107	N	5	0	0.510
1265	N	4	0	0.510
1368	N	3	0	0.510
1537	N	2	0	0.510
1550	N	1	0	0.510

To compute the Gehan test statistic, we make use of the following table, where R_1, R_2, h are as defined in Chapter 14.

Treatment	Observation	Relapse (Y) or Not (N)	R_1	R_2	h
1	128	Y	0	21	−21
1	137	Y	1	20	−19
2	148	Y	2	19	−17
2	254	Y	3	18	−15
1	281	Y	4	17	−13
2	323	Y	5	16	−11
2	552	Y	6	15	−9
1	625	Y	7	14	−7
1	730	Y	8	13	−5
2	816	N	9	0	9
2	839	N	9	0	9
2	876	N	9	0	9
2	898	Y	9	9	0
2	995	Y	10	8	2
1	1025	Y	11	7	4
2	1107	N	12	0	12
2	1265	N	12	0	12
1	1351	Y	12	4	8
2	1368	N	13	0	13
2	1537	N	13	0	13
2	1550	N	13	0	13
1	1776	N	13	0	13

Now

$$V = \text{sum of } h \text{ values for sample } 2 = 40.$$

$$\sum h^2 = \text{sum of the squares of the } h \text{ values for both samples} = 3,052.$$

$$SD(V) = \sqrt{\frac{mn \sum h^2}{(m+n)(m+n-1)}} = \sqrt{\frac{8(14)(3,052)}{22(21)}}$$

$$= \sqrt{739.88} = 27.2.$$

Then

$$V^* = \frac{V}{SD(V)} = \frac{40}{27.2} = 1.47.$$

Referring $V^* = 1.47$ to Table C2 we find the one-tailed P value is .0708 (and the two-tailed P value is .1416). Thus at $\alpha = .05$ we cannot conclude that μ_2 (the mean of time to manifestation of arteriosclerosis for the conceptual population in the arteriosclerosis prevention program) is greater than μ_1 (the mean of time to manifestation of arteriosclerosis for the conceptual population not in the prevention program). However, since the one-sided P value is relatively low, we might well take the position that the data suggest that the prevention program is useful for delaying the development of arteriosclerosis.

3. We let $Z_{(1)} < Z_{(2)} < \cdots < Z_{(n)}$ denote the ordered uncensored survival times. Then the estimated probability of death at time $Z_{(i)}$, given survival up to time $Z_{(i)}$, is $1/(n-i+1)$. Therefore, the estimated probability of surviving past time $Z_{(i)}$ is

$$\left(1-\frac{1}{n}\right) \times \left(1-\frac{1}{n-1}\right) \times \left(1-\frac{1}{n-2}\right) \times \cdots \times \left(1-\frac{1}{n-i+1}\right).$$

We can rewrite this product as

$$\left(\frac{n-1}{n}\right) \times \left(\frac{n-2}{n-1}\right) \times \left(\frac{n-3}{n-2}\right) \times \cdots \times \left(\frac{n-i}{n-i+1}\right).$$

Then observe that the numerator of a given term in the product cancels with the denominator of the next term so that the product "telescopes" to $(n-i)/n$.

APPENDIX C

Mathematical and Statistical Tables

TABLE C1. Square Roots

For each positive integer n between 1 and 500, the table contains the values of $\sqrt{n}$ and $\sqrt{10n}$. Moving the decimal point two places in n is equivalent to moving it one place in $\sqrt{n}$. As examples of the use of the table, note $\sqrt{22.5} = 4.74$, $\sqrt{225} = 15$, $\sqrt{2250} = 47.4$.

n	$\sqrt{n}$	$\sqrt{10n}$	n	$\sqrt{n}$	$\sqrt{10n}$	n	$\sqrt{n}$	$\sqrt{10n}$
1	1.000	3.162	44	6.633	20.976	87	9.327	29.496
2	1.414	4.472	45	6.708	21.213	88	9.381	29.665
3	1.732	5.477	46	6.782	21.448	89	9.434	29.833
4	2.000	6.325	47	6.856	21.679	90	9.487	30.000
5	2.236	7.071	48	6.928	21.909	91	9.539	30.166
6	2.449	7.746	49	7.000	22.136	92	9.592	30.332
7	2.646	8.367	50	7.071	22.361	93	9.644	30.496
8	2.828	8.944	51	7.141	22.583	94	9.695	30.659
9	3.000	9.487	52	7.211	22.804	95	9.747	30.822
10	3.162	10.000	53	7.280	23.022	96	9.798	30.984
11	3.317	10.488	54	7.348	23.238	97	9.849	31.145
12	3.464	10.954	55	7.416	23.452	98	9.899	31.305
13	3.606	11.402	56	7.483	23.664	99	9.950	31.464
14	3.742	11.832	57	7.550	23.875	100	10.000	31.623
15	3.873	12.247	58	7.616	24.083	101	10.050	31.780
16	4.000	12.649	59	7.681	24.290	102	10.100	31.937
17	4.123	13.038	60	7.746	24.495	103	10.149	32.094
18	4.243	13.416	61	7.810	24.698	104	10.198	32.249
19	4.359	13.784	62	7.874	24.900	105	10.247	32.404
20	4.472	14.142	63	7.937	25.100	106	10.296	32.558
21	4.583	14.491	64	8.000	25.298	107	10.344	32.711
22	4.690	14.832	65	8.062	25.495	108	10.392	32.863
23	4.796	15.166	66	8.124	25.690	109	10.440	33.015
24	4.899	15.492	67	8.185	25.884	110	10.488	33.166
25	5.000	15.811	68	8.246	26.077	111	10.536	33.317
26	5.099	16.125	69	8.307	26.268	112	10.583	33.466
27	5.196	16.432	70	8.367	26.458	113	10.630	33.615
28	5.292	16.733	71	8.426	26.646	114	10.677	33.764
29	5.385	17.029	72	8.485	26.833	115	10.724	33.912
30	5.477	17.321	73	8.544	27.019	116	10.770	34.059
31	5.568	17.607	74	8.602	27.203	117	10.817	34.205
32	5.657	17.889	75	8.660	27.386	118	10.863	34.351
33	5.745	18.166	76	8.718	27.568	119	10.909	34.496
34	5.831	18.439	77	8.775	27.749	120	10.954	34.641
35	5.916	18.708	78	8.832	27.928	121	11.000	34.785
36	6.000	18.974	79	8.888	28.107	122	11.045	34.928
37	6.083	19.235	80	8.944	28.284	123	11.091	35.071
38	6.164	19.494	81	9.000	28.460	124	11.136	35.214
39	6.245	19.748	82	9.055	28.636	125	11.180	35.355
40	6.325	20.000	83	9.110	28.810	126	11.225	35.496
41	6.403	20.248	84	9.165	28.983	127	11.269	35.637
42	6.481	20.494	85	9.220	29.155	128	11.314	35.777
43	6.557	20.736	86	9.274	29.326	129	11.358	35.917

n	$\sqrt{n}$	$\sqrt{10n}$	n	$\sqrt{n}$	$\sqrt{10n}$	n	$\sqrt{n}$	$\sqrt{10n}$
130	11.402	36.056	176	13.266	41.952	222	14.900	47.117
131	11.446	36.194	177	13.304	42.071	223	14.933	47.223
132	11.489	36.332	178	13.342	42.190	224	14.967	47.329
133	11.533	36.469	179	13.379	42.308	225	15.000	47.434
134	11.576	36.606	180	13.416	42.426	226	15.033	47.539
135	11.619	36.742	181	13.454	42.544	227	15.067	47.645
136	11.662	36.878	182	13.491	42.661	228	15.100	47.749
137	11.705	37.014	183	13.528	42.778	229	15.133	47.854
138	11.747	37.148	184	13.565	42.895	230	15.166	47.958
139	11.790	37.283	185	13.601	43.012	231	15.199	48.062
140	11.832	37.417	186	13.638	43.128	232	15.232	48.166
141	11.874	37.550	187	13.675	43.243	233	15.264	48.270
142	11.916	37.683	188	13.711	43.359	234	15.297	48.374
143	11.958	37.815	189	13.748	43.474	235	15.330	48.477
144	12.000	37.947	190	13.784	43.589	236	15.362	48.580
145	12.042	38.079	191	13.820	43.704	237	15.395	48.683
146	12.083	38.210	192	13.856	43.818	238	15.427	48.785
147	12.124	38.341	193	13.892	43.932	239	15.460	48.888
148	12.166	38.471	194	13.928	44.045	240	15.492	48.990
149	12.207	38.601	195	13.964	44.159	241	15.524	49.092
150	12.247	38.730	196	14.000	44.272	242	15.556	49.193
151	12.288	38.859	197	14.036	44.385	243	15.588	49.295
152	12.329	38.987	198	14.071	44.497	244	15.620	49.396
153	12.369	39.115	199	14.107	44.609	245	15.652	49.497
154	12.410	39.243	200	14.142	44.721	246	15.684	49.598
155	12.450	39.370	201	14.177	44.833	247	15.716	49.699
156	12.490	39.497	202	14.213	44.944	248	15.748	49.800
157	12.530	39.623	203	14.248	45.056	249	15.780	49.900
158	12.570	39.749	204	14.283	45.166	250	15.811	50.000
159	12.610	39.875	205	14.318	45.277	251	15.843	50.100
160	12.649	40.000	206	14.353	45.387	252	15.875	50.200
161	12.689	40.125	207	14.387	45.497	253	15.906	50.299
162	12.728	40.249	208	14.422	45.607	254	15.937	50.398
163	12.767	40.373	209	14.457	45.717	255	15.969	50.498
164	12.806	40.497	210	14.491	45.826	256	16.000	50.596
165	12.845	40.620	211	14.526	45.935	257	16.031	50.695
166	12.884	40.743	212	14.560	46.043	258	16.062	50.794
167	12.923	40.866	213	14.595	46.152	259	16.093	50.892
168	12.961	40.988	214	14.629	46.260	260	16.125	50.990
169	13.000	41.110	215	14.663	46.368	261	16.155	51.088
170	13.038	41.231	216	14.697	46.476	262	16.186	51.186
171	13.077	41.352	217	14.731	46.583	263	16.217	51.284
172	13.115	41.473	218	14.765	46.690	264	16.248	51.381
173	13.153	41.593	219	14.799	46.797	265	16.279	51.478
174	13.191	41.713	220	14.832	46.904	266	16.310	51.575
175	13.229	41.833	221	14.866	47.011	267	16.340	51.672

n	$\sqrt{n}$	$\sqrt{10n}$	n	$\sqrt{n}$	$\sqrt{10n}$	n	$\sqrt{n}$	$\sqrt{10n}$
268	16.371	51.769	314	17.720	56.036	360	18.974	60.000
269	16.401	51.865	315	17.748	56.125	361	19.000	60.083
270	16.432	51.962	316	17.776	56.214	362	19.026	60.166
271	16.462	52.058	317	17.804	56.303	363	19.053	60.249
272	16.492	52.154	318	17.833	56.391	364	19.079	60.332
273	16.523	52.249	319	17.861	56.480	365	19.105	60.415
274	16.553	52.345	320	17.889	56.569	366	19.131	60.498
275	16.583	52.440	321	17.916	56.657	367	19.157	60.581
276	16.613	52.536	322	17.944	56.745	368	19.183	60.663
277	16.643	52.631	323	17.972	56.833	369	19.209	60.745
278	16.673	52.726	324	18.000	56.921	370	19.235	60.828
279	16.703	52.820	325	18.028	57.009	371	19.261	60.910
280	16.733	52.915	326	18.055	57.096	372	19.287	60.992
281	16.763	53.009	327	18.083	57.184	373	19.313	61.074
282	16.793	53.104	328	18.111	57.271	374	19.339	61.156
283	16.823	53.198	329	18.138	57.359	375	19.365	61.237
284	16.852	53.292	330	18.166	57.446	376	19.391	61.319
285	16.882	53.385	331	18.193	57.533	377	19.416	61.400
286	16.912	53.479	332	18.221	57.619	378	19.442	61.482
287	16.941	53.572	333	18.248	57.706	379	19.468	61.563
288	16.971	53.666	334	18.276	57.793	380	19.494	61.644
289	17.000	53.759	335	18.303	57.879	381	19.519	61.725
290	17.029	53.852	336	18.330	57.966	382	19.545	61.806
291	17.059	53.944	337	18.358	58.052	383	19.570	61.887
292	17.088	54.037	338	18.385	58.138	384	19.596	61.968
293	17.117	54.129	339	18.412	58.224	385	19.621	62.048
294	17.146	54.222	340	18.439	58.310	386	19.647	62.129
295	17.176	54.314	341	18.466	58.395	387	19.672	62.209
296	17.205	54.406	342	18.493	58.481	388	19.698	62.290
297	17.234	54.498	343	18.520	58.566	389	19.723	62.370
298	17.263	54.589	344	18.547	58.652	390	19.748	62.450
299	17.292	54.681	345	18.574	58.737	391	19.774	62.530
300	17.321	54.772	346	18.601	58.822	392	19.799	62.610
301	17.349	54.863	347	18.628	58.907	393	19.824	62.690
302	17.378	54.955	348	18.655	58.992	394	19.849	62.769
303	17.407	55.045	349	18.682	59.076	395	19.875	62.849
304	17.436	55.136	350	18.708	59.161	396	19.900	62.929
305	17.464	55.227	351	18.735	59.245	397	19.925	63.008
306	17.493	55.317	352	18.762	59.330	398	19.950	63.087
307	17.521	55.408	353	18.788	59.414	399	19.975	63.166
308	17.550	55.498	354	18.815	59.498	400	20.000	63.246
309	17.578	55.588	355	18.841	59.582	401	20.025	63.325
310	17.607	55.678	356	18.868	59.666	402	20.050	63.403
311	17.635	55.767	357	18.894	59.749	403	20.075	63.482
312	17.664	55.857	358	18.921	59.833	404	20.100	63.561
313	17.692	55.946	359	18.947	59.917	405	20.125	63.640

n	$\sqrt{n}$	$\sqrt{10n}$	n	$\sqrt{n}$	$\sqrt{10n}$	n	$\sqrt{n}$	$\sqrt{10n}$
406	20.149	63.718	438	20.928	66.182	470	21.679	68.557
407	20.174	63.797	439	20.952	66.257	471	21.703	68.629
408	20.199	63.875	440	20.976	66.332	472	21.726	68.702
409	20.224	63.953	441	21.000	66.408	473	21.749	68.775
410	20.248	64.031	442	21.024	66.483	474	21.772	68.848
411	20.273	64.109	443	21.048	66.558	475	21.794	68.920
412	20.298	64.187	444	21.071	66.633	476	21.817	68.993
413	20.322	64.265	445	21.095	66.708	477	21.840	69.065
414	20.347	64.343	446	21.119	66.783	478	21.863	69.138
415	20.372	64.420	447	21.142	66.858	479	21.886	69.210
416	20.396	64.498	448	21.166	66.933	480	21.909	69.282
417	20.421	64.576	449	21.190	67.007	481	21.932	69.354
418	20.445	64.653	450	21.213	67.082	482	21.954	69.426
419	20.469	64.730	451	21.237	67.157	483	21.977	69.498
420	20.494	64.807	452	21.260	67.231	484	22.000	69.570
421	20.518	64.885	453	21.284	67.305	485	22.023	69.642
422	20.543	64.962	454	21.307	67.380	486	22.045	69.714
423	20.567	65.038	455	21.331	67.454	487	22.068	69.785
424	20.591	65.115	456	21.354	67.528	488	22.091	69.857
425	20.616	65.192	457	21.378	67.602	489	22.113	69.929
426	20.640	65.269	458	21.401	67.676	490	22.136	70.000
427	20.664	65.345	459	21.424	67.750	491	22.159	70.071
428	20.688	65.422	460	21.448	67.823	492	22.181	70.143
429	20.712	65.498	461	21.471	67.897	493	22.204	70.214
430	20.736	65.574	462	21.494	67.971	494	22.226	70.285
431	20.761	65.651	463	21.517	68.044	495	22.249	70.356
432	20.785	65.727	464	21.541	68.118	496	22.271	70.427
433	20.809	65.803	465	21.564	68.191	497	22.293	70.498
434	20.833	65.879	466	21.587	68.264	498	22.316	70.569
435	20.857	65.955	467	21.610	68.337	499	22.338	70.640
436	20.881	66.030	468	21.633	68.411	500	22.361	70.711
437	20.905	66.106	469	21.656	68.484			

TABLE C2.   Areas for the Standard Normal Distribution

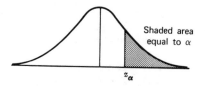

Shaded area equal to α

z_α

For each z, the corresponding entry in the table is the area to the right of z under the standard normal curve. For example, the area to the right of $z = 2.23$ is .0129.

z	.00	.01	.02	.03	.04	.05	.06	.07	.08	.09
.0	.5000	.4960	.4920	.4880	.4840	.4801	.4761	.4721	.4681	.4641
.1	.4602	.4562	.4522	.4483	.4443	.4404	.4364	.4325	.4286	.4247
.2	.4207	.4168	.4129	.4090	.4052	.4013	.3974	.3936	.3897	.3859
.3	.3821	.3783	.3745	.3707	.3669	.3632	.3594	.3557	.3520	.3483
.4	.3446	.3409	.3372	.3336	.3300	.3264	.3228	.3192	.3156	.3121
.5	.3085	.3050	.3015	.2981	.2946	.2912	.2877	.2843	.2810	.2776
.6	.2743	.2709	.2676	.2643	.2611	.2578	.2546	.2514	.2483	.2451
.7	.2420	.2389	.2358	.2327	.2296	.2266	.2236	.2206	.2177	.2148
.8	.2119	.2090	.2061	.2033	.2005	.1977	.1949	.1922	.1894	.1867
.9	.1841	.1814	.1788	.1762	.1736	.1711	.1685	.1660	.1635	.1611
1.0	.1587	.1562	.1539	.1515	.1492	.1469	.1446	.1423	.1401	.1379
1.1	.1357	.1335	.1314	.1292	.1271	.1251	.1230	.1210	.1190	.1170
1.2	.1151	.1131	.1112	.1093	.1075	.1056	.1038	.1020	.1003	.0985
1.3	.0968	.0951	.0934	.0918	.0901	.0885	.0869	.0853	.0838	.0823
1.4	.0808	.0793	.0778	.0764	.0749	.0735	.0721	.0708	.0694	.0681
1.5	.0668	.0655	.0643	.0630	.0618	.0606	.0594	.0582	.0571	.0559
1.6	.0548	.0537	.0526	.0516	.0505	.0495	.0485	.0475	.0465	.0455
1.7	.0446	.0436	.0427	.0418	.0409	.0401	.0392	.0384	.0375	.0367
1.8	.0359	.0351	.0344	.0336	.0329	.0322	.0314	.0307	.0301	.0294
1.9	.0287	.0281	.0274	.0268	.0262	.0256	.0250	.0244	.0239	.0233
2.0	.0228	.0222	.0217	.0212	.0207	.0202	.0197	.0192	.0188	.0183
2.1	.0179	.0174	.0170	.0166	.0162	.0158	.0154	.0150	.0146	.0143
2.2	.0139	.0136	.0132	.0129	.0125	.0122	.0119	.0116	.0113	.0110
2.3	.0107	.0104	.0102	.0099	.0096	.0094	.0091	.0089	.0087	.0084
2.4	.0082	.0080	.0078	.0075	.0073	.0071	.0069	.0068	.0066	.0064
2.5	.0062	.0060	.0059	.0057	.0055	.0054	.0052	.0051	.0049	.0048
2.6	.0047	.0045	.0044	.0043	.0041	.0040	.0039	.0038	.0037	.0036

Table C2. (*continued*)

z	.00	.01	.02	.03	.04	.05	.06	.07	.08	.09
2.7	.0035	.0034	.0033	.0032	.0031	.0030	.0029	.0028	.0027	.0026
2.8	.0026	.0025	.0024	.0023	.0023	.0022	.0021	.0021	.0020	.0019
2.9	.0019	.0018	.0018	.0017	.0016	.0016	.0015	.0015	.0014	.0014
3.0	.0013	.0013	.0013	.0012	.0012	.0011	.0011	.0011	.0010	.0010
3.1	.0010	.0009	.0009	.0009	.0008	.0008	.0008	.0008	.0007	.0007
3.2	.0007	.0007	.0006	.0006	.0006	.0006	.0006	.0005	.0005	.0005
3.3	.0005	.0005	.0005	.0004	.0004	.0004	.0004	.0004	.0004	.0003
3.4	.0003	.0003	.0003	.0003	.0003	.0003	.0003	.0003	.0003	.0002

TABLE C3. Number of Combinations $\binom{N}{n}$ of N Things Taken n at a Time

Examples: $\binom{17}{4} = 2380$, $\binom{21}{11} = 352{,}716$

N\n	2	3	4	5	6	7	8	9	10	11	12	13	14	15
2	1													
3	3	1												
4	6	4	1											
5	10	10	5	1										
6	15	20	15	6	1									
7	21	35	35	21	7	1								
8	28	56	70	56	28	8	1							
9	36	84	126	126	84	36	9	1						
10	45	120	210	252	210	120	45	10	1					
11	55	165	330	462	462	330	165	55	11	1				
12	66	220	495	792	924	792	495	220	66	12	1			
13	78	286	715	1,287	1,716	1,716	1,287	715	286	78	13	1		
14	91	364	1,001	2,002	3,003	3,432	3,003	2,002	1,001	364	91	14	1	
15	105	455	1,365	3,003	5,005	6,435	6,435	5,005	3,003	1,365	455	105	15	1
16	120	560	1,820	4,368	8,008	11,440	12,870	11,440	8,008	4,368	1,820	560	120	16
17	136	680	2,380	6,188	12,376	19,448	24,310	24,310	19,448	12,376	6,188	2,380	680	136
18	153	816	3,060	8,568	18,564	31,824	43,758	48,620	43,758	31,824	18,564	8,568	3,060	816
19	171	969	3,876	11,628	27,132	50,388	75,582	92,378	92,378	75,582	50,388	27,132	11,628	3,876
20	190	1,140	4,845	15,504	38,760	77,520	125,970	167,960	184,756	167,960	125,970	77,520	38,760	15,504
21	210	1,330	5,985	20,349	54,264	116,280	203,490	293,930	352,716	352,716	293,930	203,490	116,280	54,264
22	231	1,540	7,315	26,334	74,613	170,544	319,770	497,420	646,646	705,432	646,646	497,420	319,770	170,544
23	253	1,771	8,855	33,649	100,947	245,157	490,314	817,190	1,144,066	1,352,078	1,352,078	1,144,066	817,190	490,314
24	276	2,024	10,626	42,504	134,596	346,104	735,471	1,307,504	1,961,256	2,496,144	2,704,156	2,496,144	1,961,256	1,307,504
25	300	2,300	12,650	53,130	177,100	480,700	1,081,575	2,042,975	3,268,760	4,457,400	5,200,300	5,200,300	4,457,400	3,268,760
26	325	2,600	14,950	65,780	230,230	657,800	1,562,275	3,124,550	5,311,735	7,726,160	9,657,700	10,400,600	9,657,700	7,726,160
27	351	2,925	17,550	80,730	296,010	888,030	2,220,075	4,686,825	8,436,285	13,037,895	17,383,860	20,058,300	20,058,300	17,383,860
28	378	3,276	20,475	98,280	376,740	1,184,040	3,108,105	6,906,900	13,123,110	21,474,180	30,421,755	37,442,160	40,116,600	37,442,160
29	406	3,654	23,751	118,755	475,020	1,560,780	4,292,145	10,015,005	20,030,010	34,597,290	51,895,935	67,863,915	77,558,760	77,558,760
30	435	4,060	27,405	142,506	593,775	2,035,800	5,852,925	14,307,150	30,045,015	54,627,300	86,493,225	119,759,850	145,422,675	155,117,520

TABLE C4. 1000 Random Digits

07048	52841	54969	87057	30570	50494	29936	93967	10641	79871
09165	56926	17294	03803	31755	11321	33681	12997	17625	25954
35654	69761	83791	63371	28189	19944	04514	56533	89108	27861
79065	63956	39443	30373	55571	00919	15377	36851	28318	40846
27969	74368	77782	88616	06368	07345	00725	81221	78417	37992
47528	70548	25078	80729	27806	42877	80287	21759	61980	52447
65694	95760	64031	24046	77606	91163	51492	20958	18384	49840
24253	39427	80642	36718	92164	77732	69754	01291	53704	33054
34302	60309	27186	22418	59962	13934	67591	17476	21559	73437
76809	84341	74012	50947	83214	19967	44219	75929	13182	34858
85183	35958	04301	49628	91493	66103	65699	04241	82441	38112
27541	79187	99777	22894	83283	56218	86183	74497	21070	78935
74188	09083	54938	79920	27158	24864	31116	33173	43032	52000
13270	57457	30968	65978	67679	91216	47969	39204	46030	93954
89150	53922	40537	23169	46948	05519	72171	85417	31580	98102
49980	44551	99908	46115	92508	77184	44556	69725	42878	60298
26810	40280	15387	30976	15478	77703	34109	02682	52877	36755
35056	23942	42645	67063	44118	46433	83172	95689	60923	32769
09873	65959	77912	70059	07704	16015	57527	09818	84379	35903
40806	30051	54251	73489	47215	90651	90083	21019	63860	41369

TABLE C5. Upper α Percentile Points of the t Distribution with ν Degrees of Freedom

The entries in the table are the values $t_{\alpha,\nu}$ such that the area, to the right of $t_{\alpha,\nu}$ under the t curve for ν degrees of freedom, is equal to α.

Examples: $t_{.10,13} = 1.350$, $t_{.005,21} = 2.831$.

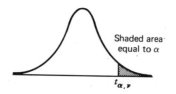

Shaded area equal to α

$t_{\alpha,\nu}$

ν	0.25	0.10	0.05	0.025	0.01	0.005	0.0005
1	1.000	3.078	6.314	12.706	31.821	63.657	636.619
2	0.816	1.886	2.920	4.303	6.965	9.925	31.598
3	0.765	1.638	2.353	3.182	4.541	5.841	12.924
4	0.741	1.533	2.132	2.776	3.747	4.604	8.610
5	0.727	1.476	2.015	2.571	3.365	4.032	6.869
6	0.718	1.440	1.943	2.447	3.143	3.707	5.959
7	0.711	1.415	1.895	2.365	2.998	3.499	5.408
8	0.706	1.397	1.860	2.306	2.896	3.355	5.041
9	0.703	1.383	1.833	2.262	2.821	3.250	4.781
10	0.700	1.372	1.812	2.228	2.764	3.169	4.587
11	0.697	1.363	1.796	2.201	2.718	3.106	4.437
12	0.695	1.356	1.782	2.179	2.681	3.055	4.318
13	0.694	1.350	1.771	2.160	2.650	3.012	4.221
14	0.692	1.345	1.761	2.145	2.624	2.977	4.140
15	0.691	1.341	1.753	2.131	2.602	2.947	4.073
16	0.690	1.337	1.746	2.120	2.583	2.921	4.015
17	0.689	1.333	1.740	2.110	2.567	2.898	3.965
18	0.688	1.330	1.734	2.101	2.552	2.878	3.922
19	0.688	1.328	1.729	2.093	2.539	2.861	3.883
20	0.687	1.325	1.725	2.086	2.528	2.845	3.850

Table C5. (continued)

ν	α						
	0.25	0.10	0.05	0.025	0.01	0.005	0.0005
21	0.686	1.323	1.721	2.080	2.518	2.831	3.819
22	0.686	1.321	1.717	2.074	2.508	2.819	3.792
23	0.685	1.319	1.714	2.069	2.500	2.807	3.768
24	0.685	1.318	1.711	2.064	2.492	2.797	3.745
25	0.684	1.316	1.708	2.060	2.485	2.787	3.725
26	0.684	1.315	1.706	2.056	2.479	2.779	3.707
27	0.684	1.314	1.703	2.052	2.473	2.771	3.690
28	0.683	1.313	1.701	2.048	2.467	2.763	3.674
29	0.683	1.311	1.699	2.045	2.462	2.756	3.659
30	0.683	1.310	1.697	2.042	2.457	2.750	3.646
40	0.681	1.303	1.684	2.021	2.423	2.704	3.551
60	0.679	1.296	1.671	2.000	2.390	2.660	3.460
100	0.677	1.290	1.660	1.984	2.364	2.626	3.390
∞	0.674	1.282	1.645	1.960	2.326	2.576	3.291

TABLE C6. Confidence Limits for a Success Probability p

For given n, α, and number of successes B, the tabled entries are p_L and p_U, lower and upper end points, respectively, of a confidence interval for the true probability of success p, with confidence coefficient at least $100(1-\alpha)\%$.

Example: For $n=3$, $B=2$, and $\alpha=.10$ (so that the confidence coefficient is at least 90%), $p_L=.1353$ and $p_U=.9830$.

$n = 1$

B	α	p_L	p_U
0	.010	.0000	.9950
	.020	.0000	.9900
	.050	.0000	.9750
	.100	.0000	.9500
	.200	.0000	.9000
1	.010	.0050	1.0000
	.020	.0100	1.0000
	.050	.0250	1.0000
	.100	.0500	1.0000
	.200	.1000	1.0000

$n = 2$

B	α	p_L	p_U
0	.010	.0000	.9293
	.020	.0000	.9000
	.050	.0000	.8419
	.100	.0000	.7764
	.200	.0000	.6838
1	.010	.0025	.9975
	.020	.0050	.9950
	.050	.0126	.9874
	.100	.0253	.9747
	.200	.0513	.9487
2	.010	.0707	1.0000
	.020	.1000	1.0000
	.050	.1581	1.0000
	.100	.2236	1.0000
	.200	.3162	1.0000

$n = 3$

B	α	p_L	p_U
0	.010	.0000	.8290
	.020	.0000	.7846
	.050	.0000	.7076
	.100	.0000	.6316
	.200	.0000	.5358
1	.010	.0017	.9586
	.020	.0033	.9411
	.050	.0084	.9057
	.100	.0170	.8647
	.200	.0345	.8042
2	.010	.0414	.9983
	.020	.0589	.9967
	.050	.0943	.9916
	.100	.1353	.9830
	.200	.1958	.9655
3	.010	.1710	1.0000
	.020	.2154	1.0000
	.050	.2924	1.0000
	.100	.3684	1.0000
	.200	.4642	1.0000

$n = 4$

B	α	p_L	p_U
0	.010	.0000	.7341
	.020	.0000	.6838
	.050	.0000	.6024
	.100	.0000	.5271
	.200	.0000	.4377

$n = 4$

B	α	p_L	p_U
1	.010	.0013	.8891
	.020	.0025	.8591
	.050	.0063	.8059
	.100	.0127	.7514
	.200	.0260	.6795
2	.010	.0294	.9706
	.020	.0420	.9580
	.050	.0676	.9324
	.100	.0976	.9024
	.200	.1426	.8574
3	.010	.1109	.9987
	.020	.1409	.9975
	.050	.1941	.9937
	.100	.2486	.9873
	.200	.3205	.9740
4	.010	.2659	1.0000
	.020	.3162	1.0000
	.050	.3976	1.0000
	.100	.4729	1.0000
	.200	.5623	1.0000

$n = 5$

B	α	p_L	p_U
0	.010	.0000	.6534
	.020	.0000	.6019
	.050	.0000	.5218
	.100	.0000	.4507
	.200	.0000	.3690
1	.010	.0010	.8149
	.020	.0020	.7779
	.050	.0051	.7164
	.100	.0102	.6574
	.200	.0209	.5839
2	.010	.0229	.9172
	.020	.0327	.8944
	.050	.0527	.8534
	.100	.0764	.8107
	.200	.1122	.7534

$n = 5$

B	α	p_L	p_U
3	.010	.0828	.9771
	.020	.1056	.9673
	.050	.1466	.9473
	.100	.1893	.9236
	.200	.2466	.8878
4	.010	.1851	.9990
	.020	.2221	.9980
	.050	.2836	.9949
	.100	.3426	.9898
	.200	.4161	.9791
5	.010	.3466	1.0000
	.020	.3981	1.0000
	.050	.4782	1.0000
	.100	.5493	1.0000
	.200	.6310	1.0000

$n = 6$

B	α	p_L	p_U
0	.010	.0000	.5865
	.020	.0000	.5358
	.050	.0000	.4593
	.100	.0000	.3930
	.200	.0000	.3187
1	.010	.0008	.7460
	.020	.0017	.7057
	.050	.0042	.6412
	.100	.0085	.5818
	.200	.0174	.5103
2	.010	.0187	.8564
	.020	.0268	.8269
	.050	.0433	.7772
	.100	.0628	.7287
	.200	.0926	.6668
3	.010	.0663	.9337
	.020	.0847	.9153
	.050	.1181	.8819
	.100	.1532	.8468
	.200	.2009	.7991

Table C6. (*continued*)

n = 6			
B	α	p_L	p_U
4	.010	.1436	.9813
	.020	.1731	.9732
	.050	.2228	.9567
	.100	.2713	.9372
	.200	.3332	.9074
5	.010	.2540	.9992
	.020	.2943	.9983
	.050	.3588	.9958
	.100	.4182	.9915
	.200	.4897	.9826
6	.010	.4135	1.0000
	.020	.4642	1.0000
	.050	.5407	1.0000
	.100	.6070	1.0000
	.200	.6813	1.0000

n = 7			
B	α	p_L	p_U
0	.010	.0000	.5309
	.020	.0000	.4821
	.050	.0000	.4096
	.100	.0000	.3482
	.200	.0000	.2803
1	.010	.0007	.6849
	.020	.0014	.6434
	.050	.0036	.5787
	.100	.0073	.5207
	.200	.0149	.4526
2	.010	.0158	.7970
	.020	.0227	.7637
	.050	.0367	.7096
	.100	.0534	.6587
	.200	.0788	.5962
3	.010	.0553	.8823
	.020	.0708	.8577
	.050	.0990	.8159
	.100	.1288	.7747
	.200	.1696	.7214

n = 7			
B	α	p_L	p_U
4	.010	.1177	.9447
	.020	.1423	.9292
	.050	.1841	.9010
	.100	.2253	.8712
	.200	.2786	.8304
5	.010	.2030	.9842
	.020	.2363	.9773
	.050	.2904	.9633
	.100	.3413	.9466
	.200	.4038	.9212
6	.010	.3151	.9993
	.020	.3566	.9986
	.050	.4213	.9964
	.100	.4793	.9927
	.200	.5474	.9851
7	.010	.4691	1.0000
	.020	.5179	1.0000
	.050	.5904	1.0000
	.100	.6518	1.0000
	.200	.7197	1.0000

n = 8			
B	α	p_L	p_U
0	.010	.0000	.4843
	.020	.0000	.4377
	.050	.0000	.3694
	.100	.0000	.3123
	.200	.0000	.2501
1	.010	.0006	.6315
	.020	.0013	.5899
	.050	.0032	.5265
	.100	.0064	.4707
	.200	.0131	.4062
2	.010	.0137	.7422
	.020	.0197	.7068
	.050	.0319	.6509
	.100	.0464	.5997
	.200	.0686	.5382

	n = 8				n = 9		
B	α	p_L	p_U	B	α	p_L	p_U
3	.010	.0475	.8303	1	.010	.0006	.5850
	.020	.0608	.8018		.020	.0011	.5440
	.050	.0852	.7551		.050	.0028	.4825
	.100	.1111	.7108		.100	.0057	.4291
	.200	.1469	.6554		.200	.0116	.3684
4	.010	.0999	.9001	2	.010	.0121	.6926
	.020	.1210	.8790		.020	.0174	.6563
	.050	.1570	.8430		.050	.0281	.6001
	.100	.1929	.8071		.100	.0410	.5496
	.200	.2397	.7603		.200	.0608	.4901
5	.010	.1697	.9525	3	.010	.0416	.7809
	.020	.1982	.9392		.020	.0534	.7500
	.050	.2449	.9148		.050	.0749	.7007
	.100	.2892	.8889		.100	.0978	.6551
	.200	.3446	.8531		.200	.1295	.5994
6	.010	.2578	.9863	4	.010	.0868	.8539
	.020	.2932	.9803		.020	.1053	.8290
	.050	.3491	.9681		.050	.1370	.7880
	.100	.4003	.9536		.100	.1687	.7486
	.200	.4618	.9314		.200	.2104	.6990
7	.010	.3685	.9994	5	.010	.1461	.9132
	.020	.4101	.9987		.020	.1710	.8947
	.050	.4735	.9968		.050	.2120	.8630
	.100	.5293	.9936		.100	.2514	.8313
	.200	.5938	.9869		.200	.3010	.7896
8	.010	.5157	1.0000	6	.010	.2191	.9584
	.020	.5623	1.0000		.020	.2500	.9466
	.050	.6306	1.0000		.050	.2993	.9251
	.100	.6877	1.0000		.100	.3449	.9022
	.200	.7499	1.0000		.200	.4006	.8705

	n = 9		
B	α	p_L	p_U
0	.010	.0000	.4450
	.020	.0000	.4005
	.050	.0000	.3363
	.100	.0000	.2831
	.200	.0000	.2257

B	α	p_L	p_U
7	.010	.3074	.9879
	.020	.3437	.9826
	.050	.3999	.9719
	.100	.4504	.9590
	.200	.5099	.9392
8	.010	.4150	.9994
	.020	.4560	.9989
	.050	.5175	.9972
	.100	.5709	.9943
	.200	.6316	.9884

421

Table C6. (*continued*)

n = 9

B	α	p_L	p_U
9	.010	.5550	1.0000
	.020	.5995	1.0000
	.050	.6637	1.0000
	.100	.7169	1.0000
	.200	.7743	1.0000

n = 10

B	α	p_L	p_U
0	.010	.0000	.4113
	.020	.0000	.3690
	.050	.0000	.3085
	.100	.0000	.2589
	.200	.0000	.2057
1	.010	.0005	.5443
	.020	.0010	.5044
	.050	.0025	.4450
	.100	.0051	.3942
	.200	.0105	.3369
2	.010	.0109	.6482
	.020	.0155	.6117
	.050	.0252	.5561
	.100	.0368	.5069
	.200	.0545	.4496
3	.010	.0370	.7351
	.020	.0475	.7029
	.050	.0667	.6525
	.100	.0873	.6066
	.200	.1158	.5517
4	.010	.0768	.8091
	.020	.0932	.7817
	.050	.1216	.7376
	.100	.1500	.6965
	.200	.1876	.6458

n = 10

B	α	p_L	p_U
5	.010	.1283	.8717
	.020	.1504	.8496
	.050	.1871	.8129
	.100	.2224	.7776
	.200	.2673	.7327
6	.010	.1909	.9232
	.020	.2183	.9068
	.050	.2624	.8784
	.100	.3035	.8500
	.200	.3542	.8124
7	.010	.2649	.9630
	.020	.2971	.9525
	.050	.3475	.9333
	.100	.3934	.9127
	.200	.4483	.8842
8	.010	.3518	.9891
	.020	.3883	.9845
	.050	.4439	.9748
	.100	.4931	.9632
	.200	.5504	.9455
9	.010	.4557	.9995
	.020	.4956	.9990
	.050	.5550	.9975
	.100	.6058	.9949
	.200	.6631	.9895
10	.010	.5887	1.0000
	.020	.6310	1.0000
	.050	.6915	1.0000
	.100	.7411	1.0000
	.200	.7943	1.0000

TABLE C7. Binomial probabilities $Pr(B = x) = \binom{n}{x} p^x (1-p)^{n-x}$

Example: When $n = 6$ and $p = .30$, $Pr(B = 4) = .0595$. Also, $Pr(B \geq 4) = .0595 + .0102 + .0007 = .0704$.

n	x\\p	.01	.05	.10	.15	.20	.25	.30	$\frac{1}{3}$	.35	.40	.45	.50
4	0	.9606	.8145	.6561	.5220	.4096	.3164	.2401	.1975	.1785	.1296	.0915	.0625
	1	.0388	.1715	.2916	.3685	.4096	.4219	.4116	.3951	.3845	.3456	.2995	.2500
	2	.0006	.0135	.0486	.0975	.1536	.2109	.2646	.2963	.3105	.3456	.3675	.3750
	3	.0000[a]	.0005	.0036	.0115	.0256	.0469	.0756	.0988	.1115	.1536	.2005	.2500
	4	.0000	.0000	.0001	.0005	.0016	.0039	.0081	.0123	.0150	.0256	.0410	.0625
5	0	.9510	.7738	.5905	.4437	.3277	.2373	.1681	.1317	.1160	.0778	.0503	.0312
	1	.0480	.2036	.3280	.3915	.4096	.3955	.3601	.3292	.3124	.2592	.2059	.1562
	2	.0010	.0214	.0729	.1382	.2048	.2637	.3087	.3292	.3364	.3456	.3369	.3125
	3	.0000	.0011	.0081	.0244	.0512	.0879	.1323	.1646	.1811	.2304	.2757	.3125
	4	.0000	.0000	.0004	.0022	.0064	.0146	.0283	.0412	.0488	.0768	.1128	.1562
	5	.0000	.0000	.0000	.0001	.0003	.0010	.0024	.0041	.0053	.0102	.0185	.0312
6	0	.9415	.7351	.5314	.3771	.2621	.1780	.1176	.0878	.0754	.0467	.0277	.0156
	1	.0571	.2321	.3543	.3993	.3932	.3560	.3025	.2634	.2437	.1866	.1359	.0937
	2	.0014	.0305	.0984	.1762	.2458	.2966	.3241	.3292	.3280	.3110	.2780	.2344
	3	.0000	.0021	.0146	.0415	.0819	.1318	.1852	.2195	.2355	.2765	.3032	.3125
	4	.0000	.0001	.0012	.0055	.0154	.0330	.0595	.0823	.0951	.1382	.1861	.2344
	5	.0000	.0000	.0001	.0004	.0015	.0044	.0102	.0165	.0205	.0369	.0609	.0937
	6	.0000	.0000	.0000	.0000	.0001	.0002	.0007	.0014	.0018	.0041	.0083	.0156
7	0	.9321	.6983	.4783	.3206	.2097	.1335	.0824	.0585	.0490	.0280	.0152	.0078
	1	.0659	.2573	.3720	.3960	.3670	.3115	.2471	.2048	.1848	.1306	.0872	.0547
	2	.0020	.0406	.1240	.2097	.2753	.3115	.3177	.3073	.2985	.2613	.2140	.1641
	3	.0000	.0036	.0230	.0617	.1147	.1730	.2269	.2561	.2679	.2903	.2918	.2734
	4	.0000	.0002	.0026	.0109	.0287	.0577	.0972	.1280	.1442	.1935	.2388	.2734
	5	.0000	.0000	.0002	.0012	.0043	.0115	.0250	.0384	.0466	.0774	.1172	.1641
	6	.0000	.0000	.0000	.0001	.0004	.0013	.0036	.0064	.0084	.0172	.0320	.0547
	7	.0000	.0000	.0000	.0000	.0000	.0001	.0002	.0005	.0006	.0016	.0037	.0078
8	0	.9227	.6634	.4305	.2725	.1678	.1001	.0576	.0390	.0319	.0168	.0084	.0039
	1	.0746	.2793	.3826	.3847	.3355	.2670	.1977	.1561	.1373	.0896	.0548	.0312
	2	.0026	.0515	.1488	.2376	.2936	.3115	.2965	.2731	.2587	.2090	.1569	.1094
	3	.0001	.0054	.0331	.0839	.1468	.2076	.2541	.2731	.2786	.2787	.2568	.2187
	4	.0000	.0004	.0046	.0185	.0459	.0865	.1361	.1707	.1875	.2322	.2627	.2734
	5	.0000	.0000	.0004	.0026	.0092	.0231	.0467	.0683	.0808	.1239	.1719	.2187
	6	.0000	.0000	.0000	.0002	.0011	.0038	.0100	.0171	.0217	.0413	.0703	.1094
	7	.0000	.0000	.0000	.0000	.0001	.0004	.0012	.0024	.0033	.0079	.0164	.0312
	8	.0000	.0000	.0000	.0000	.0000	.0000	.0001	.0002	.0002	.0007	.0017	.0039
9	0	.9135	.6302	.3874	.2316	.1342	.0751	.0404	.0260	.0207	.0101	.0046	.0020
	1	.0830	.2985	.3874	.3679	.3020	.2253	.1556	.1171	.1004	.0605	.0339	.0176
	2	.0034	.0629	.1722	.2597	.3020	.3003	.2668	.2341	.2162	.1612	.1110	.0703
	3	.0001	.0077	.0446	.1069	.1762	.2336	.2668	.2731	.2716	.2508	.2119	.1641
	4	.0000	.0006	.0074	.0283	.0661	.1168	.1715	.2048	.2194	.2508	.2600	.2461
	5	.0000	.0000	.0008	.0050	.0165	.0389	.0735	.1024	.1181	.1672	.2128	.2461
	6	.0000	.0000	.0001	.0006	.0028	.0087	.0210	.0341	.0424	.0743	.1160	.1641

n	x\p	.01	.05	.10	.15	.20	.25	.30	$\frac{1}{3}$	.35	.40	.45	.50
	7	.0000	.0000	.0000	.0000	.0003	.0012	.0039	.0073	.0098	.0212	.0407	.0703
	8	.0000	.0000	.0000	.0000	.0000	.0001	.0004	.0009	.0013	.0035	.0083	.0176
	9	.0000	.0000	.0000	.0000	.0000	.0000	.0000	.0001	.0001	.0003	.0008	.0020
10	0	.9044	.5987	.3487	.1969	.1074	.0563	.0282	.0173	.0135	.0060	.0025	.0010
	1	.0914	.3151	.3874	.3474	.2684	.1877	.1211	.0867	.0725	.0403	.0207	.0098
	2	.0042	.0746	.1937	.2759	.3020	.2816	.2335	.1951	.1757	.1209	.0763	.0439
	3	.0001	.0105	.0574	.1298	.2013	.2503	.2668	.2601	.2522	.2150	.1665	.1172
	4	.0000	.0010	.0112	.0401	.0881	.1460	.2001	.2276	.2377	.2508	.2384	.2051
	5	.0000	.0001	.0015	.0085	.0264	.0584	.1029	.1366	.1536	.2007	.2340	.2461
	6	.0000	.0000	.0001	.0012	.0055	.0162	.0368	.0569	.0689	.1115	.1596	.2051
	7	.0000	.0000	.0000	.0001	.0008	.0031	.0090	.0163	.0212	.0425	.0746	.1172
	8	.0000	.0000	.0000	.0000	.0001	.0004	.0014	.0030	.0043	.0106	.0229	.0439
	9	.0000	.0000	.0000	.0000	.0000	.0000	.0001	.0003	.0005	.0016	.0042	.0098
	10	.0000	.0000	.0000	.0000	.0000	.0000	.0000	.0000	.0000	.0001	.0003	.0010

[a] An entry of .0000 does not mean that the probability is exactly zero, rather that it is zero to four decimal places.

TABLE C8. Upper α Percentile Points of the Chi-Squared Distribution with ν Degrees of Freedom

The entries in the table are the values $\chi^2_{\alpha,\nu}$ such that the area, to the right of $\chi^2_{\alpha,\nu}$ under the chi-squared curve for ν degrees of freedom, is equal to α.

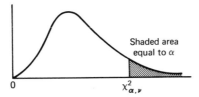

Shaded area equal to α

Examples: $\chi^2_{.10,7} = 12.02$, $\chi^2_{.05,15} = 25.00$.

Degrees of freedom ν	0.99	0.95	0.90	α 0.80	0.70	0.60	0.50
1	0.00	0.00	0.02	0.06	0.15	0.27	0.45
2	0.02	0.10	0.21	0.45	0.71	1.02	1.39
3	0.11	0.35	0.58	1.01	1.42	1.87	2.37
4	0.30	0.71	1.06	1.65	2.19	2.75	3.36
5	0.55	1.15	1.61	2.34	3.00	3.66	4.35
6	0.87	1.64	2.20	3.07	3.83	4.57	5.35
7	1.24	2.17	2.83	3.82	4.67	5.49	6.35
8	1.65	2.73	3.49	4.59	5.53	6.42	7.34
9	2.09	3.33	4.17	5.38	6.39	7.36	8.34
10	2.56	3.94	4.87	6.18	7.27	8.30	9.34
11	3.05	4.57	5.58	6.99	8.15	9.24	10.34
12	3.57	5.23	6.30	7.81	9.03	10.18	11.34
13	4.11	5.89	7.04	8.63	9.93	11.13	12.34
14	4.66	6.57	7.79	9.47	10.82	12.08	13.34
15	5.23	7.26	8.55	10.31	11.72	13.03	14.34
16	5.81	7.96	9.31	11.15	12.62	13.98	15.34
17	6.41	8.67	10.09	12.00	13.53	14.94	16.34
18	7.01	9.39	10.86	12.86	14.44	15.89	17.34
19	7.63	10.12	11.65	13.72	15.35	16.85	18.34
20	8.26	10.85	12.44	14.58	16.27	17.81	19.34
21	8.90	11.59	13.24	15.44	17.18	18.77	20.34
22	9.54	12.34	14.04	16.31	18.10	19.73	21.34
23	10.20	13.09	14.85	17.19	19.02	20.69	22.34
24	10.86	13.85	15.66	18.06	19.94	21.65	23.34
25	11.52	14.61	16.47	18.94	20.87	22.62	24.34
26	12.20	15.38	17.29	19.82	21.79	23.58	25.34
27	12.88	16.15	18.11	20.70	22.72	24.54	26.34
28	13.56	16.93	18.94	21.59	23.65	25.51	27.34
29	14.26	17.71	19.77	22.48	24.58	26.48	28.34
30	14.95	18.49	20.60	23.36	25.51	27.44	29.34

Table C8. *(continued)*

Degrees of freedom ν	0.40	0.30	0.20	α 0.10	0.05	0.01	0.001
1	0.71	1.07	1.64	2.71	3.84	6.63	10.83
2	1.83	2.41	3.22	4.61	5.99	9.21	13.82
3	2.95	3.66	4.64	6.25	7.81	11.34	16.27
4	4.04	4.88	5.99	7.78	9.49	13.28	18.47
5	5.13	6.06	7.29	9.24	11.07	15.09	20.52
6	6.21	7.23	8.56	10.64	12.59	16.81	22.46
7	7.28	8.38	9.80	12.02	14.07	18.48	24.32
8	8.35	9.52	11.03	13.36	15.51	20.09	26.12
9	9.41	10.66	12.24	14.68	16.92	21.67	27.88
10	10.47	11.78	13.44	15.99	18.31	23.21	29.59
11	11.53	12.90	14.63	17.28	19.68	24.73	31.26
12	12.58	14.01	15.81	18.55	21.03	26.22	32.91
13	13.64	15.12	16.98	19.81	22.36	27.69	34.53
14	14.69	16.22	18.15	21.06	23.68	29.14	36.12
15	15.73	17.32	19.31	22.31	25.00	30.58	37.70
16	16.78	18.42	20.47	23.54	26.30	32.00	39.25
17	17.82	19.51	21.61	24.77	27.59	33.41	40.79
18	18.87	20.60	22.76	25.99	28.87	34.81	42.31
19	19.91	21.69	23.90	27.20	30.14	36.19	43.82
20	20.95	22.77	25.04	28.41	31.41	37.57	45.31
21	21.99	23.86	26.17	29.62	32.67	38.93	46.80
22	23.03	24.94	27.30	30.81	33.92	40.29	48.27
23	24.07	26.02	28.43	32.01	35.17	41.64	49.73
24	25.11	27.10	29.55	33.20	36.42	42.98	51.18
25	26.14	28.17	30.68	34.38	37.65	44.31	52.62
26	27.18	29.25	31.79	35.56	38.89	45.64	54.05
27	28.21	30.32	32.91	36.74	40.11	46.96	55.48
28	29.25	31.39	34.03	37.92	41.34	48.28	56.89
29	30.28	32.46	35.14	39.09	42.56	49.59	58.30
30	31.32	33.53	36.25	40.26	43.77	50.89	59.70

Adapted in part from Table 2 of *New Tables of the Incomplete Gamma-function Ratio and of Percentage Points of the Chi-square and Beta Distributions,* by H. L. Harter, Aerospace Research Laboratories, United States Air Force. Washington, D.C., Government Printing Office, 1964.

TABLE C9. Upper α Percentile Points of the F Distribution with ν_1 Degrees of Freedom in the Numerator and ν_2 Degrees of Freedom in the Denominator

The entries in the table are the values F_{α,ν_1,ν_2} such that the area, to the right of F_{α,ν_1,ν_2} under the F curve for ν_1 degrees of freedom in the numerator and ν_2 degrees of freedom in the denominator, is equal to α.

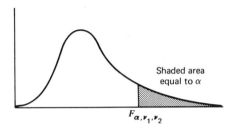

Shaded area equal to α

F_{α,ν_1,ν_2}

Examples: $F_{.01,3,5}=12.1$, $F_{.05,12,6}=4.00$.

ν_2	α	1	2	3	4	5	6	7	8	9	10	11	12
1	0.975	$0.0^2 15$	0.026	0.057	0.082	0.100	0.113	0.124	0.132	0.139	0.144	0.149	0.153
	0.05	161	200	216	225	230	234	237	239	241	242	243	244
	0.025	648	800	864	900	922	937	948	957	963	969	973	977
	0.01	405^1	500^1	540^1	562^1	576^1	586^1	593^1	598^1	602^1	606^1	608^1	611^1
	0.001	406^3	500^3	540^3	562^3	576^3	586^3	593^3	598^3	602^3	606^3	609^3	611^3
	.0005	162^4	200^4	216^4	225^4	231^4	234^4	237^4	239^4	241^4	242^4	243^4	244^4
2	0.975	$0.0^2 13$	0.026	0.062	0.094	0.119	0.138	0.153	0.165	0.175	0.183	0.190	0.196
	0.05	18.5	19.0	19.2	19.2	19.3	19.3	19.4	19.4	19.4	19.4	19.4	19.4
	0.025	38.5	39.0	39.2	39.2	39.3	39.3	39.4	39.4	39.4	39.4	39.4	39.4
	0.01	98.5	99.0	99.2	99.2	99.3	99.3	99.4	99.4	99.4	99.4	99.4	99.4
	0.001	998	999	999	999	999	999	999	999	999	999	999	999
	0.0005	200^1	200^1	200^1	200^1	200^1	200^1	200^1	200^1	200^1	200^1	200^1	200^1
3	0.975	$0.0^2 12$	0.026	0.065	0.100	0.129	0.152	0.170	0.185	0.197	0.207	0.216	0.224
	0.05	10.1	9.55	9.28	9.12	9.01	8.94	8.89	8.85	8.81	8.79	8.76	8.74
	0.025	17.4	16.0	15.4	15.1	14.9	14.7	14.6	14.5	14.5	14.4	14.4	14.3
	0.01	34.1	30.8	29.5	28.7	28.2	27.9	27.7	27.5	27.3	27.2	27.1	27.1
	0.001	167	149	141	137	135	133	132	131	130	129	129	128
	0.0005	266	237	225	218	214	211	209	208	207	206	204	204
4	0.975	$0.0^2 11$	0.026	0.066	0.104	0.135	0.161	0.181	0.198	0.212	0.224	0.234	0.243
	0.05	7.71	6.94	6.59	6.39	6.26	6.16	6.09	6.04	6.00	5.96	5.94	5.91
	0.025	12.2	10.6	9.98	9.60	9.36	9.20	9.07	8.98	8.90	8.84	8.79	8.75
	0.01	21.2	18.0	16.7	16.0	15.5	15.2	15.0	14.8	14.7	14.5	14.4	14.4
	0.001	74.1	61.2	56.2	53.4	51.7	50.5	49.7	49.0	48.5	48.0	47.7	47.4
	0.0005	106	87.4	80.1	76.1	73.6	71.9	70.6	69.7	68.9	68.3	67.8	67.4
5	0.975	$0.0^2 11$	0.025	0.067	0.107	0.140	0.167	0.189	0.208	0.223	0.236	0.248	0.257
	0.05	6.61	5.79	5.41	5.19	5.05	4.95	4.88	4.82	4.77	4.74	4.71	4.68

Table C9. (continued)

ν_2	α \ ν_1	1	2	3	4	5	6	7	8	9	10	11	12
	0.025	10.0	8.43	7.76	7.39	7.15	6.98	6.85	6.76	6.68	6.62	6.57	6.52
	0.01	16.3	13.3	12.1	11.4	11.0	10.7	10.5	10.3	10.2	10.1	9.96	9.89
	0.001	47.2	37.1	33.2	31.1	29.7	28.8	28.2	27.6	27.2	26.9	26.6	26.4
	0.0005	63.6	49.8	44.4	41.5	39.7	38.5	37.6	36.9	36.4	35.9	35.6	35.2
6	0.975	$0.0^2 11$	0.025	0.068	0.109	0.143	0.172	0.195	0.215	0.231	0.246	0.258	0.268
	0.05	5.99	5.14	4.76	4.53	4.39	4.28	4.21	4.15	4.10	4.06	4.03	4.00
	0.025	8.81	7.26	6.60	6.23	5.99	5.82	5.70	5.60	5.52	5.46	5.41	5.37
	0.01	13.7	10.9	9.78	9.15	8.75	8.47	8.26	8.10	7.98	7.87	7.79	7.72
	0.001	35.5	27.0	23.7	21.9	20.8	20.0	19.5	19.0	18.7	18.4	18.2	18.0
	0.0005	46.1	34.8	30.4	28.1	26.6	25.6	24.9	24.3	23.9	23.5	23.2	23.0
7	0.975	$0.0^2 10$	0.025	0.068	0.110	0.146	0.176	0.200	0.221	0.238	0.253	0.266	0.277
	0.05	5.59	4.74	4.35	4.12	3.97	3.87	3.79	3.73	3.68	3.64	3.60	3.57
	0.025	8.07	6.54	5.89	5.52	5.29	5.12	4.99	4.90	4.82	4.76	4.71	4.67
	0.01	12.2	9.55	8.45	7.85	7.46	7.19	6.99	6.84	6.72	6.62	6.54	6.47
	0.001	29.2	21.7	18.8	17.2	16.2	15.5	15.0	14.6	14.3	14.1	13.9	13.7
	0.0005	37.0	27.2	23.5	21.4	20.2	19.3	18.7	18.2	17.8	17.5	17.2	17.0

ν_2	α \ ν_1	15	20	24	30	40	50	60	100	120	200	500	∞
1	0.975	0.161	0.170	0.175	0.180	0.184	0.187	0.189	0.193	0.194	0.196	0.198	0.199
	0.05	246	248	249	250	251	252	252	253	253	254	254	254
	0.025	985	993	997	100^1	101^1	101^1	101^1	101^1	101^1	102^1	102^1	102^1
	0.01	616^1	621^1	623^1	626^1	629^1	630^1	631^1	633^1	634^1	635^1	636^1	637^1
	0.001	616^3	621^3	623^3	626^3	629^3	630^3	631^3	633^3	634^3	635^3	636^3	637^3
	0.0005	246^4	248^4	249^4	250^4	251^4	252^4	252^4	253^4	253^4	253^4	254^4	254^4
2	0.975	0.210	0.224	0.232	0.239	0.247	0.251	0.255	0.261	0.263	0.266	0.269	0.271
	0.05	19.4	19.4	19.5	19.5	19.5	19.5	19.5	19.5	19.5	19.5	19.5	19.5
	0.025	39.4	39.4	39.5	39.5	39.5	39.5	39.5	39.5	39.5	39.5	39.5	39.5
	0.01	99.4	99.4	99.5	99.5	99.5	99.5	99.5	99.5	99.5	99.5	99.5	99.5
	0.001	999	999	999	999	999	999	999	999	999	999	999	999
	0.0005	200^1	200^1	200^1	200^1	200^1	200^1	200^1	200^1	200^1	200^1	200^1	200^1
3	0.975	0.241	0.259	0.269	0.279	0.289	0.295	0.299	0.308	0.310	0.314	0.318	0.321
	0.05	8.70	8.66	8.63	8.62	8.59	8.58	8.57	8.55	8.55	8.54	8.53	8.53
	0.025	14.3	14.2	14.1	14.1	14.0	14.0	14.0	14.0	13.9	13.9	13.9	13.9
	0.01	26.9	26.7	26.6	26.5	26.4	26.4	26.3	26.2	26.2	26.2	26.1	26.1
	0.001	127	126	126	125	125	125	124	124	124	124	124	123
	0.0005	203	201	200	199	199	198	198	197	197	197	196	196
4	0.975	0.263	0.284	0.296	0.308	0.320	0.327	0.332	0.342	0.346	0.351	0.356	0.359
	0.05	5.86	5.80	5.77	5.75	5.72	5.70	5.69	5.66	5.66	5.65	5.64	5.63
	0.025	8.66	8.56	8.51	8.46	8.41	8.38	8.36	8.32	8.31	8.29	8.27	8.26
	0.01	14.2	14.0	13.9	13.8	13.7	13.7	13.7	13.6	13.6	13.5	13.5	13.5
	0.001	46.8	46.1	45.8	45.4	45.1	44.9	44.7	44.5	44.4	44.3	44.1	44.0
	0.0005	66.5	65.5	65.1	64.6	64.1	63.8	63.6	63.2	63.1	62.9	62.7	62.6

Table C9. (continued)

ν_2	α \ ν_1	15	20	24	30	40	50	60	100	120	200	500	∞
5	0.975	0.280	0.304	0.317	0.330	0.344	0.353	0.359	0.370	0.374	0.380	0.386	0.390
	0.05	4.62	4.56	4.53	4.50	4.46	4.44	4.43	4.41	4.40	4.39	4.37	4.36
	0.025	6.43	6.33	6.28	6.23	6.18	6.14	6.12	6.08	6.07	6.05	6.03	6.02
	0.01	9.72	9.55	9.47	9.38	9.29	9.24	9.20	9.13	9.11	9.08	9.04	9.02
	0.001	25.9	25.4	25.1	24.9	24.6	24.4	24.3	24.1	24.1	23.9	23.8	23.8
	0.0005	34.6	33.9	33.5	33.1	32.7	32.5	32.3	32.1	32.0	31.8	31.7	31.6
6	0.975	0.293	0.320	0.334	0.349	0.364	0.375	0.381	0.394	0.398	0.405	0.412	0.415
	0.05	3.94	3.87	3.84	3.81	3.77	3.75	3.74	3.71	3.70	3.69	3.68	3.67
	0.025	5.27	5.17	5.12	5.07	5.01	4.98	4.96	4.92	4.90	4.88	4.86	4.85
	0.01	7.56	7.40	7.31	7.23	7.14	7.09	7.06	6.99	6.97	6.93	6.90	6.88
	0.001	17.6	17.1	16.9	16.7	16.4	16.3	16.2	16.0	16.0	15.9	15.8	15.7
	0.0005	22.4	21.9	21.7	21.4	21.1	20.9	20.7	20.5	20.4	20.3	20.2	20.1
7	0.975	0.304	0.333	0.348	0.364	0.381	0.392	0.399	0.413	0.418	0.426	0.433	0.437
	0.05	3.51	3.44	3.41	3.38	3.34	3.32	3.30	3.27	3.27	3.25	3.24	3.23
	0.025	4.57	4.47	4.42	4.36	4.31	4.28	4.25	4.21	4.20	4.18	4.16	4.14
	0.01	6.31	6.16	6.07	5.99	5.91	5.86	5.82	5.75	5.74	5.70	5.67	5.65
	0.001	13.3	12.9	12.7	12.5	12.3	12.2	12.1	11.9	11.9	11.8	11.7	11.7
	0.0005	16.5	16.0	15.7	15.5	15.2	15.1	15.0	14.7	14.7	14.6	14.5	14.4

ν_2	α \ ν_1	1	2	3	4	5	6	7	8	9	10	11	12
8	0.975	$0.0^2 10$	0.025	0.069	0.111	0.148	0.179	0.204	0.226	0.244	0.259	0.273	0.285
	0.05	5.32	4.46	4.07	3.84	3.69	3.58	3.50	3.44	3.39	3.35	3.31	3.28
	0.025	7.57	6.06	5.42	5.05	4.82	4.65	4.53	4.43	4.36	4.30	4.24	4.20
	0.01	11.3	8.65	7.59	7.01	6.63	6.37	6.18	6.03	5.91	5.81	5.73	5.67
	0.001	25.4	18.5	15.8	14.4	13.5	12.9	12.4	12.0	11.8	11.5	11.4	11.2
	0.0005	31.6	22.8	19.4	17.6	16.4	15.7	15.1	14.6	14.3	14.0	13.8	13.6
9	0.975	$0.0^2 10$	0.025	0.069	0.112	0.150	0.181	0.207	0.230	0.248	0.265	0.279	0.291
	0.05	5.12	4.26	3.86	3.63	3.48	3.37	3.29	3.23	3.18	3.14	3.10	3.07
	0.025	7.21	5.71	5.08	4.72	4.48	4.32	4.20	4.10	4.03	3.96	3.91	3.87
	0.01	10.6	8.02	6.99	6.42	6.06	5.80	5.61	5.47	5.35	5.26	5.18	5.11
	0.001	22.9	16.4	13.9	12.6	11.7	11.1	10.7	10.4	10.1	9.89	9.71	9.57
	0.0005	28.0	19.9	16.8	15.1	14.1	13.3	12.8	12.4	12.1	11.8	11.6	11.4
10	0.975	$0.0^2 10$	0.025	0.069	0.113	0.151	0.183	0.210	0.233	0.252	0.269	0.283	0.296
	0.05	4.96	4.10	3.71	3.48	3.33	3.22	3.14	3.07	3.02	2.98	2.94	2.91
	0.025	6.94	5.46	4.83	4.47	4.24	4.07	3.95	3.85	3.78	3.72	3.66	3.62
	0.01	10.0	7.56	6.55	5.99	5.64	5.39	5.20	5.06	4.94	4.85	4.77	4.71
	0.001	21.0	14.9	12.6	11.3	10.5	9.92	9.52	9.20	8.96	8.75	8.58	8.44
	0.0005	25.5	17.9	15.0	13.4	12.4	11.8	11.3	10.9	10.6	10.3	10.1	9.93
11	0.975	$0.0^2 10$	0.025	0.069	0.114	0.152	0.185	0.212	0.236	0.256	0.273	0.288	0.301
	0.05	4.84	3.98	3.59	3.36	3.20	3.09	3.01	2.95	2.90	2.85	2.82	2.79
	0.025	6.72	5.26	4.63	4.28	4.04	3.88	3.76	3.66	3.59	3.53	3.47	3.43

ν_2	α \ ν_1	1	2	3	4	5	6	7	8	9	10	11	12
	0.01	9.65	7.21	6.22	5.67	5.32	5.07	4.89	4.74	4.63	4.54	4.46	4.40
	0.001	19.7	13.8	11.6	10.3	9.58	9.05	8.66	8.35	8.12	7.92	7.76	7.62
	0.0005	23.6	16.4	13.6	12.2	11.2	10.6	10.1	9.76	9.48	9.24	9.04	8.88
12	0.975	$0.0^2 10$	0.025	0.070	0.114	0.153	0.186	0.214	0.238	0.259	0.276	0.292	0.305
	0.05	4.75	3.89	3.49	3.26	3.11	3.00	2.91	2.85	2.80	2.75	2.72	2.69
	0.025	6.55	5.10	4.47	4.12	3.89	3.73	3.61	3.51	3.44	3.37	3.32	3.28
	0.01	9.33	6.93	5.95	5.41	5.06	4.82	4.64	4.50	4.39	4.30	4.22	4.16
	0.001	18.6	13.0	10.8	9.63	8.89	8.38	8.00	7.71	7.48	7.29	7.14	7.01
	0.0005	22.2	15.3	12.7	11.2	10.4	9.74	9.28	8.94	8.66	8.43	8.24	8.08
15	0.975	$0.0^2 10$	0.025	0.070	0.116	0.156	0.190	0.219	0.244	0.265	0.284	0.300	0.315
	0.05	4.54	3.68	3.29	3.06	2.90	2.79	2.71	2.64	2.59	2.54	2.51	2.48
	0.025	6.20	4.76	4.15	3.80	3.58	3.41	3.29	3.20	3.12	3.06	3.01	2.96
	0.01	8.68	6.36	5.42	4.89	4.56	4.32	4.14	4.00	3.89	3.80	3.73	3.67
	0.001	16.6	11.3	9.34	8.25	7.57	7.09	6.74	6.47	6.26	6.08	5.93	5.81
	0.0005	19.5	13.2	10.8	9.48	8.66	8.10	7.68	7.36	7.11	6.91	6.75	6.60
20	0.975	$0.0^2 10$	0.025	0.071	0.117	0.158	0.193	0.224	0.250	0.273	0.292	0.310	0.325
	0.05	4.35	3.49	3.10	2.87	2.71	2.60	2.51	2.45	2.39	2.35	2.31	2.28
	0.025	5.87	4.46	3.86	3.51	3.29	3.13	3.01	2.91	2.84	2.77	2.72	2.68
	0.01	8.10	5.85	4.94	4.43	4.10	3.87	3.70	3.56	3.46	3.37	3.29	3.23
	0.001	14.8	9.95	8.10	7.10	6.46	6.02	5.69	5.44	5.24	5.08	4.94	4.82
	0.0005	17.2	11.4	9.20	8.02	7.28	6.76	6.38	6.08	5.85	5.66	5.51	5.38

ν_2	α \ ν_1	15	20	24	30	40	50	60	100	120	200	500	∞
8	0.975	0.313	0.343	0.360	0.377	0.395	0.407	0.415	0.431	0.435	0.442	0.450	0.456
	0.05	3.22	3.15	3.12	3.08	3.04	3.02	3.01	2.97	2.97	2.95	2.94	2.93
	0.025	4.10	4.00	3.95	3.89	3.84	3.81	3.78	3.74	3.73	3.70	3.68	3.67
	0.01	5.52	5.36	5.28	5.20	5.12	5.07	5.03	4.96	4.95	4.91	4.88	4.86
	0.001	10.8	10.5	10.3	10.1	9.92	9.80	9.73	9.57	9.54	9.46	9.39	9.34
	0.0005	13.1	12.7	12.5	12.2	12.0	11.8	11.8	11.6	11.5	11.4	11.4	11.3
9	0.975	0.320	0.352	0.370	0.388	0.408	0.420	0.428	0.446	0.450	0.459	0.467	0.473
	0.05	3.01	2.94	2.90	2.86	2.83	2.80	2.79	2.76	2.75	2.73	2.72	2.71
	0.025	3.77	3.67	3.61	3.56	3.51	3.47	3.45	3.40	3.39	3.37	3.35	3.33
	0.01	4.96	4.81	4.73	4.65	4.57	4.52	4.48	4.42	4.40	4.36	4.33	4.31
	0.001	9.24	8.90	8.72	8.55	8.37	8.26	8.19	8.04	8.00	7.93	7.86	7.81
	0.0005	11.0	10.6	10.4	10.2	9.94	9.80	9.71	9.53	9.49	9.40	9.32	9.26
10	0.975	0.327	0.360	0.379	0.398	0.419	0.431	0.441	0.459	0.464	0.474	0.483	0.488
	0.05	2.85	2.77	2.74	2.70	2.66	2.64	2.62	2.59	2.58	2.56	2.55	2.54
	0.025	3.52	3.42	3.37	3.31	3.26	3.22	3.20	3.15	3.14	3.12	3.09	3.08
	0.01	4.56	4.41	4.33	4.25	4.17	4.12	4.08	4.01	4.00	3.96	3.93	3.91
	0.001	8.13	7.80	7.64	7.47	7.30	7.19	7.12	6.98	6.94	6.87	6.81	6.76
	0.0005	9.56	9.16	8.96	8.75	8.54	8.42	8.33	8.16	8.12	8.04	7.96	7.90

ν_2	α \ ν_1	15	20	24	30	40	50	60	100	120	200	500	∞
11	0.975	0.332	0.368	0.386	0.407	0.429	0.442	0.450	0.472	0.476	0.485	0.495	0.503
	0.05	2.72	2.65	2.61	2.57	2.53	2.51	2.49	2.46	2.45	2.43	2.42	2.40
	0.025	3.33	3.23	3.17	3.12	3.06	3.03	3.00	2.96	2.94	2.92	2.90	2.88
	0.01	4.25	4.10	4.02	3.94	3.86	3.81	3.78	3.71	3.69	3.66	3.62	3.60
	0.001	7.32	7.01	6.85	6.68	6.52	6.41	6.35	6.21	6.17	6.10	6.04	6.00
	0.0005	8.52	8.14	7.94	7.75	7.55	7.43	7.35	7.18	7.14	7.06	6.98	6.93
12	0.975	0.337	0.374	0.394	0.416	0.437	0.450	0.461	0.481	0.487	0.498	0.508	0.514
	0.05	2.62	2.54	2.51	2.47	2.43	2.40	2.38	2.35	2.34	2.32	2.31	2.30
	0.025	3.18	3.07	3.02	2.96	2.91	2.87	2.85	2.80	2.79	2.76	2.74	2.72
	0.01	4.01	3.86	3.78	3.70	3.62	3.57	3.54	3.47	3.45	3.41	3.38	3.36
	0.001	6.71	6.40	6.25	6.09	5.93	5.83	5.76	5.63	5.59	5.52	5.46	5.42
	0.0005	7.74	7.37	7.18	7.00	6.80	6.68	6.61	6.45	6.41	6.33	6.25	6.20
15	0.975	0.349	0.389	0.410	0.433	0.458	0.474	0.485	0.508	0.514	0.526	0.538	0.546
	0.05	2.40	2.33	2.29	2.25	2.20	2.18	2.16	2.12	2.11	2.10	2.08	2.07
	0.025	2.86	2.76	2.70	2.64	2.59	2.55	2.52	2.47	2.46	2.44	2.41	2.40
	0.01	3.52	3.37	3.29	3.21	3.13	3.08	3.05	2.98	2.96	2.92	2.89	2.87
	0.001	5.54	5.25	5.10	4.95	4.80	4.70	4.64	4.51	4.47	4.41	4.35	4.31
	0.0005	6.27	5.93	5.75	5.58	5.40	5.29	5.21	5.06	5.02	4.94	4.87	4.83
20	0.975	0.363	0.406	0.430	0.456	0.484	0.503	0.514	0.541	0.548	0.562	0.575	0.585
	0.05	2.20	2.12	2.08	2.04	1.99	1.97	1.95	1.91	1.90	1.88	1.86	1.84
	0.025	2.57	2.46	2.41	2.35	2.29	2.25	2.22	2.17	2.16	2.13	2.10	2.09
	0.01	3.09	2.94	2.86	2.78	2.69	2.64	2.61	2.54	2.52	2.48	2.44	2.42
	0.001	4.56	4.29	4.15	4.01	3.86	3.77	3.70	3.58	3.54	3.48	3.42	3.38
	0.0005	5.07	4.75	4.58	4.42	4.24	4.15	4.07	3.93	3.90	3.82	3.75	3.70

ν_2	α \ ν_1	1	2	3	4	5	6	7	8	9	10	11	12
24	0.975	$0.0^2$10	0.025	0.071	0.117	0.159	0.195	0.227	0.253	0.277	0.297	0.315	0.331
	0.05	4.26	3.40	3.01	2.78	2.62	2.51	2.42	2.36	2.30	2.25	2.21	2.18
	0.025	5.72	4.32	3.72	3.38	3.15	2.99	2.87	2.78	2.70	2.64	2.59	2.54
	0.01	7.82	5.61	4.72	4.22	3.90	3.67	3.50	3.36	3.26	3.17	3.09	3.03
	0.001	14.0	9.34	7.55	6.59	5.98	5.55	5.23	4.99	4.80	4.64	4.50	4.39
	0.0005	16.2	10.6	8.52	7.39	6.68	6.18	5.82	5.54	5.31	5.13	4.98	4.85
30	0.975	$0.0^2$10	0.025	0.071	0.118	0.161	0.197	0.229	0.257	0.281	0.302	0.321	0.327
	0.05	4.17	3.32	2.92	2.69	2.53	2.42	2.33	2.27	2.21	2.16	2.13	2.09
	0.025	5.57	4.18	3.59	3.25	3.03	2.87	2.75	2.65	2.57	2.51	2.46	2.41
	0.01	7.56	5.39	4.51	4.02	3.70	3.47	3.30	3.17	3.07	2.98	2.91	2.84
	0.001	13.3	8.77	7.05	6.12	5.53	5.12	4.82	4.58	4.39	4.24	4.11	4.00
	0.0005	15.2	9.90	7.90	6.82	6.14	5.66	5.31	5.04	4.82	4.65	4.51	4.38
40	0.975	$0.0^3$99	0.025	0.071	0.119	0.162	0.199	0.232	0.260	0.285	0.307	0.327	0.344
	0.05	4.08	3.23	2.84	2.61	2.45	2.34	2.25	2.18	2.12	2.08	2.04	2.00
	0.025	5.42	4.05	3.46	3.13	2.90	2.74	2.62	2.53	2.45	2.39	2.33	2.29

Table C9. (*continued*)

ν_2	α \ ν_1	1	2	3	4	5	6	7	8	9	10	11	12
	0.01	7.31	5.18	4.31	3.83	3.51	3.29	3.12	2.99	2.89	2.80	2.73	2.66
	0.001	12.6	8.25	6.60	5.70	5.13	4.73	4.44	4.21	4.02	3.87	3.75	3.64
	0.0005	14.4	9.25	7.33	6.30	5.64	5.19	4.85	4.59	4.38	4.21	4.07	3.95
60	0.975	0.0^399	0.025	0.071	0.120	0.163	0.202	0.235	0.264	0.290	0.313	0.333	0.351
	0.05	4.00	3.15	2.76	2.53	2.37	2.25	2.17	2.10	2.04	1.99	1.95	1.92
	0.025	5.29	3.93	3.34	3.01	2.79	2.63	2.51	2.41	2.33	2.27	2.22	2.17
	0.01	7.08	4.98	4.13	3.65	3.34	3.12	2.95	2.82	2.72	2.63	2.56	2.50
	0.001	12.0	7.76	6.17	5.31	4.76	4.37	4.09	3.87	3.69	3.54	3.43	3.31
	0.0005	13.6	8.65	6.81	5.82	5.20	4.76	4.44	4.18	3.98	3.82	3.69	3.57
120	0.975	0.0^399	0.025	0.072	0.120	0.165	0.204	0.238	0.268	0.295	0.318	0.340	0.359
	0.05	3.92	3.07	2.68	2.45	2.29	2.18	2.09	2.02	1.96	1.91	1.87	1.83
	0.025	5.15	3.80	3.23	2.89	2.67	2.52	2.39	2.30	2.22	2.16	2.10	2.05
	0.01	6.85	4.79	3.95	3.48	3.17	2.96	2.79	2.66	2.56	2.47	2.40	2.34
	0.001	11.4	7.32	5.79	4.95	4.42	4.04	3.77	3.55	3.38	3.24	3.12	3.02
	0.0005	12.8	8.10	6.34	5.39	4.79	4.37	4.07	3.82	3.63	3.47	3.34	3.22
∞	0.975	0.0^398	0.025	0.072	0.121	0.166	0.206	0.241	0.272	0.300	0.325	0.347	0.367
	0.05	3.84	3.00	2.60	2.37	2.21	2.10	2.01	1.94	1.88	1.83	1.79	1.75
	0.025	5.02	3.69	3.12	2.79	2.57	2.41	2.29	2.19	2.11	2.05	1.99	1.94
	0.01	6.63	4.61	3.78	3.32	3.02	2.80	2.64	2.51	2.41	2.32	2.25	2.18
	0.001	10.8	6.91	5.42	4.62	4.10	3.74	3.47	3.27	3.10	2.96	2.84	2.74
	0.0005	12.1	7.60	5.91	5.00	4.42	4.02	3.72	3.48	3.30	3.14	3.02	2.90

ν_2	α \ ν_1	15	20	24	30	40	50	60	100	120	200	500	∞
24	0.975	0.370	0.415	0.441	0.468	0.498	0.518	0.531	0.562	0.568	0.585	0.599	0.610
	0.05	2.11	2.03	1.98	1.94	1.89	1.86	1.84	1.80	1.79	1.77	1.75	1.73
	0.025	2.44	2.33	2.27	2.21	2.15	2.11	2.08	2.02	2.01	1.98	1.95	1.94
	0.01	2.89	2.74	2.66	2.58	2.49	2.44	2.40	2.33	2.31	2.27	2.24	2.21
	0.001	4.14	3.87	3.74	3.59	3.45	3.35	3.29	3.16	3.14	3.07	3.01	2.97
	0.0005	4.55	4.25	4.09	3.93	3.76	3.66	3.59	3.44	3.41	3.33	3.27	3.22
30	0.975	0.378	0.426	0.453	0.482	0.515	0.535	0.551	0.585	0.592	0.610	0.625	0.639
	0.05	2.01	1.93	1.89	1.84	1.79	1.76	1.74	1.70	1.68	1.66	1.64	1.62
	0.025	2.31	2.20	2.14	2.07	2.01	1.97	1.94	1.88	1.87	1.84	1.81	1.79
	0.01	2.70	2.55	2.47	2.39	2.30	2.25	2.21	2.13	2.11	2.07	2.03	2.01
	0.001	3.75	3.49	3.36	3.22	3.07	2.98	2.92	2.79	2.76	2.69	2.63	2.59
	0.0005	4.10	3.80	3.65	3.48	3.32	3.22	3.15	3.00	2.97	2.89	2.82	2.78
40	0.975	0.387	0.437	0.466	0.498	0.533	0.556	0.573	0.610	0.620	0.641	0.662	0.674
	0.05	1.92	1.84	1.79	1.74	1.69	1.66	1.64	1.59	1.58	1.55	1.53	1.51
	0.025	2.18	2.07	2.01	1.94	1.88	1.83	1.80	1.74	1.72	1.69	1.66	1.64
	0.01	2.52	2.37	2.29	2.20	2.11	2.06	2.02	1.94	1.92	1.87	1.83	1.80
	0.001	3.40	3.15	3.01	2.87	2.73	2.64	2.57	2.44	2.41	2.34	2.28	2.23
	0.0005	3.68	3.39	3.24	3.08	2.92	2.82	2.74	2.60	2.57	2.49	2.41	2.37

Table C9. (*continued*)

ν_2	α	15	20	24	30	40	50	60	100	120	200	500	∞
60	0.975	0.396	0.450	0.481	0.515	0.555	0.581	0.600	0.641	0.654	0.680	0.704	0.720
	0.05	1.84	1.75	1.70	1.65	1.59	1.56	1.53	1.48	1.47	1.44	1.41	1.39
	0.025	2.06	1.94	1.88	1.82	1.74	1.70	1.67	1.60	1.58	1.54	1.51	1.48
	0.01	2.35	2.20	2.12	2.03	1.94	1.88	1.84	1.75	1.73	1.68	1.63	1.60
	0.001	3.08	2.83	2.69	2.56	2.41	2.31	2.25	2.11	2.09	2.01	1.93	1.89
	0.0005	3.30	3.02	2.87	2.71	2.55	2.45	2.38	2.23	2.19	2.11	2.03	1.98
120	0.975	0.406	0.464	0.498	0.536	0.580	0.611	0.633	0.684	0.698	0.729	0.762	0.789
	0.05	1.75	1.66	1.61	1.55	1.50	1.46	1.43	1.37	1.35	1.32	1.28	1.25
	0.025	1.95	1.82	1.76	1.69	1.61	1.56	1.53	1.45	1.43	1.39	1.34	1.31
	0.01	2.19	2.03	1.95	1.86	1.76	1.70	1.66	1.56	1.53	1.48	1.42	1.38
	0.001	2.78	2.53	2.40	2.26	2.11	2.02	1.95	1.82	1.76	1.70	1.62	1.54
	0.0005	2.96	2.67	2.53	2.38	2.21	2.11	2.01	1.88	1.84	1.75	1.67	1.60
∞	0.975	0.418	0.480	0.517	0.560	0.611	0.645	0.675	0.741	0.763	0.813	0.878	1.00
	0.05	1.67	1.57	1.52	1.46	1.39	1.35	1.32	1.24	1.22	1.17	1.11	1.00
	0.025	1.83	1.71	1.64	1.57	1.48	1.43	1.39	1.30	1.27	1.21	1.13	1.00
	0.01	2.04	1.88	1.79	1.70	1.59	1.52	1.47	1.36	1.32	1.25	1.15	1.00
	0.001	2.51	2.27	2.13	1.99	1.84	1.73	1.66	1.49	1.45	1.34	1.21	1.00
	0.0005	2.65	2.37	2.22	2.07	1.91	1.79	1.71	1.53	1.48	1.36	1.22	1.00

notation: $593^3 = 593 \times 10^3$
$0.0^2 11 = 0.11 \times 10^{-2}$

TABLE C10. Null Distributions of Spearman's Rank Correlation Coefficient r_s

A lower-tail probability P, corresponding to a value c, is the probability, under the null hypothesis of independence, that r_s will be less than or equal to c. An upper-tail probability P, corresponding to a value c, is the probability, under the null hypothesis of independence, that r_s will be greater than or equal to c. (For $n=4$, 5, and 6, the table is arranged so that in each subtable the lower-tail values are presented first, with the upper-tail values beneath—and separated by a space from—the lower-tail values.)

Examples: $n = 5$: $Pr(r_s \leq -.400) = .258$, $Pr(r_s \geq .400) = .258$.

$n = 8$: $Pr(r_s \leq -.690) = .035$, $Pr(r_s \geq .690) = .035$.

	$n = 4$			$n = 5$			$n = 6$	
c	Σd^2	P	c	Σd^2	P	c	Σd^2	P
-1.000	20	.042	-1.000	40	.008	-1.000	70	.001
$-.800$	18	.167	$-.900$	38	.042	$-.943$	68	.008
$-.600$	16	.208	$-.800$	36	.067	$-.886$	66	.017
$-.400$	14	.375	$-.700$	34	.117	$-.829$	64	.029
$-.200$	12	.458	$-.600$	32	.175	$-.771$	62	.051
.000	10	.542	$-.500$	30	.225	$-.714$	60	.068
			$-.400$	28	.258	$-.657$	58	.088
.000	10	.542	$-.300$	26	.342	$-.600$	56	.121
.200	8	.458	$-.200$	24	.392	$-.543$	54	.149
.400	6	.375	$-.100$	22	.475	$-.486$	52	.178
.600	4	.208	.000	20	.525	$-.429$	50	.210
.800	2	.167				$-.371$	48	.249
1.000	0	.042	.000	20	.525	$-.314$	46	.282
			.100	18	.475	$-.257$	44	.329
			.200	16	.392	$-.200$	42	.357
			.300	14	.342	$-.143$	40	.401
			.400	12	.258	$-.086$	38	.460
			.500	10	.225	$-.029$	36	.500
			.600	8	.175			
			.700	6	.117	.029	34	.500
			.800	4	.067	.086	32	.460
			.900	2	.042	.143	30	.401
			1.000	0	.008	.200	28	.357
						.257	26	.329
						.314	24	.282
						.371	22	.249
						.429	20	.210
						.486	18	.178
						.543	16	.149
						.600	14	.121
						.657	12	.088
						.714	10	.068
						.771	8	.051
						.829	6	.029
						.886	4	.017
						.943	2	.008
						1.000	0	.001

Table C10. (continued)

(Lower Tail)

n = 7			n = 8			n = 9					
c	Σd^2	P	c	Σd^2	P	c	Σd^2	P	c	Σd^2	P
−1.000	112	.000	−1.000	168	.000	−1.000	240	.000	−.483	178	.097
−.964	110	.001	−.976	166	.000	−.983	238	.000	−.467	176	.106
−.929	108	.003	−.952	164	.001	−.967	236	.000	−.450	174	.115
−.893	106	.006	−.929	162	.001	−.950	234	.000	−.433	172	.125
−.857	104	.012	−.905	160	.002	−.933	232	.000	−.417	170	.135
−.821	102	.017	−.881	158	.004	−.917	230	.001	−.400	168	.146
−.786	100	.024	−.857	156	.005	−.900	228	.001	−.383	166	.156
−.750	98	.033	−.833	154	.008	−.883	226	.002	−.367	164	.168
−.714	96	.044	−.810	152	.011	−.867	224	.002	−.350	162	.179
−.679	94	.055	−.786	150	.014	−.850	222	.003	−.333	160	.193
−.643	92	.069	−.762	148	.018	−.833	220	.004	−.317	158	.205
−.607	90	.083	−.738	146	.023	−.817	218	.005	−.300	156	.218
−.571	88	.100	−.714	144	.029	−.800	216	.007	−.283	154	.231
−.536	86	.118	−.690	142	.035	−.783	214	.009	−.267	152	.247
−.500	84	.133	−.667	140	.042	−.767	212	.011	−.250	150	.260
−.464	82	.151	−.643	138	.048	−.750	210	.013	−.233	148	.276
−.429	80	.177	−.619	136	.057	−.733	208	.016	−.217	146	.290
−.393	78	.198	−.595	134	.066	−.717	206	.018	−.200	144	.307
−.357	76	.222	−.571	132	.076	−.700	204	.022	−.183	142	.322
−.321	74	.249	−.548	130	.085	−.683	202	.025	−.167	140	.339
−.286	72	.278	−.524	128	.098	−.667	200	.029	−.150	138	.354
−.250	70	.297	−.500	126	.108	−.650	198	.033	−.133	136	.372
−.214	68	.331	−.476	124	.122	−.633	196	.038	−.117	134	.388
−.179	66	.357	−.452	122	.134	−.617	194	.043	−.100	132	.405
−.143	64	.391	−.429	120	.150	−.600	192	.048	−.083	130	.422
−.107	62	.420	−.405	118	.163	−.583	190	.054	−.067	128	.440
−.071	60	.453	−.381	116	.180	−.567	188	.060	−.050	126	.456
−.036	58	.482	−.357	114	.195	−.550	186	.066	−.033	124	.474
.000	56	.518	−.333	112	.214	−.533	184	.074	−.017	122	.491
			−.310	110	.231	−.517	182	.081	.000	120	.509
			−.286	108	.250	−.500	180	.089			
			−.262	106	.268						
			−.238	104	.291						
			−.214	102	.310						
			−.190	100	.332						
			−.167	98	.352						
			−.143	96	.376						
			−.119	94	.397						
			−.095	92	.420						
			−.071	90	.441						
			−.048	88	.467						
			−.024	86	.488						
			.000	84	.512						

(Upper Tail)

	n = 7			n = 8			n = 9			n = 9	
c	Σd^2	P	c	Σd^2	P	c	Σd^2	P	c	Σd^2	P
.000	56	.518	.000	84	.512	.000	120	.509	.517	58	.081
.036	54	.482	.024	82	.488	.017	118	.491	.533	56	.074
.071	52	.453	.048	80	.467	.033	116	.474	.550	54	.066
.107	50	.420	.071	78	.441	.050	114	.456	.567	52	.060
.143	48	.391	.095	76	.420	.067	112	.440	.583	50	.054
.179	46	.357	.119	74	.397	.083	110	.422	.600	48	.048
.214	44	.331	.143	72	.376	.100	108	.405	.617	46	.043
.250	42	.297	.167	70	.352	.117	106	.388	.633	44	.038
.286	40	.278	.190	68	.332	.133	104	.372	.650	42	.033
.321	38	.249	.214	66	.310	.150	102	.354	.667	40	.029
.357	36	.222	.238	64	.291	.167	100	.339	.683	38	.025
.393	34	.198	.262	62	.268	.183	98	.322	.700	36	.022
.429	32	.177	.286	60	.250	.200	96	.307	.717	34	.018
.464	30	.151	.310	58	.231	.217	94	.290	.733	32	.016
.500	28	.133	.333	56	.214	.233	92	.276	.750	30	.013
.536	26	.118	.357	54	.195	.250	90	.260	.767	28	.011
.571	24	.100	.381	52	.180	.267	88	.247	.783	26	.009
.607	22	.083	.405	50	.163	.283	86	.231	.800	24	.007
.643	20	.069	.429	48	.150	.300	84	.218	.817	22	.005
.679	18	.055	.452	46	.134	.317	82	.205	.833	20	.004
.714	16	.044	.476	44	.122	.333	80	.193	.850	18	.003
.750	14	.033	.500	42	.108	.350	78	.179	.867	16	.002
.786	12	.024	.524	40	.098	.367	76	.168	.883	14	.002
.821	10	.017	.548	38	.085	.383	74	.156	.900	12	.001
.857	8	.012	.571	36	.076	.400	72	.146	.917	10	.001
.893	6	.006	.595	34	.066	.417	70	.135	.933	8	.000
.929	4	.003	.619	32	.057	.433	68	.125	.950	6	.000
.964	2	.001	.643	30	.048	.450	66	.115	.967	4	.000
1.000	0	.000	.667	28	.042	.467	64	.106	.983	2	.000
			.690	26	.035	.483	62	.097	1.000	0	.000
			.714	24	.029	.500	60	.089			
			.738	22	.023						
			.762	20	.018						
			.786	18	.014						
			.810	16	.011						
			.833	14	.008						
			.857	12	.005						
			.881	10	.004						
			.905	8	.002						
			.929	6	.001						
			.952	4	.001						
			.976	2	.000						
			1.000	0	.000						

(Lower Tail)

n = 10						n = 11					
c	Σd^2	P	c	Σd^2	P	c	Σd^2	P	c	Σd^2	P
−1.000	330	.000	−.491	246	.077	−1.000	440	.000	−.491	328	.065
−.988	328	.000	−.479	244	.083	−.991	438	.000	−.482	326	.069
−.976	326	.000	−.467	242	.089	−.982	436	.000	−.473	324	.073
−.964	324	.000	−.455	240	.096	−.973	434	.000	−.464	322	.077
−.952	322	.000	−.442	238	.102	−.964	432	.000	−.455	320	.082
−.939	320	.000	−.430	236	.109	−.955	430	.000	−.445	318	.087
−.927	318	.000	−.418	234	.116	−.945	428	.000	−.436	316	.091
−.915	316	.000	−.406	232	.124	−.936	426	.000	−.427	314	.096
−.903	314	.000	−.394	230	.132	−.927	424	.000	−.418	312	.102
−.891	312	.001	−.382	228	.139	−.918	422	.000	−.409	310	.107
−.879	310	.001	−.370	226	.148	−.909	420	.000	−.400	308	.112
−.867	308	.001	−.358	224	.156	−.900	418	.000	−.391	306	.118
−.855	306	.001	−.345	222	.165	−.891	416	.000	−.382	304	.124
−.842	304	.002	−.333	220	.174	−.882	414	.000	−.373	302	.130
−.830	302	.002	−.321	218	.184	−.873	412	.000	−.364	300	.137
−.818	300	.003	−.309	216	.193	−.864	410	.001	−.355	298	.143
−.806	298	.004	−.297	214	.203	−.855	408	.001	−.345	296	.150
−.794	296	.004	−.285	212	.214	−.845	406	.001	−.336	294	.157
−.782	294	.005	−.273	210	.224	−.836	404	.001	−.327	292	.163
−.770	292	.006	−.261	208	.235	−.827	402	.001	−.318	290	.171
−.758	290	.008	−.248	206	.246	−.818	400	.002	−.309	288	.178
−.745	288	.009	−.236	204	.257	−.809	398	.002	−.300	286	.186
−.733	286	.010	−.224	202	.268	−.800	396	.002	−.291	284	.193
−.721	284	.012	−.212	200	.280	−.791	394	.003	−.282	282	.201
−.709	282	.013	−.200	198	.292	−.782	392	.003	−.273	280	.209
−.697	280	.015	−.188	196	.304	−.773	390	.004	−.264	278	.217
−.685	278	.017	−.176	194	.316	−.764	388	.004	−.255	276	.226
−.673	276	.019	−.164	192	.328	−.755	386	.005	−.245	274	.234
−.661	274	.022	−.152	190	.341	−.745	384	.006	−.236	272	.243
−.648	272	.024	−.139	188	.354	−.736	382	.006	−.227	270	.252
−.636	270	.027	−.127	186	.367	−.727	380	.007	−.218	268	.260
−.624	268	.030	−.115	184	.379	−.718	378	.008	−.209	266	.270
−.612	266	.033	−.103	182	.393	−.709	376	.009	−.200	264	.279
−.600	264	.037	−.091	180	.406	−.700	374	.010	−.191	262	.288
−.588	262	.040	−.079	178	.419	−.691	372	.011	−.182	260	.298
−.576	260	.044	−.067	176	.433	−.682	370	.013	−.173	258	.307
−.564	258	.048	−.055	174	.446	−.673	368	.014	−.164	256	.317
−.552	256	.052	−.042	172	.459	−.664	366	.015	−.155	254	.327
−.539	254	.057	−.030	170	.473	−.655	364	.017	−.145	252	.337
−.527	252	.062	−.018	168	.486	−.645	362	.018	−.136	250	.347
−.515	250	.067	−.006	166	.500	−.636	360	.020	−.127	248	.357
−.503	248	.072				−.627	358	.022	−.118	246	.367
						−.618	356	.024	−.109	244	.377
						−.609	354	.026	−.100	242	.388
						−.600	352	.028	−.091	240	.398
						−.591	350	.030	−.082	238	.409
						−.582	348	.033	−.073	236	.419
						−.573	346	.035	−.064	234	.430
						−.564	344	.038	−.055	232	.441
						−.555	342	.041	−.045	230	.452
						−.545	340	.044	−.036	228	.462
						−.536	338	.047	−.027	226	.473
						−.527	336	.050	−.018	224	.484
						−.518	334	.054	−.009	222	.495
						−.509	332	.057	.000	220	.505
						−.500	330	.061			

Table C10. (continued)

(Upper Tail)

$n = 10$

c	Σd^2	P	c	Σd^2	P
.006	164	.500	.515	80	.067
.018	162	.486	.527	78	.062
.030	160	.473	.539	76	.057
.042	158	.459	.552	74	.052
.055	156	.446	.564	72	.048
.067	154	.433	.576	70	.044
.079	152	.419	.588	68	.040
.091	150	.406	.600	66	.037
.103	148	.393	.612	64	.033
.115	146	.379	.624	62	.030
.127	144	.367	.636	60	.027
.139	142	.354	.648	58	.024
.152	140	.341	.661	56	.022
.164	138	.328	.673	54	.019
.176	136	.316	.685	52	.017
.188	134	.304	.697	50	.015
.200	132	.292	.709	48	.013
.212	130	.280	.721	46	.012
.224	128	.268	.733	44	.010
.236	126	.257	.745	42	.009
.248	124	.246	.758	40	.008
.261	122	.235	.770	38	.006
.273	120	.224	.782	36	.005
.285	118	.214	.794	34	.004
.297	116	.203	.806	32	.004
.309	114	.193	.818	30	.003
.321	112	.184	.830	28	.002
.333	110	.174	.842	26	.002
.345	108	.165	.855	24	.001
.358	106	.156	.867	22	.001
.370	104	.148	.879	20	.001
.382	102	.139	.891	18	.001
.394	100	.132	.903	16	.000
.406	98	.124	.915	14	.000
.418	96	.116	.927	12	.000
.430	94	.109	.939	10	.000
.442	92	.102	.952	8	.000
.455	90	.096	.964	6	.000
.467	88	.089	.976	4	.000
.479	86	.083	.988	2	.000
.491	84	.077	1.000	0	.000
.503	82	.072			

$n = 11$

c	Σd^2	P	c	Σd^2	P
.000	220	.505	.509	108	.057
.009	218	.495	.518	106	.054
.018	216	.484	.527	104	.050
.027	214	.473	.536	102	.047
.036	212	.462	.545	100	.044
.045	210	.452	.555	98	.041
.055	208	.441	.564	96	.038
.064	206	.430	.573	94	.035
.073	204	.419	.582	92	.033
.082	202	.409	.591	90	.030
.091	200	.398	.600	88	.028
.100	198	.388	.609	86	.026
.109	196	.377	.618	84	.024
.118	194	.367	.627	82	.022
.127	192	.357	.636	80	.020
.136	190	.347	.645	78	.018
.145	188	.337	.655	76	.017
.155	186	.327	.664	74	.015
.164	184	.317	.673	72	.014
.173	182	.307	.682	70	.013
.182	180	.298	.691	68	.011
.191	178	.288	.700	66	.010
.200	176	.279	.709	64	.009
.209	174	.270	.718	62	.008
.218	172	.260	.727	60	.007
.227	170	.252	.736	58	.006
.236	168	.243	.745	56	.006
.245	166	.234	.755	54	.005
.255	164	.226	.764	52	.004
.264	162	.217	.773	50	.004
.273	160	.209	.782	48	.003
.282	158	.201	.791	46	.003
.291	156	.193	.800	44	.002
.300	154	.186	.809	42	.002
.309	152	.178	.818	40	.002
.318	150	.171	.827	38	.001
.327	148	.163	.836	36	.001
.336	146	.157	.845	34	.001
.345	144	.150	.855	32	.001
.355	142	.143	.864	30	.001
.364	140	.137	.873	28	.000
.373	138	.130	.882	26	.000
.382	136	.124	.891	24	.000
.391	134	.118	.900	22	.000
.400	132	.112	.909	20	.000
.409	130	.107	.918	18	.000
.418	128	.102	.927	16	.000
.427	126	.096	.936	14	.000
.436	124	.091	945	12	.000
.445	122	.087	.955	10	.000
.455	120	.082	.964	8	.000
.464	118	.077	.973	6	.000
.473	116	.073	.982	4	.000
.482	114	.069	.991	2	.000
.491	112	.065	1.000	0	.000
.500	110	.061			

Adapted from Table J of *A Nonparametric Introduction to Statistics*, by C. H. Kraft and C. van Eeden, Macmillan, New York, 1968, with the permission of the authors and the publisher. Copyright © 1968, by the Macmillan Company.

TABLE C11. Null Distributions of Wilcoxon's Signed Rank Statistic T^+

A lower-tail probability P, corresponding to a value c, is the probability, under the null hypothesis that $\mu = 0$, that T^+ will be less than or equal to c. An upper-tail probability P, corresponding to a value c, is the probability, under the null hypothesis that $\mu = 0$, that T^+ will be greater than or equal to c. (For $n = 3, 4, \ldots, 9$, the table is arranged so that in each subtable the lower-tail values are presented first, with the upper-tail values beneath—and separated by a space from—the lower-tail values.)

Examples: $n = 5$: $Pr(T^+ \le 4) = .219$, $Pr(T^+ \ge 11) = .219$.

$n = 13$: $Pr(T^+ \le 17) = .024$, $Pr(T^+ \ge 74) = .024$.

Sample Size n

$n = 3$

c	P
0	.125
1	.250
2	.375
3	.625
3	.625
4	.375
5	.250
6	.125

$n = 4$

c	P
0	.062
1	.125
2	.188
3	.312
4	.438
5	.562
5	.562
6	.438
7	.312
8	.188
9	.125
10	.062

$n = 5$

c	P
0	.031
1	.062
2	.094
3	.156
4	.219
5	.312
6	.406
7	.500
8	.500
9	.406
10	.312
11	.219
12	.156
13	.094
14	.062
15	.031

$n = 6$

c	P
0	.016
1	.031
2	.047
3	.078
4	.109
5	.156
6	.219
7	.281
8	.344
9	.422
10	.500
11	.500
12	.422
13	.344
14	.281
15	.219
16	.156
17	.109
18	.078
19	.047
20	.031
21	.016

$n = 7$

c	P
0	.008
1	.016
2	.023
3	.039
4	.055
5	.078
6	.109
7	.148
8	.188
9	.234
10	.289
11	.344
12	.406
13	.469
14	.531
14	.531
15	.469
16	.406
17	.344
18	.289
19	.234
20	.188
21	.148
22	.109
23	.078
24	.055
25	.039
26	.023
27	.016
28	.008

$n = 8$

c	P
0	.004
1	.008
2	.012
3	.020
4	.027
5	.039
6	.055
7	.074
8	.098
9	.125
10	.156
11	.191
12	.230
13	.273
14	.320
15	.371
16	.422
17	.473
18	.527
18	.527
19	.473
20	.422
21	.371
22	.320
23	.273
24	.230
25	.191
26	.156
27	.125
28	.098
29	.074
30	.055
31	.039
32	.027
33	.020
34	.012
35	.008
36	.004

$n = 9$

c	P
0	.002
1	.004
2	.006
3	.010
4	.014
5	.020
6	.027
7	.037
8	.049
9	.064
10	.082
11	.102
12	.125
13	.150
14	.180
15	.213
16	.248
17	.285
18	.326
19	.367
20	.410
21	.455
22	.500
23	.500
24	.455
25	.410
26	.367
27	.326
28	.285
29	.248
30	.213
31	.180
32	.150
33	.125
34	.102
35	.082
36	.064
37	.049
38	.037
39	.027
40	.020
41	.014
42	.010
43	.006
44	.004
45	.002

(Lower Tail)

Sample Size n													
10		11		12		13		14		15			
c	P	c	P	c	P	c	P	c	P	c	P	c	P
0	.001	0	.000	0	.000	0	.000	0	.000	0	.000	31	.053
1	.002	1	.001	1	.000	1	.000	1	.000	1	.000	32	.060
2	.003	2	.001	2	.001	2	.000	2	.000	2	.000	33	.068
3	.005	3	.002	3	.001	3	.001	3	.000	3	.000	34	.076
4	.007	4	.003	4	.002	4	.001	4	.000	4	.000	35	.084
5	.010	5	.005	5	.002	5	.001	5	.001	5	.000	36	.094
6	.014	6	.007	6	.003	6	.002	6	.001	6	.000	37	.104
7	.019	7	.009	7	.005	7	.002	7	.001	7	.001	38	.115
8	.024	8	.012	8	.006	8	.003	8	.002	8	.001	39	.126
9	.032	9	.016	9	.008	9	.004	9	.002	9	.001	40	.138
10	.042	10	.021	10	.010	10	.005	10	.003	10	.001	41	.151
11	.053	11	.027	11	.013	11	.007	11	.003	11	.002	42	.165
12	.065	12	.034	12	.017	12	.009	12	.004	12	.002	43	.180
13	.080	13	.042	13	.021	13	.011	13	.005	13	.003	44	.195
14	.097	14	.051	14	.026	14	.013	14	.007	14	.003	45	.211
15	.116	15	.062	15	.032	15	.016	15	.008	15	.004	46	.227
16	.138	16	.074	16	.039	16	.020	16	.010	16	.005	47	.244
17	.161	17	.087	17	.046	17	.024	17	.012	17	.006	48	.262
18	.188	18	.103	18	.055	18	.029	18	.015	18	.008	49	.281
19	.216	19	.120	19	.065	19	.034	19	.018	19	.009	50	.300
20	.246	20	.139	20	.076	20	.040	20	.021	20	.011	51	.319
21	.278	21	.160	21	.088	21	.047	21	.025	21	.013	52	.339
22	.312	22	.183	22	.102	22	.055	22	.029	22	.015	53	.360
23	.348	23	.207	23	.117	23	.064	23	.034	23	.018	54	.381
24	.385	24	.232	24	.133	24	.073	24	.039	24	.021	55	.402
25	.423	25	.260	25	.151	25	.084	25	.045	25	.024	56	.423
26	.461	26	.289	26	.170	26	.095	26	.052	26	.028	57	.445
27	.500	27	.319	27	.190	27	.108	27	.059	27	.032	58	.467
		28	.350	28	.212	28	.122	28	.068	28	.036	59	.489
		29	.382	29	.235	29	.137	29	.077	29	.042	60	.511
		30	.416	30	.259	30	.153	30	.086	30	.047		
		31	.449	31	.285	31	.170	31	.097				
		32	.483	32	.311	32	.188	32	.108				
		33	.517	33	.339	33	.207	33	.121				
				34	.367	34	.227	34	.134				
				35	.396	35	.249	35	.148				
				36	.425	36	.271	36	.163				
				37	.455	37	.294	37	.179				
				38	.485	38	.318	38	.196				
				39	.515	39	.342	39	.213				
						40	.368	40	.232				
						41	.393	41	.251				
						42	.420	42	.271				
						43	.446	43	.292				
						44	.473	44	.313				
						45	.500	45	.335				
								46	.357				
								47	.380				
								48	.404				
								49	.428				
								50	.452				
								51	.476				
								52	.500				

Table C11. (continued)

(Upper Tail)

Sample Size n													
10		11		12		13		14		15			
c	P	c	P	c	P	c	P	c	P	c	P	c	P
28	.500	33	.517	39	.515	46	.500	53	.500	60	.511	91	.042
29	.461	34	.483	40	.485	47	.473	54	.476	61	.489	92	.036
30	.423	35	.449	41	.455	48	.446	55	.452	62	.467	93	.032
31	.385	36	.416	42	.425	49	.420	56	.428	63	.445	94	.028
32	.348	37	.382	43	.396	50	.393	57	.404	64	.423	95	.024
33	.312	38	.350	44	.367	51	.368	58	.380	65	.402	96	.021
34	.278	39	.319	45	.339	52	.342	59	.357	66	.381	97	.018
35	.246	40	.289	46	.311	53	.318	60	.335	67	.360	98	.015
36	.216	41	.260	47	.285	54	.294	61	.313	68	.339	99	.013
37	.188	42	.232	48	.259	55	.271	62	.292	69	.319	100	.011
38	.161	43	.207	49	.235	56	.249	63	.271	70	.300	101	.009
39	.138	44	.183	50	.212	57	.227	64	.251	71	.281	102	.008
40	.116	45	.160	51	.190	58	.207	65	.232	72	.262	103	.006
41	.097	46	.139	52	.170	59	.188	66	.213	73	.244	104	.005
42	.080	47	.120	53	.151	60	.170	67	.196	74	.227	105	.004
43	.065	48	.103	54	.133	61	.153	68	.179	75	.211	106	.003
44	.053	49	.087	55	.117	62	.137	69	.163	76	.195	107	.003
45.	.042	50	.074	56	.102	63	.122	70	.148	77	.180	108	.002
46	.032	51	.062	57	.088	64	.108	71	.134	78	.165	109	.002
47	.024	52	.051	58	.076	65	.095	72	.121	79	.151	110	.001
48	.019	53	.042	59	.065	66	.084	73	.108	80	.138	111	.001
49	.014	54	.034	60	.055	67	.073	74	.097	81	.126	112	.001
50	.010	55	.027	61	.046	68	.064	75	.086	82	.115	113	.001
51	.007	56	.021	62	.039	69	.055	76	.077	83	.104	114	.000
52	.005	57	.016	63	.032	70	.047	77	.068	84	.094	115	.000
53	.003	58	.012	64	.026	71	.040	78	.059	85	.084	116	.000
54	.002	59	.009	65	.021	72	.034	79	.052	86	.076	117	.000
55	.001	60	.007	66	.017	73	.029	80	.045	87	.068	118	.000
		61	.005	67	.013	74	.024	81	.039	88	.060	119	.000
		62	.003	68	.010	75	.020	82	.034	89	.053	120	.000
		63	.002	69	.008	76	.016	83	.029	90	.047		
		64	.001	70	.006	77	.013	84	.025				
		65	.001	71	.005	78	.011	85	.021				
		66	.000	72	.003	79	.009	86	.018				
				73	.002	80	.007	87	.015				
				74	.002	81	.005	88	.012				
				75	.001	82	.004	89	.010				
				76	.001	83	.003	90	.008				
				77	.000	84	.002	91	.007				
				78	.000	85	.002	92	.005				
						86	.001	93	.004				
						87	.001	94	.003				
						88	.001	95	.003				
						89	.000	96	.002				
						90	.000	97	.002				
						91	.000	98	.001				
								99	.001				
								100	.001				
								101	.000				
								102	.000				
								103	.000				
								104	.000				
								105	.000				

Adapted from Table C of *A Nonparametric Introduction to Statistics*, by C. H. Kraft and C. van Eeden, Macmillan, New York, 1968, with the permission of the authors and the publisher Copyright © 1968, by the Macmillan Company.

TABLE C12. Null Distributions of Wilcoxon's Rank Sum Statistic W

A lower-tail probability P, corresponding to a value c, is the probability, under the null hypothesis that $\mu_1 = \mu_2$, that W will be less than or equal to c. An upper-tail probability P, corresponding to a value c, is the probability, under the null hypothesis that $\mu_1 = \mu_2$, that W will be greater than or equal to c. (For $m = 3, 4, \ldots, 7$, the table is arranged so that in each subtable the lower-tail values are presented first, with the upper-tail values beneath—and separated by a space from—the lower-tail values.)

Examples: $m = 5$, $n = 4$: $Pr(W \le 12) = .032$, $Pr(W \ge 28) = .032$.
$m = 8$, $n = 6$: $Pr(W \le 33) = .071$, $Pr(W \ge 57) = .071$.

Larger Sample Size $m = 3$

Smaller Sample Size n					
1		2		3	
c	P	c	P	c	P
1	.250	3	.100	6	.050
2	.500	4	.200	7	.100
		5	.400	8	.200
		6	.600	9	.350
				10	.500
3	.500	6	.600	11	.500
4	.250	7	.400	12	.350
		8	.200	13	.200
		9	.100	14	.100
				15	.050

Larger Sample Size $m = 4$

Smaller Sample Size n							
1		2		3		4	
c	P	c	P	c	P	c	P
1	.200	3	.067	6	.029	10	.014
2	.400	4	.133	7	.057	11	.029
3	.600	5	.267	8	.114	12	.057
		6	.400	9	.200	13	.100
		7	.600	10	.314	14	.171
				11	.429	15	.243
				12	.571	16	.343
						17	.443
						18	.557
3	.600	7	.600	12	.571	18	.557
4	.400	8	.400	13	.429	19	.443
5	.200	9	.267	14	.314	20	.343
		10	.133	15	.200	21	.243
		11	.067	16	.114	22	.171
				17	.057	23	.100
				18	.029	24	.057
						25	.029
						26	.014

Larger Sample Size $m = 5$

| Smaller Sample Size n | | | | | | | | | |
| 1 | | 2 | | 3 | | 4 | | 5 | |
c	P	c	P	c	P	c	P	c	P
1	.167	3	.048	6	.018	10	.008	15	.004
2	.333	4	.095	7	.036	11	.016	16	.008
3	.500	5	.190	8	.071	12	.032	17	.016
		6	.286	9	.125	13	.056	18	.028
4	.500	7	.429	10	.196	14	.095	19	.048
5	.333	8	.571	11	.286	15	.143	20	.075
6	.167			12	.393	16	.206	21	.111
		8	.571	13	.500	17	.278	22	.155
		9	.429			18	.365	23	.210
		10	.286	14	.500	19	.452	24	.274
		11	.190	15	.393	20	.548	25	.345
		12	.095	16	.286			26	.421
		13	.048	17	.196	20	.548	27	.500
				18	.125	21	.452		
				19	.071	22	.365	28	.500
				20	.036	23	.278	29	.421
				21	.018	24	.206	30	.345
						25	.143	31	.274
						26	.095	32	.210
						27	.056	33	.155
						28	.032	34	.111
						29	.016	35	.075
						30	.008	36	.048
								37	.028
								38	.016
								39	.008
								40	.004

Larger Sample Size $m = 6$

| Smaller Sample Size n | | | | | | | | | | | |
| 1 | | 2 | | 3 | | 4 | | 5 | | 6 | |
c	P	c	P	c	P	c	P	c	P	c	P
1	.143	3	.036	6	.012	10	.005	15	.002	21	.001
2	.286	4	.071	7	.024	11	.010	16	.004	22	.002
3	.429	5	.143	8	.048	12	.019	17	.009	23	.004
4	.571	6	.214	9	.083	13	.033	18	.015	24	.008
		7	.321	10	.131	14	.057	19	.026	25	.013
4	.571	8	.429	11	.190	15	.086	20	.041	26	.021
5	.429	9	.571	12	.274	16	.129	21	.063	27	.032
6	.286			13	.357	17	.176	22	.089	28	.047
7	.143	9	.571	14	.452	18	.238	23	.123	29	.066
		10	.429	15	.548	19	.305	24	.165	30	.090
		11	.321			20	.381	25	.214	31	.120
		12	.214	15	.548	21	.457	26	.268	32	.155
		13	.143	16	.452	22	.543	27	.331	33	.197
		14	.071	17	.357			28	.396	34	.242
		15	.036	18	.274	22	.543	29	.465	35	.294
				19	.190	23	.457	30	.535	36	.350
				20	.131	24	.381			37	.409
				21	.083	25	.305	30	.535	38	.469
				22	.048	26	.238	31	.465	39	.531
				23	.024	27	.176	32	.396		
				24	.012	28	.129	33	.331	39	.531
						29	.086	34	.268	40	.469
						30	.057	35	.214	41	.409
						31	.033	36	.165	42	.350
						32	.019	37	.123	43	.294
						33	.010	38	.089	44	.242
						34	.005	39	.063	45	.197
								40	.041	46	.155
								41	.026	47	.120
								42	.015	48	.090
								43	.009	49	.066
								44	.004	50	.047
								45	.002	51	.032
										52	.021
										53	.013
										54	.008
										55	.004
										56	.002
										57	.001

Table C12. (continued)

Larger Sample Size $m = 7$

Smaller Sample Size n													
1		**2**		**3**		**4**		**5**		**6**		**7**	
c	P	c	P	c	P	c	P	c	P	c	P	c	P
1	.125	3	.028	6	.008	10	.003	15	.001	21	.001	28	.000
2	.250	4	.056	7	.017	11	.006	16	.003	22	.001	29	.001
3	.375	5	.111	8	.033	12	.012	17	.005	23	.002	30	.001
4	.500	6	.167	9	.058	13	.021	18	.009	24	.004	31	.002
		7	.250	10	.092	14	.036	19	.015	25	.007	32	.003
5	.500	8	.333	11	.133	15	.055	20	.024	26	.011	33	.006
6	.375	9	.444	12	.192	16	.082	21	.037	27	.017	34	.009
7	.250	10	.556	13	.258	17	.115	22	.053	28	.026	35	.013
8	.125			14	.333	18	.158	23	.074	29	.037	36	.019
		10	.556	15	.417	19	.206	24	.101	30	.051	37	.027
		11	.444	16	.500	20	.264	25	.134	31	.069	38	.036
		12	.333			21	.324	26	.172	32	.090	39	.049
		13	.250	17	.500	22	.394	27	.216	33	.117	40	.064
		14	.167	18	.417	23	.464	28	.265	34	.147	41	.082
		15	.111	19	.333	24	.536	29	.319	35	.183	42	.104
		16	.056	20	.258			30	.378	36	.223	43	.130
		17	.028	21	.192	24	.536	31	.438	37	.267	44	.159
				22	.133	25	.464	32	.500	38	.314	45	.191
				23	.092	26	.394			39	.365	46	.228
				24	.058	27	.324	33	.500	40	.418	47	.267
				25	.033	28	.264	34	.438	41	.473	48	.310
				26	.017	29	.206	35	.378	42	.527	49	.355
				27	.008	30	.158	36	.319			50	.402
						31	.115	37	.265	42	.527	51	.451
						32	.082	38	.216	43	.473	52	.500
						33	.055	39	.172	44	.418		
						34	.036	40	.134	45	.365	53	.500
						35	.021	41	.101	46	.314	54	.451
						36	.012	42	.074	47	.267	55	.402
						37	.006	43	.053	48	.223	56	.355
						38	.003	44	.037	49	.183	57	.310
								45	.024	50	.147	58	.267
								46	.015	51	.117	59	.228
								47	.009	52	.090	60	.191
								48	.005	53	.069	61	.159
								49	.003	54	.051	62	.130
								50	.001	55	.037	63	.104
										56	.026	64	.082
										57	.017	65	.064
										58	.011	66	.049
										59	.007	67	.036
										60	.004	68	.027
										61	.002	69	.019
										62	.001	70	.013
										63	.001	71	.009
												72	.006
												73	.003
												74	.002
												75	.001
												76	.001
												77	.000

Larger Sample Size *m* = 8, (Lower Tail)

Smaller Sample Size *n*															
1		2		3		4		5		6		7		8	
c	*P*	*c*	*P*	*c*	*P*	*c*	*P*	*c*	*P*	*c*	*P*	*c*	*P*	*c*	*P*
1	.111	3	.022	6	.006	10	.002	15	.001	21	.000	28	.000	36	.000
2	.222	4	.044	7	.012	11	.004	16	.002	22	.001	29	.000	37	.000
3	.333	5	.089	8	.024	12	.008	17	.003	23	.001	30	.001	38	.000
4	.444	6	.133	9	.042	13	.014	18	.005	24	.002	31	.001	39	.001
5	.556	7	.200	10	.067	14	.024	19	.009	25	.004	32	.002	40	.001
		8	.267	11	.097	15	.036	20	.015	26	.006	33	.003	41	.001
		9	.356	12	.139	16	.055	21	.023	27	.010	34	.005	42	.002
		10	.444	13	.188	17	.077	22	.033	28	.015	35	.007	43	.003
		11	.556	14	.248	18	.107	23	.047	29	.021	36	.010	44	.005
				15	.315	19	.141	24	.064	30	.030	37	.014	45	.007
				16	.388	20	.184	25	.085	31	.041	38	.020	46	.010
				17	.461	21	.230	26	.111	32	.054	39	.027	47	.014
				18	.539	22	.285	27	.142	33	.071	40	.036	48	.019
						23	.341	28	.177	34	.091	41	.047	49	.025
						24	.404	29	.218	35	.114	42	.060	50	.032
						25	.467	30	.262	36	.141	43	.076	51	.041
						26	.533	31	.311	37	.172	44	.095	52	.052
								32	.362	38	.207	45	.116	53	.065
								33	.416	39	.245	46	.140	54	.080
								34	.472	40	.286	47	.168	55	.097
								35	.528	41	.331	48	.198	56	.117
										42	.377	49	.232	57	.139
										43	.426	50	.268	58	.164
										44	.475	51	.306	59	.191
										45	.525	52	.347	60	.221
												53	.389	61	.253
												54	.433	62	.287
												55	.478	63	.323
												56	.522	64	.360
														65	.399
														66	.439
														67	.480
														68	.520

Table C12. (continued)

Larger Sample Size $m = 8$, (Upper Tail)

Smaller Sample Size n															
1		2		3		4		5		6		7		8	
c	P	c	P	c	P	c	P	c	P	c	P	c	P	c	P
5	.556	11	.556	18	.539	26	.533	35	.528	45	.525	56	.522	68	.520
6	.444	12	.444	19	.461	27	.467	36	.472	46	.475	57	.478	69	.480
7	.333	13	.356	20	.388	28	.404	37	.416	47	.426	58	.433	70	.439
8	.222	14	.267	21	.315	29	.341	38	.362	48	.377	59	.389	71	.399
9	.111	15	.200	22	.248	30	.285	39	.311	49	.331	60	.347	72	.360
		16	.133	23	.188	31	.230	40	.262	50	.286	61	.306	73	.323
		17	.089	24	.139	32	.184	41	.218	51	.245	62	.268	74	.287
		18	.044	25	.097	33	.141	42	.177	52	.207	63	.232	75	.253
		19	.022	26	.067	34	.107	43	.142	53	.172	64	.198	76	.221
				27	.042	35	.077	44	.111	54	.141	65	.168	77	.191
				28	.024	36	.055	45	.085	55	.114	66	.140	78	.164
				29	.012	37	.036	46	.064	56	.091	67	.116	79	.139
				30	.006	38	.024	47	.047	57	.071	68	.095	80	.117
						39	.014	48	.033	58	.054	69	.076	81	.097
						40	.008	49	.023	59	.041	70	.060	82	.080
						41	.004	50	.015	60	.030	71	.047	83	.065
						42	.002	51	.009	61	.021	72	.036	84	.052
								52	.005	62	.015	73	.027	85	.041
								53	.003	63	.010	74	.020	86	.032
								54	.002	64	.006	75	.014	87	.025
								55	.001	65	.004	76	.010	88	.019
										66	.002	77	.007	89	.014
										67	.001	78	.005	90	.010
										68	.001	79	.003	91	.007
										69	.000	80	.002	92	.005
												81	.001	93	.003
												82	.001	94	.002
												83	.000	95	.001
												84	.000	96	.001
														97	.001
														98	.000
														99	.000
														100	.000

Larger Sample Size m = 9, (Lower Tail)

Smaller Sample Size n

1		2		3		4		5		6		7		8		9	
c	P	c	P	c	P	c	P	c	P	c	P	c	P	c	P	c	P
1	.100	3	.018	6	.005	10	.001	15	.000	21	.000	28	.000	36	.000	45	.000
2	.200	4	.036	7	.009	11	.003	16	.001	22	.000	29	.000	37	.000	46	.000
3	.300	5	.073	8	.018	12	.006	17	.002	23	.001	30	.000	38	.000	47	.000
4	.400	6	.109	9	.032	13	.010	18	.003	24	.001	31	.001	39	.000	48	.000
5	.500	7	.164	10	.050	14	.017	19	.006	25	.002	32	.001	40	.000	49	.000
		8	.218	11	.073	15	.025	20	.009	26	.004	33	.002	41	.001	50	.000
		9	.291	12	.105	16	.038	21	.014	27	.006	34	.003	42	.001	51	.001
		10	.364	13	.141	17	.053	22	.021	28	.009	35	.004	43	.002	52	.001
		11	.455	14	.186	18	.074	23	.030	29	.013	36	.006	44	.003	53	.001
		12	.545	15	.241	19	.099	24	.041	30	.018	37	.008	45	.004	54	.002
				16	.300	20	.130	25	.056	31	.025	38	.011	46	.006	55	.003
				17	.364	21	.165	26	.073	32	.033	39	.016	47	.008	56	.004
				18	.432	22	.207	27	.095	33	.044	40	.021	48	.010	57	.005
				19	.500	23	.252	28	.120	34	.057	41	.027	49	.014	58	.007
						24	.302	29	.149	35	.072	42	.036	50	.018	59	.009
						25	.355	30	.182	36	.091	43	.045	51	.023	60	.012
						26	.413	31	.219	37	.112	44	.057	52	.030	61	.016
						27	.470	32	.259	38	.136	45	.071	53	.037	62	.020
						28	.530	33	.303	39	.164	46	.087	54	.046	63	.025
								34	.350	40	.194	47	.105	55	.057	64	.031
								35	.399	41	.228	48	.126	56	.069	65	.039
								36	.449	42	.264	49	.150	57	.084	66	.047
								37	.500	43	.303	50	.176	58	.100	67	.057
										44	.344	51	.204	59	.118	68	.068
										45	.388	52	.235	60	.138	69	.081
										46	.432	53	.268	61	.161	70	.095
										47	.477	54	.303	62	.185	71	.111
										48	.523	55	.340	63	.212	72	.129
												56	.379	64	.240	73	.149
												57	.419	65	.271	74	.170
												58	.459	66	.303	75	.193
												59	.500	67	.336	76	.218
														68	.371	77	.245
														69	.407	78	.273
														70	.444	79	.302
														71	.481	80	.333
														72	.519	81	.365
																82	.398
																83	.432
																84	.466
																85	.500

Table C12. (continued)

Larger Sample Size $m = 9$, (Upper Tail)

Smaller Sample Size n																	
1		2		3		4		5		6		7		8		9	
c	P	c	P	c	P	c	P	c	P	c	P	c	P	c	P	c	P
6	.500	12	.545	20	.500	28	.530	38	.500	48	.523	60	.500	72	.519	86	.500
7	.400	13	.455	21	.432	29	.470	39	.449	49	.477	61	.459	73	.481	87	.466
8	.300	14	.364	22	.364	30	.413	40	.399	50	.432	62	.419	74	.444	88	.432
9	.200	15	.291	23	.300	31	.355	41	.350	51	.388	63	.379	75	.407	89	.398
10	.100	16	.218	24	.241	32	.302	42	.303	52	.344	64	.340	76	.371	90	.365
		17	.164	25	.186	33	.252	43	.259	53	.303	65	.303	77	.336	91	.333
		18	.109	26	.141	34	.207	44	.219	54	.264	66	.268	78	.303	92	.302
		19	.073	27	.105	35	.165	45	.182	55	.228	67	.235	79	.271	93	.273
		20	.036	28	.073	36	.130	46	.149	56	.194	68	.204	80	.240	94	.245
		21	.018	29	.050	37	.099	47	.120	57	.164	69	.176	81	.212	95	.218
				30	.032	38	.074	48	.095	58	.136	70	.150	82	.185	96	.193
				31	.018	39	.053	49	.073	59	.112	71	.126	83	.161	97	.170
				32	.009	40	.038	50	.056	60	.091	72	.105	84	.138	98	.149
				33	.005	41	.025	51	.041	61	.072	73	.087	85	.118	99	.129
						42	.017	52	.030	62	.057	74	.071	86	.100	100	.111
						43	.010	53	.021	63	.044	75	.057	87	.084	101	.095
						44	.006	54	.014	64	.033	76	.045	88	.069	102	.081
						45	.003	55	.009	65	.025	77	.036	89	.057	103	.068
						46	.001	56	.006	66	.018	78	.027	90	.046	104	.057
								57	.003	67	.013	79	.021	91	.037	105	.047
								58	.002	68	.009	80	.016	92	.030	106	.039
								59	.001	69	.006	81	.011	93	.023	107	.031
								60	.000	70	.004	82	.008	94	.018	108	.025
										71	.002	83	.006	95	.014	109	.020
										72	.001	84	.004	96	.010	110	.016
										73	.001	85	.003	97	.008	111	.012
										74	.000	86	.002	98	.006	112	.009
										75	.000	87	.001	99	.004	113	.007
												88	.001	100	.003	114	.005
												89	.000	101	.002	115	.004
												90	.000	102	.001	116	.003
												91	.000	103	.001	117	.002
														104	.000	118	.001
														105	.000	119	.001
														106	.000	120	.001
														107	.000	121	.000
														108	.000	122	.000
																123	.000
																124	.000
																125	.000
																126	.000

Larger Sample Size $m = 10$, (Lower Tail)

Smaller Sample Size n																			
1		2		3		4		5		6		7		8		9		10	
c	P	c	P	c	P	c	P	c	P	c	P	c	P	c	P	c	P	c	P
1	.091	3	.015	6	.003	10	.001	15	.000	21	.000	28	.000	36	.000	45	.000	55	.000
2	.182	4	.030	7	.007	11	.002	16	.001	22	.000	29	.000	37	.000	46	.000	56	.000
3	.273	5	.061	8	.014	12	.004	17	.001	23	.000	30	.000	38	.000	47	.000	57	.000
4	.364	6	.091	9	.024	13	.007	18	.002	24	.001	31	.000	39	.000	48	.000	58	.000
5	.455	7	.136	10	.038	14	.012	19	.004	25	.001	32	.001	40	.000	49	.000	59	.000
6	.545	8	.182	11	.056	15	.018	20	.006	26	.002	33	.001	41	.000	50	.000	60	.000
		9	.242	12	.080	16	.027	21	.010	27	.004	34	.002	42	.001	51	.000	61	.000
		10	.303	13	.108	17	.038	22	.014	28	.005	35	.002	43	.001	52	.000	62	.000
		11	.379	14	.143	18	.053	23	.020	29	.008	36	.003	44	.002	53	.001	63	.000
		12	.455	15	.185	19	.071	24	.028	30	.011	37	.005	45	.002	54	.001	64	.001
		13	.545	16	.234	20	.094	25	.038	31	.016	38	.007	46	.003	55	.001	65	.001
				17	.287	21	.120	26	.050	32	.021	39	.009	47	.004	56	.002	66	.001
				18	.346	22	.152	27	.065	33	.028	40	.012	48	.006	57	.003	67	.001
				19	.406	23	.187	28	.082	34	.036	41	.017	49	.008	58	.004	68	.002
				20	.469	24	.227	29	.103	35	.047	42	.022	50	.010	59	.005	69	.003
				21	.531	25	.270	30	.127	36	.059	43	.028	51	.013	60	.007	70	.003
						26	.318	31	.155	37	.074	44	.035	52	.017	61	.009	71	.004
						27	.367	32	.185	38	.090	45	.044	53	.022	62	.011	72	.006
						28	.420	33	.220	39	.110	46	.054	54	.027	63	.014	73	.007
						29	.473	34	.257	40	.132	47	.067	55	.034	64	.017	74	.009
						30	.527	35	.297	41	.157	48	.081	56	.042	65	.022	75	.012
								36	.339	42	.184	49	.097	57	.051	66	.027	76	.014
								37	.384	43	.214	50	.115	58	.061	67	.033	77	.018
								38	.430	44	.246	51	.135	59	.073	68	.039	78	.022
								39	.477	45	.281	52	.157	60	.086	69	.047	79	.026
								40	.523	46	.318	53	.182	61	.102	70	.056	80	.032
										47	.356	54	.209	62	.118	71	.067	81	.038
										48	.396	55	.237	63	.137	72	.078	82	.045
										49	.437	56	.268	64	.158	73	.091	83	.053
										50	.479	57	.300	65	.180	74	.106	84	.062
										51	.521	58	.335	66	.204	75	.121	85	.072
												59	.370	67	.230	76	.139	86	.083
												60	.406	68	.257	77	.158	87	.095
												61	.443	69	.286	78	.178	88	.109
												62	.481	70	.317	79	.200	89	.124
												63	.519	71	.348	80	.223	90	.140
														72	.381	81	.248	91	.157
														73	.414	82	.274	92	.176
														74	.448	83	.302	93	.197
														75	.483	84	.330	94	.218
														76	.517	85	.360	95	.241
																86	.390	96	.264
																87	.421	97	.289
																88	.452	98	.315
																89	.484	99	.342
																90	.516	100	.370
																		101	.398
																		102	.427
																		103	.456
																		104	.485
																		105	.515

Table C12. (continued)

Larger Sample Size m = 10, (Upper Tail)

Smaller Sample Size n

1		2		3		4		5		6		7		8		9		10	
c	P	c	P	c	P	c	P	c	P	c	P	c	P	c	P	c	P	c	P
6	.545	13	.545	21	.531	30	.527	40	.523	51	.521	63	.519	76	.517	90	.516	105	.515
7	.455	14	.455	22	.469	31	.473	41	.477	52	.479	64	.481	77	.483	91	.484	106	.485
8	.364	15	.379	23	.406	32	.420	42	.430	53	.437	65	.443	78	.448	92	.452	107	.456
9	.273	16	.303	24	.346	33	.367	43	.384	54	.396	66	.406	79	.414	93	.421	108	.427
10	.182	17	.242	25	.287	34	.318	44	.339	55	.356	67	.370	80	.381	94	.390	109	.398
11	.091	18	.182	26	.234	35	.270	45	.297	56	.318	68	.335	81	.348	95	.360	110	.370
		19	.136	27	.185	36	.227	46	.257	57	.281	69	.300	82	.317	96	.330	111	.342
		20	.091	28	.143	37	.187	47	.220	58	.246	70	.268	83	.286	97	.302	112	.315
		21	.061	29	.108	38	.152	48	.185	59	.214	71	.237	84	.257	98	.274	113	.289
		22	.030	30	.080	39	.120	49	.155	60	.184	72	.209	85	.230	99	.248	114	.264
		23	.015	31	.056	40	.094	50	.127	61	.157	73	.182	86	.204	100	.223	115	.241
				32	.038	41	.071	51	.103	62	.132	74	.157	87	.180	101	.200	116	.218
				33	.024	42	.053	52	.082	63	.110	75	.135	88	.158	102	.178	117	.197
				34	.014	43	.038	53	.065	64	.090	76	.115	89	.137	103	.158	118	.176
				35	.007	44	.027	54	.050	65	.074	77	.097	90	.118	104	.139	119	.157
				36	.003	45	.018	55	.038	66	.059	78	.081	91	.102	105	.121	120	.140
						46	.012	56	.028	67	.047	79	.067	92	.086	106	.106	121	.124
						47	.007	57	.020	68	.036	80	.054	93	.073	107	.091	122	.109
						48	.004	58	.014	69	.028	81	.044	94	.061	108	.078	123	.095
						49	.002	59	.010	70	.021	82	.035	95	.051	109	.067	124	.083
						50	.001	60	.006	71	.016	83	.028	96	.042	110	.056	125	.072
								61	.004	72	.011	84	.022	97	.034	111	.047	126	.062
								62	.002	73	.008	85	.017	98	.027	112	.039	127	.053
								63	.001	74	.005	86	.012	99	.022	113	.033	128	.045
								64	.001	75	.004	87	.009	100	.017	114	.027	129	.038
								65	.000	76	.002	88	.007	101	.013	115	.022	130	.032
										77	.001	89	.005	102	.010	116	.017	131	.026
										78	.001	90	.003	103	.008	117	.014	132	.022
										79	.000	91	.002	104	.006	118	.011	133	.018
										80	.000	92	.002	105	.004	119	.009	134	.014
										81	.000	93	.001	106	.003	120	.007	135	.012
												94	.001	107	.002	121	.005	136	.009
												95	.000	108	.002	122	.004	137	.007
												96	.000	109	.001	123	.003	138	.006
												97	.000	110	.001	124	.002	139	.004
												98	.000	111	.000	125	.001	140	.003
														112	.000	126	.001	141	.003
														113	.000	127	.001	142	.002
														114	.000	128	.000	143	.001
														115	.000	129	.000	144	.001
														116	.000	130	.000	145	.001
																131	.000	146	.001
																132	.000	147	.000
																133	.000	148	.000
																134	.000	149	.000
																135	.000	150	.000
																		151	.000
																		152	.000
																		153	.000
																		154	.000
																		155	.000

Adapted from Table B of *A Nonparametric Introduction to Statistics*, by C. H. Kraft and C. van Eeden, Macmillan, New York, 1968, with the permission of the authors and the publisher. Copyright © 1968, by the Macmillan Company.

Index

451